Advances in
IMMUNOLOGY

VOLUME **98**

Advances in
IMMUNOLOGY

VOLUME **98**

Edited by

FREDERICK W. ALT

Howard Hughes Medical Institute, Boston, Massachusetts

Associate Editors

K. FRANK AUSTEN

Harvard Medical School, Boston, Massachusetts

TASUKU HONJO

Kyoto University, Kyoto, Japan

FRITZ MELCHERS

University of Basel, Basel, Switzerland

JONATHAN W. UHR

University of Texas, Dallas, Texas

EMIL R. UNANUE

Washington University, St. Louis, Missouri

AMSTERDAM • BOSTON • HEIDELBERG • LONDON
NEW YORK • OXFORD • PARIS • SAN DIEGO
SAN FRANCISCO • SINGAPORE • SYDNEY • TOKYO
Academic Press is an imprint of Elsevier

Academic Press is an imprint of Elsevier
84 Theobald's Road, London WC1X 8RR, UK
Radarweg 29, PO Box 211, 1000 AE Amsterdam, The Netherlands
Linacre House, Jordan Hill, Oxford OX2 8DP, UK
30 Corporate Drive, Suite 400, Burlington, MA 01803, USA
525 B Street, Suite 1900, San Diego, CA 92101-4495, USA

First edition 2008

ISBN: 978-0-12-374331-2
ISSN: 0065-2776

For information on all Academic Press publications
visit our website at elsevierdirect.com

Printed and bound in USA
08 09 10 11   10 9 8 7 6 5 4 3 2 1

Working together to grow
libraries in developing countries

www.elsevier.com | www.bookaid.org | www.sabre.org

ELSEVIER   BOOK AID
             International   Sabre Foundation

CONTENTS

CONTRIBUTORS

Numbers in parentheses indicate the pages on which the authors' contributions begin.

Stephen M. Anderton
University of Edinburgh, Institute of Immunology and Infection Research, School of Biological Sciences, Edinburgh EH9 3JT, United Kingdom (1)

Elisabeth Calderon-Gomez
Laboratory of immune regulation, Deutsches Rheuma-Forschungszentrum, Charitéplatz 1, Berlin, Germany (1)

J. Donald Capra
Oklahoma Medical Research Foundation, Oklahoma City, Oklahoma 73104 (151)

Tse Wen Chang
Genomics Research Center, Academia Sinica, Nankang, Taipei 11529, Taiwan (39)

Nora A. Fierro
Laboratory of Immune Cell Signaling, National Institute of Arthritis and Musculoskeletal and Skin Diseases, National Institutes of Health, Bethesda, Maryland (85)

Simon Fillatreau
Laboratory of immune regulation, Deutsches Rheuma-Forschungszentrum, Charitéplatz 1, Berlin, Germany (1)

Kai Hoehlig
Laboratory of immune regulation, Deutsches Rheuma-Forschungszentrum, Charitéplatz 1, Berlin, Germany (1)

Stephen M. Jackson
Oklahoma Medical Research Foundation, Oklahoma City, Oklahoma 73104 (151)

Judith A. James
Oklahoma Medical Research Foundation, Oklahoma City, Oklahoma 73104 (151)

Vicky Lampropoulou
Laboratory of immune regulation, Deutsches Rheuma-Forschungszentrum, Charitéplatz 1, Berlin, Germany (1)

Katherine A. McLaughlin
Program in Immunology, Harvard Medical School, Boston, Massachusetts 02115; *and* Department of Cancer Immunology and AIDS, Dana-Farber Cancer Institute, Boston, Massachusetts 02115 (*121*)

Patricia Neves
Laboratory of immune regulation, Deutsches Rheuma-Forschungszentrum, Charitéplatz 1, Berlin, Germany (*1*)

Ana Olivera
Laboratory of Immune Cell Signaling, National Institute of Arthritis and Musculoskeletal and Skin Diseases, National Institutes of Health, Bethesda, Maryland (*85*)

Ariel Y. Pan
Genomics Research Center, Academia Sinica, Nankang, Taipei 11529, Taiwan (*39*)

Juan Rivera
Laboratory of Immune Cell Signaling, National Institute of Arthritis and Musculoskeletal and Skin Diseases, National Institutes of Health, Bethesda, Maryland (*85*)

Toralf Roch
Laboratory of immune regulation, Deutsches Rheuma-Forschungszentrum, Charitéplatz 1, Berlin, Germany (*1*)

Ulrich Steinhoff
Max Planck Institute for Infection Biology, Charitéplatz 1, Berlin, Germany (*1*)

Ryo Suzuki
Laboratory of Immune Cell Signaling, National Institute of Arthritis and Musculoskeletal and Skin Diseases, National Institutes of Health, Bethesda, Maryland (*85*)

Patrick C. Wilson
Oklahoma Medical Research Foundation, Oklahoma City, Oklahoma 73104 (*151*)

Kai W. Wucherpfennig
Department of Neurology and Program in Immunology, Harvard Medical School, Boston, Massachusetts 02115; *and* Department of Cancer Immunology and AIDS, Dana-Farber Cancer Institute, Boston, Massachusetts 02115 (*121*)

CHAPTER 1

Immune Regulation by B Cells and Antibodies: A View Towards the Clinic

Kai Hoehlig,* Vicky Lampropoulou,* Toralf Roch,*
Patricia Neves,* Elisabeth Calderon-Gomez,*
Stephen M. Anderton,† Ulrich Steinhoff,‡ and
Simon Fillatreau*

Contents		

* Laboratory of immune regulation, Deutsches Rheuma-Forschungszentrum, Charitéplatz 1, Berlin, Germany
† University of Edinburgh, Institute of Immunology and Infection Research, School of Biological Sciences, Edinburgh EH9 3JT, United Kingdom
‡ Max Planck Institute for Infection Biology, Charitéplatz 1, Berlin, Germany

Advances in Immunology, Volume 98
ISSN 0065-2776, DOI: 10.1016/S0065-2776(08)00401-X

Abstract   B lymphocytes contribute to immunity in multiple ways, including production of antibodies, presentation of antigen to T cells, organogenesis of secondary lymphoid organs, and secretion of cytokines. Recent clinical trials have shown that depleting B cells can be highly beneficial for patients with autoimmune diseases, implicating B cells and antibodies as key drivers of pathology. However, it should be kept in mind that B cell responses and antibodies also have important regulatory roles in limiting autoimmune pathology. Here, we analyze clinical examples illustrating the potential of antibodies as treatment for immune-mediated disorders and discuss the underlying mechanisms. Furthermore, we examine the regulatory functions of activated B cells, their involvement in the termination of some experimental autoimmune diseases, and their use in cell-based therapy for such pathologies. These suppressive functions of B cells and antibodies do not only open new ways for harnessing autoimmune illnesses, but they also should be taken into account when designing new strategies for vaccination against microbes and tumors.

1. INTRODUCTION

B cells are unique for their capacity to secrete antibodies, and this has long been regarded as their main function. Antibodies protect against infectious diseases, and deficiencies in antibody increase susceptibility to infections by microbes such as *Streptococcus pneumoniae*, *Haemophilus influenzae*, and *Staphylococcus aureus*, and by Gram-negative bacteria such as pseudomonas (Cunningham-Rundles and Ponda, 2005).

Antibodies are also major players in the pathogenesis of autoimmune diseases. Idiopathic thrombocytopenia purpura (ITP), a disease associated with low platelet counts and mucocutaneous bleedings is driven primarily by autoantibodies. Thus, transfer of serum from ITP patients can induce thrombocytopenia in healthy subjects (Harrington *et al.*, 1951), and there can be a loss of platelets in neonates of affected mothers (Cines and Blanchette, 2002). Other examples of antibody-mediated pathologies include hyperthyroidism and myasthenia gravis. In hyperthyroidism, autoantibodies chronically stimulate the thyroid-stimulating hormone receptor and thereby sustain an increased production of thyroid hormone. In myasthenia gravis, autoantibodies trigger the loss of acetylcholine receptors and consequently provoke failure of neuromuscular transmission and muscle weakness. Experimental animal models have confirmed the capacity of autoantibodies to induce autoimmune diseases. Typical examples are the induction of thrombocytopenia and arthritis by injection of antiplatelet antibodies or of serum from arthritic K/BxN mice, respectively (Korganow *et al.*, 1999; Siragam *et al.*, 2006).

Assuming a generally harmful function of antibodies in autoimmune diseases, one might predict that reduced antibody titers protect against these pathologies. Paradoxically, patients suffering from antibody deficiencies show a higher incidence of autoimmune disorders. Common variable immunodeficiency (CVID) is an immunodeficiency characterized by hypogammaglobulinemia (1999). Most patients have normal B cell numbers, but these fail to differentiate into antibody-producing plasma cells (Agematsu *et al.*, 2002). Remarkably, the incidence of autoimmune diseases in CVID patients is 20% (Cunningham-Rundles and Bodian, 1999). The most common pathologies are ITP and autoimmune hemolytic anemia, occasionally occurring together in the form of Evan's syndrome (Cunningham-Rundles and Bodian, 1999). Other autoimmune disorders such as rheumatoid arthritis, juvenile rheumatoid arthritis, and pernicious anemia are also found at increased frequencies in these patients. Standard treatment for CVID is the periodic injection of intravenous immunoglobulins (IVIG), which provides some protection against both infections and autoimmune diseases (Kazatchkine and Kaveri, 2001). Why do antibody deficiencies result in enhanced susceptibility to autoimmune pathologies?

One possibility is that these patients suffer from repeated infections. Microbes can trigger autoimmunity through epitope mimicry, tissue destruction, or adjuvant effects. Indeed, childhood ITP is often a consequence of viral infection (Rand and Wright, 1998). Infections can facilitate platelet destruction by stimulating production of interferon (IFN)-γ, a cytokine that strongly activates macrophages and thereby increases elimination of antibody-coated platelets (Musaji *et al.*, 2005). However, it is sometimes an autoimmune disease, but not an infection, that leads to diagnosis of CVID. Thus, CVID patients may have a natural predisposition to autoimmune diseases, independently of their susceptibility to infections. Defects in the antibody system may impair immunological tolerance if antibodies have immune regulating functions. Indeed, IgG contributes to immune homeostasis via the inhibitory Fc receptor FcγRIIB. Deficiency in FcγRIIB results in an increased production of autoantibodies and an enhanced susceptibility to autoimmune diseases (Bolland and Ravetch, 2000; Yuasa *et al.*, 1999). In addition to IgG, circulating IgM also limits production of autoantibodies (Ehrenstein *et al.*, 2000), and suppresses autoimmune diseases (Boes *et al.*, 2000). Furthermore, IgM antibodies can mask autoreactivity of circulating IgG through idiotypic interactions (Adib *et al.*, 1990). This inhibitory effect is impaired in patients with active Hashimoto's thyroiditis (Hurez *et al.*, 1993) and warm autoimmune hemolytic anemia (Stahl *et al.*, 2000). IgM also facilitates clearance of apoptotic cells that are considered a major source of autoantigens in autoimmune pathology (Botto *et al.*, 1998; Kim *et al.*, 2002; Scott *et al.*, 2001). Finally, antibodies can participate in the development and repair of certain tissues such as the central nervous system. Oligodendrocyte progenitors express FcγR, and mice lacking FcγR display a severe myelination defect characterized by a reduced production of myelin proteins and a diminished frequency of myelinated neurons (Nakahara *et al.*, 2003). IgM antibodies recognizing antigens expressed at the surface of oligodendrocytes can promote remyelination (Asakura *et al.*, 1998; Bieber *et al.*, 2001). Thus, it is clear that antibodies are involved in the maintenance of immunological tolerance so that significant alterations of the antibody repertoire might knock these homeostatic equilibriums out of balance and thereby facilitate the onset and progression of autoimmune diseases.

Like antibodies, B cells themselves have a dual role in autoimmune disorders, sometimes driving inflammatory diseases, while in other circumstances suppressing autoimmune symptoms. Evidence for a regulatory role of B cells in autoimmune disease was first provided by Wolf and Janeway who observed that mice lacking B cells develop an unusually severe form of experimental autoimmune encephalomyelitis (EAE), resulting in a chronic disease while mice with wild-type B cells recovered after a short episode of paralysis (Wolf *et al.*, 1996). Dissection of the mechanisms by which B cells control resolution of EAE revealed that B cells regulate autoimmune disease through provision of IL-10 (Fillatreau *et al.*, 2002). A beneficial function for IL-10-producing B cells has also been identified in models of ulcerative colitis (UC) (Mizoguchi *et al.*, 2002), and in collagen-induced arthritis (CIA) (Mauri *et al.*, 2003). Furthermore, IL-10-producing B cells have been identified in humans (Duddy *et al.*, 2004), and preliminary data showing that B cells from multiple sclerosis (MS) patients produce reduced amounts of IL-10 suggests that these cells possess a regulatory function in human disease as well (Duddy *et al.*, 2007). This regulatory role of B cells is not limited to autoimmune situations. For instance, chronic infection with *Schistosoma mansoni* induces IL-10 production by B cells, which can suppress anaphylaxis (Mangan *et al.*, 2004). Similarly, some viruses have developed mechanisms to stimulate IL-10-producing B cells and thereby subvert the immune system (Jude *et al.*, 2003; Velupillai *et al.*, 2006).

The dual role of B cells in pathogenesis and resolution of autoimmune inflammation is illustrated by the fact that both B cell depletion and administration of polyclonal immunoglobulins are effective treatments for some autoimmune disorders. Thus, the B-cell-depleting antibody Rituximab (Reff *et al.*, 1994) induces a significant rise in platelet counts in 33–54% of ITP patients (Bennett *et al.*, 2006; Wang *et al.*, 2005), while infusion of IVIG increases platelet counts in more than 70% of these patients (Bussel and Pham, 1987). An interpretation of these results is that depleting B lymphocytes is advantageous when these cells directly contribute to pathogenesis; while administrating antibodies can suppress disease by restoring regulatory mechanisms whose failure promoted onset and progression of disease. Here, we analyze clinical examples that document the capacity of antibodies to suppress immunopathology, and discuss possible mechanisms. We also examine the concept that activated B cells can have important regulatory activities, and highlight their potential for cell-based therapy of autoimmune diseases. Finally, we discuss what makes a B cell or an antibody protective rather than pathogenic with respect to development of autoimmune pathologies.

2. PREVENTION OF RHESUS D HEMOLYTIC DISEASE OF THE NEWBORN WITH ANTI-D ANTIBODIES

2.1. HDFN is mediated by maternal IgG

Rhesus hemolytic disease of the fetus and newborn (HDFN) occurs when a pregnant mother develops an IgG response against red blood cells (RBC) of her fetus leading to destruction of these RBC. This mostly happens when a rhesus D-negative (D^-) mother carries a rhesus-D-positive (D^+) child. The mother's immune system can be sensitized to fetal D^+ RBC as a result of transplacental hemorrhage or at delivery. The primary response is usually weak and often consists of IgM antibodies that do not cross the placenta. Thus, pathology typically starts upon re-exposure to D^+ RBC during a second pregnancy, when anti-D antibodies switch to IgG subclasses that efficiently cross the placenta. Clinical manifestations of HDFN range from asymptomatic mild anemia to still birth, hydrops fetalis, or kernicterus, the two last conditions being fatal for 25%

of affected neonates (Bowman, 1998). HDFN was a significant cause of fetal mortality and morbidity as approximately 15–17% of Caucasians are D^- (Urbaniak and Greiss, 2000). Death rate was approximately 150 per 100,000 births, corresponding to 10% of perinatal deaths (Bowman, 1998).

2.2. Anti-D prevents HDFN

The concept of antibody-mediated suppression, according to which specific antibodies can prevent humoral immunity against a relevant antigen, suggested that deliberate administration of human anti-D immunoglobulin could prevent maternal immunization against fetal D antigens and thereby prevent HDFN. Indeed, clinical trials demonstrated the therapeutic efficacy of anti-D immunoglobulin, which was licensed as treatment for HDFN in 1968 in North America (1969; Clarke *et al.*, 1963; Freda *et al.*, 1964). In North America, D^- women nowadays receive anti-D immunoglobulin between 28 and 32 weeks of gestation in all pregnancies, and an additional injection is performed at delivery if the fetus is D^+. (Bowman, 1988). This treatment offers a protection rate of 96%, and remains today the most efficient way to prevent HDFN (Bowman and Pollock, 1987). Anti-D immunoglobulin is manufactured from pools of plasma from individuals previously exposed to D^+ RBC during pregnancy, transfusion, or deliberate immunization, implying limited availability. To overcome this problem, human monoclonal anti-D antibodies have been generated (Fletcher and Thomson, 1995), and two clones (BRAD-3 and BRAD-5) that prevented rhesus sensitization in human trials are under further development (Kumpel, 2002).

Anti-D prophylaxis is robust. It is effective with only 20% of D molecules on RBC being masked (Kumpel *et al.*, 1995), and is accompanied by an inhibition of humoral immunity towards other antigens expressed by the D^+ RBC (Woodrow *et al.*, 1975). Furthermore, suppression can last up to 6–16 months after treatment (1971; Kumpel and Elson, 2001). Thus, anti-D must engage regulatory mechanisms of general interest for dampening unwanted immune reactions.

2.3. How does anti-D treatment work?

Several models have been proposed to explain the suppressive effects of anti-D. Fc receptors are thought to have a central role because the Fc portion of anti-D is required for suppression. A popular concept proposes that anti-D-coated RBC inactivate B cells by cross-linking their antigen receptor to the inhibitory Fc receptor FcγRIIB. Indeed, cross-linking of B cell receptor for antigen (BCR) and FcγRIIB strongly inhibits B cell activation *in vitro* (Phillips and Parker, 1984). This model predicts a suppression that spreads to all antigens of D^+ RBC. It may also explain why suppression lasts for several months, until new D-reactive B cells are generated. Another model proposes that anti-D inhibits B cell priming by promotion of immunological "ignorance," either by diverting D^+ RBC away from microenvironments where B cell priming takes place, or by accelerating D^+ RBC removal so that they are eliminated before B cell priming can occur (Jones *et al.*, 1957). A limitation of such "ignorance" models is their inability to explain long-lasting effects. A third hypothesis proposes that anti-D serves as Trojan horse targeting suppressive molecules to RBC-reactive B cells. For instance, the Fc portion of anti-D could carry TGF-β to RBC-reactive B cells and consequently induce their inhibition (Bouchard *et al.*, 1994, 1995; Rowley and Stach, 1998). This would require the Fc tail of anti-D but be independent of Fc receptors. The three models described predict a RBC-specific suppression. However, suppression extends beyond RBC antigens because anti-D is also used to treat ITP, an autoimmune disease that does not involve anti-RBC antibodies.

An experimental approach to dissect the mechanisms of anti-D-mediated suppression is the transfer of sheep RBC into mice. Using this model, Heyman and colleagues have shown that suppression of humoral immunity by anti-D-like antibodies is independent of FcγRIIB, FcγRI, FcγRIII, and the neonatal Fc receptor (Karlsson *et al.*, 1999, 2001). Although these experiments may not reflect perfectly the clinical setting of anti-D prophylaxis, they have challenged the long-held belief that these Fc receptors are prime mediators of anti-D inhibition. The concept that immunological ignorance is the major mechanism of suppression has also been challenged by experiments showing that T cell priming occurs although humoral immunity is suppressed (Brinc *et al.*, 2007).

2.4. Conclusions and perspectives

Anti-D prophylaxis demonstrates that antibodies can be used to prevent humoral immunity against a complex antigen in a genetically diverse population. Although the mechanism of action remains unknown (Kumpel and Elson, 2001), anti-D is also used to treat patients suffering from ITP, indicating its therapeutic potential against immune-driven pathology beyond HDFN. What does anti-D teach us about the natural functions of antibodies? It suggests that, similarly to anti-D, IgG naturally present in serum can regulate humoral immunity against viruses, bacteria, or cell-associated self-antigens. This could have important implications for host defense mechanisms and maintenance of self-tolerance. However, this notion remains difficult to test experimentally as long as the cells and receptors mediating anti-D suppressive effects remain unknown.

3. ANTIBODY-BASED TREATMENT OF IMMUNE THROMBOCYTOPENIA PURPURA

ITP is a disease characterized by bleedings and low platelet counts. Its annual incidence is about 50–100 cases per million people, with about half of cases occurring in children (Cines and Blanchette, 2002; McMillan, 1997). Natural history of ITP differs between children and adults. Childhood ITP often follows an infection, and it is usually an acute syndrome resolving spontaneously in <6 months (Bussel, 2006). In adults, ITP is often a chronic disease requiring therapeutic intervention and it is rarely preceded by an infection (Bussel, 2006). Patients with platelet counts below 20,000/μl have a significant risk of fatal bleedings, mostly intracranial hemorrhage (Cines and Bussel, 2005; Cohen *et al.*, 2000). IVIG and anti-D are two effective treatments for ITP. Before discussing how these antibody-based therapies modulate ITP, we will first examine data from B cell depletion in ITP patients that precise the pathogenic mechanisms involved in this disease.

3.1. Roles of antibodies and B cells in ITP

3.1.1. Roles of antibodies
By showing that injection of plasma from chronic ITP patients into healthy recipients can transfer thrombocytopenia, Harrington has proven that humoral factors play an important role in ITP (Harrington *et al.*, 1951). Consistently, ITP patients often have elevated titers of anti-platelet antibodies (Kiefel *et al.*, 1996). Injection of monoclonal anti-platelet antibodies into mice is sufficient to provoke thrombocytopenia (Siragam *et al.*, 2006). Thus, the main mechanism of pathogenesis in ITP is thought to be clearance of opsonized platelets by FcγR-expressing phagocytic cells in spleen and liver. This model is supported by several observations: splenectomy often ameliorates ITP (Choi *et al.*, 2001), and polymorphisms in FcγRIIA, FcγRIIIA, and FcγRIIIB influence susceptibility to that disease (Ivan and Colovai, 2006). Furthermore, blocking FcγRIII with antibodies is clinically beneficial (Clarkson *et al.*, 1986).

3.1.2. Roles of B cells
Rituximab is a genetically engineered anti-CD20 antibody initially developed for the treatment of non-Hodgkin lymphoma. The standard administration protocol consists of one injection per week during four weeks. Rituximab eliminates most circulating B cells within a few days, and depletion usually lasts for 6–12 months. Depletion is very rapid for blood B cells, and slower for lymph node and splenic B cells (Martin and Chan, 2006). Marginal zone B cells, germinal centre B cells, and B1 cells are more resistant to depletion than follicular B cells. Rituximab does not deplete pro-B cells and plasma cells, because these cells do not express

CD20. Nevertheless, short-lived plasma cells are probably lost within a few weeks because they cannot be renewed from mature B cells. In contrast, long-lived plasma cells can certainly persist for several months after Rituximab treatment (Dilillo *et al.*, 2008). Consistently, Rituximab does not alter serum IgG levels, while IgM levels decrease more significantly (Bennett *et al.*, 2006). The respective contributions of short-lived and long-lived plasma cells to anti-platelet IgG production are not known.

Rituximab induced an important elevation of platelet counts in 33–54% of cases, confirming that B cells drive ITP in a subset of patients possibly via antibodies (Bennett *et al.*, 2006; Wang *et al.*, 2005). There is also evidence that B cells contribute to pathogenesis via antibody-independent mechanisms. For instance, platelet counts usually rose within a week after the first Rituximab dose, most likely before any reduction of autoantibody levels (Stasi *et al.*, 2001). Furthermore, some untreated patients underwent remission while maintaining constant anti-platelet antibody levels (Imbach *et al.*, 1991; Ozsoylu *et al.*, 1991). Finally, Schlomchik and colleagues have shown in an experimental model of systemic lupus erythematosus that B cells can promote autoimmune pathogenesis independently of antibodies, possibly as antigen-presenting cells prompting aggressive T cell responses (Chan *et al.*, 1999). Modulation of T cell activation is certainly a mechanism worth considering since B cell-depletion modulates T cell activation (Bouaziz *et al.*, 2007), and positive response to Rituximab correlates with a reversion of T cell abnormalities in chronic ITP patients (Stasi *et al.*, 2007). It is notable that B cell depletion also reduces T cell reactivity in MS, leading to a decrease of T cell numbers in cerebrospinal fluid (Cross *et al.*, 2006). According to these data, it seems likely that T cells play an important role in ITP.

Some patients do not respond to Rituximab, suggesting that B cells are not key drivers of disease in these cases, although it cannot be excluded that long-lived plasma cells maintain ITP independently of B cells. Several observations indicate the existence of antibody-independent mechanisms of thrombocytopenia. First, anti-platelet antibodies are detected in only 50–70% of ITP patients (Brighton *et al.*, 1996; Warner *et al.*, 1999), and not serum from all ITP patients transferred thrombocytopenia in Harrington's experiment (Harrington *et al.*, 1951). Second, T cells from ITP patients can directly lyse autologous platelets, possibly via Fas, granzymes, or perforin (Olsson *et al.*, 2003; Zhang *et al.*, 2006). ITP patients have additional T cell abnormalities including an augmented anti-platelet T cell reactivity, expansion of some T cell clones, and increased T_H1/T_H2 and Tc1/Tc2 ratios (Semple, 2002; Semple and Freedman, 1991; Shimomura *et al.*, 1996). Thus, T cells certainly contribute to the destruction of platelets in ITP. On the other hand, Manz and colleagues have suggested that both short-lived and long-lived plasma cells produce pathogenic autoantibodies (Hoyer *et al.*, 2004). Long-lived plasma cells could therefore perpetuate disease in B cell-depleted individuals.

These results indicate that ITP is a complex disease that involves antibodies, T cells, and also antibody-independent functions of B cells. The dominant pathogenic process is likely to differ according to patients and disease stages. This complexity should be kept in mind when discussing how treatments suppress ITP.

3.2. Treatment of ITP by IVIG or anti-D

IVIG is a polyclonal preparation of human IgG obtained from large pools of plasma usually including more than 5,000 donors. The distribution of isotype subclasses corresponds to that of normal human serum. The half-life of IVIG is about three weeks in healthy individuals (Ephrem *et al.*, 2005).

The first use of IVIG in ITP patients was reported in 1981 by Imbach and colleagues (Imbach *et al.*, 1981). Epidemiological studies have determined that IVIG increases platelet counts in about 64% of cases (Bussel and Pham, 1987). Salama and colleagues then noticed that IVIG contains some anti-RBC IgG that cause a mild hemolysis in ITP patients treated with IVIG (Salama *et al.*, 1983). This led them to propose that the active component of IVIG acts similarly to anti-D by forming large immune complexes (ICs) of opsonized RBC, and to show that anti-D also increases platelet counts in ITP patients (Salama *et al.*, 1983). Following confirmation of these results (Salama *et al.*, 1984; Salama *et al.*, 1986), anti-D was licensed for treatment of ITP in 1995. Anti-D increases platelet numbers in 70% of patients who are D^+ and nonsplenectomized (Scaradavou *et al.*, 1997). Splenectomized patients show a reduced response, and D^- patients do not respond. It is remarkable that IVIG and anti-D seem to be effective for at least as many patients as Rituximab. However, it has to be kept in mind that these treatments are difficult to compare because they were tested in different cohorts of patients.

3.3. Does IVIG treatment work via Fc receptors, and if so how?

The therapeutic effect of IVIG primarily involves its Fc portion, because elimination of this domain strongly diminishes protection (Burdach *et al.*, 1986), while removing the antigen-binding part usually has no effect (Solal-Celigny *et al.*, 1983). IVIG does not require the complement system, implying that it works via Fc receptors. Similarly, clinical response to anti-D involves the Fc receptors FcγRIIa and FcγRIIIa (Cooper *et al.*, 2004).

3.3.1. Role of competition mechanisms

IVIG and anti-D can block access of self-reactive antibodies to the Fc receptors mediating pathogenesis, such as those expressed on macrophages, and thereby prevent the clearance of circulating IgG-coated platelets (Kimberly *et al.*, 1984). Indeed, IVIG inhibits the phagocytic activity of macrophages (Fehr *et al.*, 1982), and blocking FcγRIII with a monoclonal antibody is sufficient to increase platelet numbers in some patients (Clarkson *et al.*, 1986). IVIG could also limit pathogenesis by accelerating the elimination of pathogenic IgG, for instance by saturating the neonatal Fc receptor (FcRn) which extends the life-span of circulating IgG (Hansen and Balthasar, 2002a,b; Roopenian *et al.*, 2003; Ward *et al.*, 2003). A common feature of these two mechanisms is the involvement of competition between IVIG and pathogenic antibodies, implying that protective and pathogenic IgG have to bind to the same Fc receptors.

3.3.2. Role of immune regulating mechanisms

Injection of monoclonal anti-platelet antibody provokes in mice a transient thrombocytopenia treatable by IVIG or anti-RBC antibodies (a mimic for anti-D) as in humans. However, IVIG is not protective in mice lacking the inhibitory receptor FcγRIIB, indicating that competition between IVIG and pathogenic IgG cannot prevent pathogenesis in absence of immune regulation by FcγRIIB (Samuelsson *et al.*, 2001; Siragam *et al.*, 2005). Similarly, IVIG-mediated protection against antibody-induced arthritis requires FcγRIIB (Samuelsson *et al.*, 2001). Following administration of IVIG, FcγRIIB is up-regulated on splenic effector macrophages, which reduces the ratio between activating and inhibitory Fc receptors on these cells. Because FcγRIIB contains immunoreceptor tyrosine-based inhibitory motifs (ITIM), this leads to an inhibition of activating Fc receptors, which signal via immunoreceptor tyrosine-based activation motifs (ITAM), and to suppression of phagocytosis. The signaling pathway by which FcγRIIB mediates this suppression is not known. FcγRIIB can transmit inhibitory signals via two pathways in B cells. The first involves SHIP1 and possibly SHP-1 (Ravetch and Lanier, 2000). The second is ITIM-independent and recruits Bruton tyrosine kinase (Btk) (Pearse *et al.*, 1999; Ravetch and Lanier, 2000). However, IVIG ameliorates experimental ITP independently of SHIP1, SHP-1, and Btk (Crow *et al.*, 2003). It has to be noted that anti-D-like antibodies are protective in FcγRIIB-deficient mice, demonstrating that this treatment works via a different mechanism than IVIG (Samuelsson *et al.*, 2001; Siragam *et al.*, 2005).

3.4. How does IVIG up-regulate FcγRIIB on effector macrophages?

Recent experiments have proven that IVIG up-regulates expression of FcγRIIB indirectly by triggering an intermediate cell type that then increases the expression of FcγRIIB on effector macrophages (Bruhns *et al.*, 2003; Siragam *et al.*, 2006). This "two cells" model explains why IVIG does not up-regulate FcγRIIB on effector macrophages in colony-stimulating factor (CSF)-1-deficient mice (Bruhns *et al.*, 2003), which lack subsets of dendritic cells (DC) and macrophages (MacDonald *et al.*, 2005) and are not protected from antibody-driven inflammation (Bruhns *et al.*, 2003). Which intermediate cell type initiates the suppressive effect of IVIG? The absence of particular populations of DC and macrophages in CSF-1-deficient mice points to one of these cell-types. Indeed, injection of DC exposed to IVIG *in vitro* augments expression of FcγRIIB on effector macrophages and transfers the protective effect of IVIG against antibody-mediated inflammation (Siragam *et al.*, 2006). Thus, DC can initiate the therapeutic effect of IVIG. To mediate this function, they require signaling via activating Fcγ receptors but not FcγRIIB (Siragam *et al.*, 2006). Thus, activating Fc receptors do not only produce disease but they also suppress antibody-mediated pathology. Distinct cell types execute these opposite functions: activating Fcγ receptors from effector macrophages contribute to disease while those from DC restrain immune pathogenesis. The function of activating Fcγ receptors is highly cell-type specific because IVIG-treated macrophages, T cells, or B cells could not transfer protection (Siragam *et al.*, 2006). These data collectively show that immune regulating mechanisms are the primary mediators of IVIG therapy because they are both necessary and sufficient for protection, even in absence of any competition between IVIG and pathogenic antibodies. It is remarkable that protective and pathogenic antibodies do not need to be related in terms of antigenic specificities. Instead, these data suggest that the quantity of protective antibodies determines the threshold for an autoreactive antibody to stimulate pathogenesis, assuming that up-regulation of FcγRIIB on effector macrophages is proportional to the global amount of protective IgG. This model predicts that antibody deficiencies can facilitate antibody-mediated inflammation because a reduction of the amount of protective antibody would impair this regulatory circuit. Indeed, patients suffering from CVID have reduced antibody levels and they are highly prone to antibody-mediated autoimmune diseases such as ITP or autoimmune hemolytic anemia (Cunningham-Rundles and Bodian, 1999). This model also suggests that IVIG ameliorates antibody-driven inflammation by correcting the rupture in immune regulating mechanisms that facilitated onset of disease i.e. a weakening of antibody-mediated suppression.

3.5. Component of IVIG protecting against antibody-mediated inflammation

IVIG is a complex preparation containing monomeric IgG (>95%) and a small amount of IC. IC can form by IgG aggregation during the industrial fractionation process (Matejtschuk *et al.*, 2002), or by interaction between IVIG and endogenous serum proteins after administration (Dietrich *et al.*, 1992; Lamoureux *et al.*, 2003, 2004). Such interactions are certainly favored by the large amounts of IVIG administered (1–2 g/kg), which provokes a rapid 2–3-fold increase in plasma IgG levels. Application of IVIG in the clinic is currently limited by availability and elevated production costs. Worldwide consumption of IVIG has nearly tripled during the period 1992–2004, increasing from 19.4 to 55 tons/year (Bayry *et al.*, 2007), so that demand for IVIG nowadays largely exceeds its supply (Bayry *et al.*, 2007; Pendergrast *et al.*, 2005). Identifying the active component of IVIG could allow production of IVIG substitutes and refinement of this therapy.

3.5.1. Monomeric IgG versus ICs

Augener and colleagues have observed that the level of IVIG-mediated protection correlates with the amount of IgG aggregates, and therefore suggested that IC constitute the active component of IVIG (Augener *et al.*, 1985). Consistently, small IC prepared artificially by mixing ovalbumin with anti-ovalbumin antibodies protect from antibody-mediated arthritis and ITP as efficiently as IVIG (Siragam *et al.*, 2005). Similarly, IC formed *in vivo* by injection of monoclonal antibodies specific for endogenous proteins such as albumin or transferrin achieve protection (Siragam *et al.*, 2005). In all cases, protection is dependent on FcγRIIB, as observed for IVIG (Siragam *et al.*, 2005). These data indicate that small IC contain the active component of IVIG, and IC indeed protect from antibody-mediated inflammation at lower doses than IVIG (Bazin *et al.*, 2004; Siragam *et al.*, 2005). Thus, small IC could substitute IVIG for the treatment of autoimmune diseases. Importantly, these data suggest that antibody-mediated suppression is independent of antigen-specificity, as discussed in Section 3.4.

Salama and colleagues have proposed that the active component of IVIG consists of anti-RBC antibodies. This hypothesis was the basis for treating ITP patients with anti-D (Scaradavou *et al.*, 1997; Ware and Zimmerman, 1998). In experimental ITP the protective effect of anti-D can be mimicked by injecting either certain RBC-reactive monoclonal antibodies, or antibody-coated liposomes (Deng and Balthasar, 2005), or RBC coated *ex vivo* with ovalbumin and opsonized with anti-OVA antibodies. In all cases suppression was independent of FcγRIIB (Siragam *et al.*, 2005; Song *et al.*, 2005). The fact that IVIG and anti-D suppress immune responses via different mechanisms implies that anti-RBC antibodies are not an active component of IVIG.

The active component of anti-D is certainly made of D-reactive antibodies since this treatment is inefficient in D⁻ ITP patients. Some clinical studies have tested whether anti-D could be replaced by monoclonal anti-D antibodies. Although monoclonal anti-RBC antibodies could prevent experimental ITP in rodent, this approach was unsuccessful in humans (Godeau *et al.*, 1996). Thus, reacting with D is insufficient for an antibody to be protective. In order to identify the characteristics of suppressive anti-D antibodies, Kjaersgaard and colleagues have compared several human monoclonal antibodies for their capacity to inhibit phagocytosis of antibody-coated platelets *in vitro* (Kjaersgaard *et al.*, 2007). Human antibodies of isotype IgG3 were more suppressive than those of IgG1 subclass. A certain binding affinity between IgG and RBC may also be required for protection (Song *et al.*, 2003, 2005).

3.5.2. What makes an IgG protective or pathogenic?

IgG differ by their oligosaccharides contents. In particular, each Fc domain has a conserved N-glycosylation site at position asparagin-297 that can bind approximately 32 different glycans (Arnold *et al.*, 2007). These sugars regulate the quaternary structure of the Fc domain and thereby shape the interaction between IgG and Fc receptors (Arnold *et al.*, 2007; Mimura *et al.*, 2000, 2001). Kaneko and colleagues have examined how presence of sialic acid in these sugars affects the functional properties of anti-platelet IgG and IVIG in ITP (Kaneko *et al.*, 2006). Sialic acids are usually exposed to the environment because they are found at distal sides of oligosaccharides chains. Following elimination of sialic acids, anti-platelet antibodies became more toxic and IVIG less protective, suggesting that sialic acids determine the biological activity of antibodies, acting as a switch between pro- and anti-inflammatory functions (Kaneko *et al.*, 2006). Deregulation of IgG sialylation may be involved in autoimmune pathology since levels of asialylated IgG are increased in rheumatoid arthritis and this alteration correlates with disease activity (Parekh *et al.*, 1985; Rook *et al.*, 1991). It is notable that sialylated IgG represent only 1–2% of IVIG, which may explain why large doses of IVIG are required to achieve a significant protection.

How can sialic acids change the function of IgG? Sialic acids can interact with endogenous lectins, such as sialic acid-binding immunoglobulin superfamily lectins (Siglecs), which can contain ITIM motifs and thereby inhibit signaling via ITAM-bearing Fcγ receptors as shown for the siglec CD33 expressed by myeloid cells (Crocker, 2002; Crocker *et al.*, 2007; Paul *et al.*, 2000). Alternatively, sialic acids could alter the interaction of IgG with activating Fc receptors so that these receptors propagate an inhibitory signal. It is now clear that ITAM, which were first characterized for their role in immune activation, can also convey suppressive signals (Hamerman and Lanier, 2006). For instance, the ITAM-bearing IgA receptor FcαR can inhibit activating Fcγ receptor (Pasquier *et al.*, 2005). Similarly, some form of signaling via ITAM-bearing CD3ζ can inhibit T cell activation as shown in studies with altered peptide ligands (De Magistris *et al.*, 1992; Sloan-Lancaster *et al.*, 1994). Inhibitory signals usually follow partial ITAM phosphorylation, while stimulatory signals are associated with complete phosphorylation (Kersh *et al.*, 1999; Madrenas *et al.*, 1995; Pasquier *et al.*, 2005; Sloan-Lancaster *et al.*, 1994).

These ITAM-mediated inhibitory signals can suppress diverse types of immune diseases (Franco *et al.*, 1994; Pasquier *et al.*, 2005). Inhibitory and stimulatory ligands usually differ by their affinity for the receptor. For instance, antagonist peptides are of lower affinity for TCR than agonists (Lyons *et al.*, 1996). Similarly, monomeric IgA triggers a suppressive signal via FcαR while multimeric IgA is stimulatory (Pasquier *et al.*, 2005). Thus, it is possible that sialylated IgG, which have a lower affinity for activating Fc receptors than unsialylated IgG (Kaneko *et al.*, 2006), trigger a suppressive signal via these receptors.

The models above suggest the existence of a distinct category of protective IgG distinguishable from pathogenic IgG by its sialic acid content. In contrast, Siragam and colleagues have demonstrated that activation of Fcγ signaling in DC with agonistic antibodies to Fcγ receptors or to a receptor that does not bind IgG but signals via this chain, generates cells that transfer protection against ITP as efficiently as DC treated with IVIG, which suggests that suppression is not restricted to a certain category of IgG (Siragam *et al.*, 2006). In the study of Siragam, the function of IgG primarily depends on the cell-type with which it interacts. How can we reconcile these observations? Do sialic acids facilitate the delivery of IC to DC via soluble siglecs? Explaining the role of sialic acids in the therapeutic activity of IVIG remains an important task to understand what defines protective IgG and how suppression of antibody-mediated inflammation is achieved.

3.6. Effects of IVIG on T cell responses

T cell-mediated inflammation certainly plays an important role in ITP. T cells can stimulate production of anti-platelet antibodies, and they can lyse platelets directly (Olsson *et al.*, 2003; Zhang *et al.*, 2006). Furthermore, ITP is typically associated with heightened production of T_H1 cytokines, which are potent activators of macrophages (Ogawara *et al.*, 2003; Semple, 2002; Semple and Freedman, 1991; Shimomura *et al.*, 1996). Only few studies so far have examined the effect of IVIG on T cells in ITP. Nevertheless, it seems that IVIG can suppress the dominant T_H1 cytokine profile (Mouzaki *et al.*, 2002). Furthermore, IVIG can suppress DC function and strengthen T cell-mediated regulation.

3.6.1. IVIG suppresses DC function

DC plays a crucial role in the initiation and programming of T cell responses (Banchereau and Steinman, 1998). DC from ITP patients express higher levels of co-stimulatory molecules and stimulate stronger T cell proliferative responses than cells from healthy controls, suggesting that DC are constantly stimulated in ITP patients (Catani *et al.*, 2006). In agreement with this notion, the gene expression signature of ITP is characterized by genes induced by type 1 IFN, which is produced by activated DC (Cella *et al.*, 1999; Siegal *et al.*, 1999; Sood *et al.*, 2008). IVIG inhibits the provision by DC of co-stimulatory molecules and cytokines important for T cell activation (Bayry *et al.*, 2003a,b). For instance, IVIG limits production of IL-12 and stimulates secretion of anti-inflammatory mediators such as IL-1 receptor antagonist and IL-10 (Bayry *et al.*, 2003a). This modulation of DC function is likely to down-regulate T cell-mediated immunity, and possibly also antibody production. DC can indeed directly drive B cell differentiation into plasma cell via type I IFN and IL-6 (Jego *et al.*, 2003). The effects of IVIG on DC are not well defined with regard to the receptors and intra-cellular signaling pathways involved.

3.6.2. IVIG stimulates T cell-mediated regulation

A recent study has examined the effect of IVIG in a model of T cell-mediated inflammation (Ephrem *et al.*, 2007). The model used was EAE induced in C57BL/6 mice by the immunodominant peptide from mouse myelin oligodendrocyte glycoprotein, which is a pathology strictly driven by T cells and independent of both B cells and antibodies (Bourquin *et al.*, 2003; Fillatreau *et al.*, 2002; Kuchroo *et al.*, 1993; Park *et al.*, 2005). Mice receiving IVIG were almost completely protected from EAE due to an increased frequency of $CD4^+$ $CD25^+$ $Foxp3^+$ regulatory T cells, and an augmentation of their suppressive activity (Ephrem *et al.*, 2007). In ITP, the percentage of regulatory T cells is significantly decreased during episodes of active disease compared to phases of remission and to healthy controls (Liu *et al.*, 2007). Thus, IVIG could ameliorate ITP by strengthening T cell-mediated regulation.

3.7. Effects of IVIG on macrophage activation

The activation status of macrophages, which are the main phagocytes of opsonized platelets, determines the pathogenicity of anti-platelet antibodies. Thus, inhibition of macrophages by IVIG significantly reduces antibody-mediated disease, whereas stimulation of their phagocytic activity by the T_H1 cytokine IFN-γ, which is produced at higher levels in ITP patients, exacerbates ITP (Park-Min *et al.*, 2007; Semple, 2002). Such aggravating effect of IFN-γ could account for the causal relationship between viral infections and childhood ITP (Musaji *et al.*, 2004; Rand and Wright, 1998). It is notable that IVIG suppresses the harmful effect of IFN-γ on antibody-mediated ITP. This protective function of IVIG is independent of FcγRIIB but requires activating Fc receptor whose triggering directly decreases IFN-γ receptor expression by DC, macrophages, and NK cells (Park-Min *et al.*, 2007). Inhibition of IFN-γ signaling requires prolonged exposure to IVIG, implying a role for sustained signaling.

Transduction of suppressive signals upon prolonged ITAM-signaling has been documented for the B cell receptor that signals via the ITAM-containing molecules Igα and Igβ (Healy and Goodnow, 1998).

3.8. Conclusions and perspectives

IVIG is an effective treatment for ITP that certainly suppresses disease by limiting the pathogenic potential of self-reactive antibodies, inhibition of DC hyper-activation, strengthening of T cell-mediated immune regulation, and blocking of innate immune signaling via IFN-γ. Figure 1.1 recapitulates the suppressive mechanisms of IVIG discussed in this review. Kaneko and colleagues have proposed that sialylated IgG are the

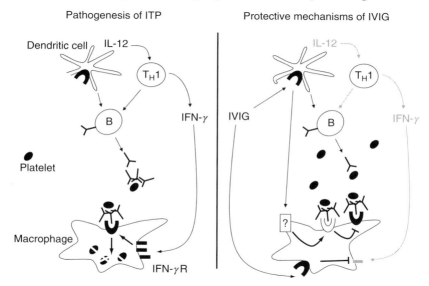

FIGURE 1.1 Protective mechanisms of IVIG in antibody-mediated ITP. Anti-platelet antibodies are a primary cause of platelet loss in ITP. Opsonized platelets are captured via activating Fc receptors (dark horseshoe) by macrophages and destroyed intra-cellularly. T cells contribute to pathogenesis by facilitating the differentiation of B cells into antibody-producing plasma cells, and by secreting the inflammatory cytokine IFN-γ, which is a potent macrophage activator capable of catalyzing the phagocytosis of osponized platelets. They can also lyse platelets directly (not shown). IVIG engages several protective mechanisms via DC and macrophages. IVIG limits production of IL-12 by DC, thereby dampening T$_H$1 inflammation. IVIG also suppresses IFN-γ signaling by down-regulating expression of IFN-γ receptor on effector macrophages. In addition, IVIG stimulates via DC expression of the inhibitory Fc receptor FcγRIIB (white horseshoe) on effector macrophages, which directly inhibits activating Fc receptors. This mechanism is essential for IVIG-mediated suppression of antibody-mediated inflammation.

components of IVIG involved in suppression of antibody-mediated inflammation. However, other protective functions of IVIG such as inhibition of macrophages activation by IFN-γ are independent of FcγRIIB and probably also of sialylated IgG. Accordingly, IVIG should be considered as a sum of modules, in which multiple components have additive effects that synergize to protect the host against infections and autoimmune pathology. Such composite nature could complicate the production of universal IVIG substitutes. A possible side-effect of IVIG substitutes could be to increase susceptibility to infections. Because IVIG inhibits IFN-γ signaling, which is crucial for anti-microbial immunity, Park-Min and colleagues have tested whether IVIG impairs host defense against the Gram-positive bacteria *Listeria monocytogenes*. Indeed, mice treated with IVIG failed to restrain *Listeria monocytogenes* infection, similarly to IFN-γ receptor-deficient mice (Park-Min *et al.*, 2007). Although infections are not a reported side-effect of IVIG in human (Katz *et al.*, 2007), these findings indicate that global immune suppression could be a risk for substitutes that only mimic the suppressive components of IVIG, while IVIG also contains IgG that protect against infections, as shown by its beneficial effects in immunodeficient patients.

IVIG is made of IgG naturally present in serum of healthy individuals, so that its immune regulating components are almost certainly present in the natural antibody repertoire. Indeed, a small amount of IC is found in normal serum. A category of antibodies susceptible of making IC is polyreactive IgG, which interact efficiently with a broad range of self-antigens (Tiller *et al.*, 2007). Polyreactive IgG could bind to serum proteins such as albumin, or transferrin, or to RBC, and thereby produce small IC having the suppressive properties of IVIG or large IC having the suppressive properties of anti-D, respectively. Can polyreactive IgG protect against autoantibody-mediated pathogenesis? Suppression of antibody-mediated inflammation by IVIG is mediated at least partly by antigen non-specific mechanisms, suggesting that suppressive IgG also inhibit humoral immunity against microbes. Interestingly, Kaneko and colleagues have shown that IgG sialylation diminishes during active immune responses (Kaneko *et al.*, 2006). Thus, this regulatory circuit may be inhibited during infections to allow effective immune defense, which may contribute to the detrimental effect of some infections on immunopathology.

4. IMMUNE REGULATING FUNCTIONS OF NAÏVE AND ACTIVATED B CELLS

Treatment of autoimmune diseases by non-specific immune suppressive drugs always carries a risk of favoring infections. Several MS patients treated with the integrin blocker Natalizumab, which impairs leukocyte infiltration into the CNS, developed progressive multifocal leukoence-phalopathy (PML), a severe disease of the brain caused by reactivation of polyomavirus (2005; Langer-Gould *et al.*, 2005). Thus, antigen-specific therapies remain highly desirable for treatment of autoimmune diseases. Provision of autoantigen in soluble form can induce antigen-specific tolerance and protection from experimental autoimmune diseases (Anderton, 2001). However, this approach can provoke fatal side-effects when applied during an ongoing disease (Pedotti *et al.*, 2001, 2003; Smith *et al.*, 2005). A peptide-based trial in MS patients was interrupted due to risk of anaphylaxis (Kappos *et al.*, 2000). Alternatively, specific tolerance can be achieved by administration of autoantigen-presenting cells, a strategy that does not bear the risks of anaphylaxis observed with soluble peptides (Smith *et al.*, 2005). B cells are particularly tolerogenic antigen-presenting cells (APC), suggesting that they could be of therapeutic value against autoimmune diseases. Here, we review data highlighting the tolerogenic properties of naïve and activated B cells, and their potential as cell-based therapy for autoimmune diseases.

4.1. Naïve B cells and tolerance

Naïve B cells express low levels of MHC and co-stimulatory molecules, and they are poor activators of naïve T cells (Lassila *et al.*, 1988; Ronchese and Hausmann, 1993). Upon stimulation by resting B cells, naïve T cells undergo a phase of minimal expansion followed by extensive cell death and induction of hypo-responsiveness in the remaining cells (Croft *et al.*, 1997). Thus, naïve B cells can purge the T cell repertoire of unwanted specificities. Mason and colleagues have designed a system to target auto-antigen specifically to naïve B cells *in vivo* and tested its ability to prevent autoimmune disease. Their strategy involves the construction of fusion proteins linking autoantigens to an antibody specific for IgD, a receptor specifically expressed on naïve B cells (Eynon and Parker, 1992). Such targeting of autoantigen to naïve B cells protects rats from EAE (Day *et al.*, 1992). This treatment is also effective when applied shortly after EAE induction. Thus, naïve B cells can prevent the differentiation of T cells into drivers of pathogenesis, and also control the pathogenic poten-tial of primed T cells. Disease suppression correlates with a switch in cytokine production by T cells from IFN-γ towards IL-4 and IL-13, and a reduction of leukocyte infiltration into the spinal cord (Saoudi *et al.*, 1995).

4.2. Cytokine-mediated regulation by activated B cells

Following activation, B cells display elevated levels of MHC and co-stimulatory molecules (Cassell and Schwartz, 1994; Kennedy *et al.*, 1994). As a result, they can stimulate proliferation, survival, and differentiation of T cells (Cassell and Schwartz, 1994; Malynn *et al.*, 1985). However, activated B cells can also act as suppressor cells driving resolution of autoimmune pathologies. For instance, they control recovery from EAE (Fillatreau *et al.*, 2002; Wolf *et al.*, 1996), and inhibit disease progression in CIA and UC (Mauri *et al.*, 2003; Mizoguchi *et al.*, 2002). How do activated B cells protect from autoimmune pathology?

Activated B cells can produce distinct arrays of cytokines (Harris *et al.*, 2000). Some B cells, called Be1 cells, secrete cytokines typically associated with T_H1 responses such as IFN-γ and IL-12, while Be2 cells secrete the cytokine IL-4 characteristic of T_H2 cells (Harris *et al.*, 2000). In addition, activated B cells can secrete large amounts of IL-6 and IL-10 that play instrumental roles in T cell-mediated immunity (Lampropoulou *et al.*, 2008). IL-6 is a crucial co-stimulator for T cell responses. It releases naïve $CD4^+$ T cells from the inhibitory activity of $CD4^+$ $CD25^+$ T regulatory cells, and protects activated $CD4^+$ T cells from apoptosis (Pasare and Medzhitov, 2003; Rochman *et al.*, 2005). Consistently, IL-6-deficient mice are resistant to T cell-dependent autoimmune pathologies such as EAE or CIA (Alonzi *et al.*, 1998; Samoilova *et al.*, 1998). On the opposite, IL-10 suppresses immune reactions. IL-10-deficient mice suffer a chronic EAE, and a more severe CIA (Bettelli *et al.*, 1998; Finnegan *et al.*, 2003). They also spontaneously develop inflammatory bowel disease (IBD) (Kuhn *et al.*, 1993). The function of B cells in autoimmune diseases could depend on the cytokines they produce. Indeed, the suppressive capacity of activated B cells in EAE, CIA, and IBD was found to typically involve their production of IL-10 (Fillatreau *et al.*, 2002; Mauri *et al.*, 2003; Mizoguchi *et al.*, 2002). Thus, mice in which only B cells cannot secrete IL-10 develop a chronic EAE while mice with wild-type B cells recover after a short episode of paralysis (Fillatreau *et al.*, 2002).

B cell-derived IL-10 suppresses autoimmune pathologies of different etiologies: EAE is primarily driven by inflammatory T cells of T_H1 and T_H17 types (Kuchroo *et al.*, 1993; Park *et al.*, 2005), UC is caused by T_H2 cells (Iijima *et al.*, 1999), and CIA involves a dominant T_H1 response and requires pathogenic antibodies (Svensson *et al.*, 1998). How do IL-10-producing B cells regulate these autoimmune diseases? IL-10 strongly inhibits DC and macrophages (Moore *et al.*, 2001), and IL-10 from B cells directly inhibits the capacity of CpG-activated DC to stimulate T cell proliferation *in vitro* (Lampropoulou *et al.*, 2008). Similarly, Byrne and colleagues have shown that B cells activated by ultraviolet irradiation produce IL-10 and inhibit DC function (Byrne and Halliday, 2005; Matsumura *et al.*, 2006). B cell-derived IL-10 also represses production of IL-6 and IL-12 by DC, which can inhibit differentiation of T_H1 and T_H17 cells, respectively (Lampropoulou *et al.*, 2008). In agreement with these data, DC from B cell-deficient mice produce higher amounts of IL-12 than DC from wild-type mice after immunization, and this deviates the T cell response towards T_H1 (Moulin *et al.*, 2000). Thus, suppression could involve paralysis of innate immunity following interactions of IL-10-producing B cells with DC and macrophages. IL-10-producing B cells may also regulate immune responses by stimulating regulatory $CD4^+$ and $CD8^+$ T cells as discussed in detail in other reviews (Fillatreau and Anderton, 2007; Mizoguchi and Bhan, 2006).

4.3. Signals driving IL-10 secretion by B cells

Identification of the receptors triggering a regulatory function in B cells is important because it could allow therapeutic manipulation of this differentiation pathway in diseases. BCR and CD40 have crucial roles because primed B cells secrete IL-10 upon re-stimulation via BCR and CD40, and B cells require activation through these receptors to allow remission from EAE (Fillatreau *et al.*, 2002). Similarly, B cells from arthritic mice manifested a suppressive activity following re-stimulation via CD40 *in vitro* (Mauri *et al.*, 2003). Co-receptors of BCR regulate the amounts of IL-10 released. For instance, B cells lacking CD19 secrete reduced quantities of IL-10 and have a reduced capacity to suppress EAE (Matsushita *et al.*, 2006). During actively induced autoimmune diseases such as EAE or CIA, the BCR agonists controlling the regulatory function of B cells could be antigens contained in the immunization cocktail or antigenic debris released as a result of tissue destruction in the target organ. Several cell types could supply B cells with CD40L. $CD4^+$ T cells are likely candidates. Alternatively, invariant T cells restricted to CD1d and MR1 non-polymorphic MHC-Ib molecules also express CD40L and suppress autoimmune diseases by increasing IL-10 secretion by B cells (Croxford *et al.*, 2006; Lehuen *et al.*, 1998; Miyamoto *et al.*, 2001).

It is notable that signals provided via BCR and CD40 are also involved in germinal centre formation and plasma cell differentiation. Considering it unlikely that all activated B cells produce IL-10, these data suggest that other pathways control the initial cell fate decision driving a naive B cell to produce IL-10. Indeed, signals delivered via CD1d initiate B cell programming for IL-10 production in IBD (Mizoguchi *et al.*, 2002). In that model, B cells also require MHC-II for suppression, but the role of this pathway is to expand the initial population of IL-10-producing B cells rather than to prime B cells for IL-10-production (Mizoguchi *et al.*, 2002). Altogether these data suggest a two-step model for the establishment of B cell-mediated regulation. In a first phase, receptors such as CD1d initiate IL-10 production by B cells. However, this probably produces too few IL-10-producing B cells for effective regulation of disease. In a second phase, engagement of receptors classically involved in B cell survival and expansion, such as CD40, amplifies the initial population of IL-10-producing B cells and boosts their cytokine secretion, thereby allowing more effective suppression. This second phase may preferentially expand B cells specific for autoantigen if autoreactive CD4$^+$ T cells are the main source of CD40L, which would establish a negative feedback circuit between pathogenic T cells and B cell-mediated regulation. Other signals can be involved in the regulatory function of B cells. For instance, apoptotic cells increase the levels of IL-10 produced by B cells, and injection of apoptotic cells has a regulatory effect on CIA that is mediated by IL-10 and B cells (Gray *et al.*, 2007). Which signals initiate IL-10 production by B cells during EAE and CIA?

4.4. TLR-activated B cells and regulation of EAE

Naïve B cells do not secrete IL-10 upon co-stimulation through BCR and CD40, implying the need of a different signal (Lampropoulou *et al.*, 2008). This signal can be provided by Toll-like receptor (TLR) agonists such as lipopolysaccharides (LPS) from Gram-negative bacteria or CpG oligonucleotides, which are potent inducers of IL-10 production by naïve B cells (Lampropoulou *et al.*, 2008). TLR control the regulatory function of B cells during EAE because mice with B cell-specific deletions of *myd88* or both *tlr-2* and *tlr-4* do not recover from disease (Lampropoulou *et al.*, 2008). Altogether, these data suggest that, similarly to CD1d, TLR can initiate a regulatory function in naïve B cells. Remarkably, MyD88-deficient B cells can form germinal centers and mount antigen-specific antibody responses, indicating that a primary role of MyD88-signalling in B cells may be initiation of their suppressive properties. This function of MyD88 is highly cell-type specific (Barr *et al.*, 2007; Lampropoulou *et al.*, 2008). Although B cells activated via TLR produce a cytokine milieu that inhibits T cell activation in an IL-10-dependent manner *in vitro*, DC activated in the same way secrete little IL-10 and provide a milieu enhancing T cell proliferation (Lampropoulou *et al.*, 2008). Furthermore, MyD88-deficient mice do not develop EAE, demonstrating that MyD88 signaling in different cells sequentially controls initiation, progression, and resolution of autoimmune disease (Lampropoulou *et al.*, 2008; Prinz *et al.*, 2006).

4.5. Do TLR-activated B cells always mediate a suppressive function?

Evidence is accumulating that only certain TLR, including TLR-2 and TLR-4, can stimulate a regulatory function in B cells, while TLR-9 seems ineffective. For instance, TLR-2 and TLR-4, but not TLR-9, are required on B cells for recovery from EAE (Lampropoulou *et al.*, 2008). Similarly, Scott and colleagues have shown that TLR-4- but not TLR-9-activated B cells are tolerogenic in an adoptive transfer model of suppression (Lei *et al.*, 2005). Thus, only specific TLR-agonists can trigger a regulatory function in B cells. Such TLR-activated B cells probably also require being in a specific cellular microenvironment to achieve suppression since supernatants from LPS-activated B cells can either suppress or stimulate T cell activation depending on the presence of DC in the culture (Lampropoulou *et al.*, 2008). This dual effect of LPS-activated B cells was related to the fact that they do not only produce IL-10, which suppresses DC, but also IL-6, which stimulates T cells (Lampropoulou *et al.*, 2008). Thus, DC may be required as a partner for suppression. Interestingly, DC exposed to IL-10 and the TLR-agonist LPS produce the B cell-attracting chemokine CXCL13 (Perrier *et al.*, 2004), suggesting that TLR-activated B cells and DC can enter a self-amplified regulatory circuit leading to suppression of DC function and further recruitment of B lymphocytes. B cells and CXCL-13-expressing myeloid cells co-localize in inflammatory lesions from patients with rheumatoid arthritis or IBD (Carlsen *et al.*, 2004). This may represent an attempt to limit local inflammation. Collectively, these observations indicate that only under specific circumstances will TLR-activated B cells have a suppressive activity, despite that these receptors are commonly involved in immune responses. Further clarification of theses aspects should help to reconcile activated B cells as effectors and regulators.

4.6. Therapeutic potential of TLR-activated B cells

It is remarkable that TLR-activated B cells can suppress an ongoing pathology such as EAE and mediate recovery from disease. TLR-stimulated B cells express multiple immune regulating molecules such as IL-10, FasL, and TGF-β (Tian *et al.*, 2001), indicating broad suppressive functions that could provide an effective cell-based therapy for autoimmune diseases. Indeed, LPS-activated B cells protect NOD mice from diabetes upon adoptive transfer (Tian *et al.*, 2001), and LPS-activated B cells retrovirally transduced to express a myelin autoantigen protect mice from EAE induced by that antigen (Chen *et al.*, 2001). Scott and colleagues have compared the tolerogenic capacities of naïve and TLR-activated B cells using a transgenic mouse that constitutively expresses a relevant antigen. Notably, they found that TLR-activated B cells suppress an ongoing immune response, which naïve B cells are unable to regulate (Zambidis *et al.*, 1997). Thus, B cells acquire a remarkable capacity to inhibit T cell-mediated inflammation upon TLR-activation. Could they be used as cell-based therapy for autoimmune disease? Preliminary results are encouraging: TLR-activated B cells transduced to express the relevant antigen abrogate EAE progression when injected after remission from the first clinical episode in a relapsing-remitting disease model (Chen *et al.*, 2004).

Scott and colleagues have optimized the form of autoantigen expressed by retrovirally transduced TLR-activated B cells to obtain better suppression. They constructed fusion proteins linking autoantigen to IgG1 because immunoglobulin carriers are potent tolerogens (Borel, 1980; Pike *et al.*, 1981; Scott *et al.*, 1979; Zambidis and Scott, 1996). LPS-activated B cells expressing an antigen-IgG1 fusion protein tolerized the relevant B and T cells more effectively than B cells expressing the non chimeric form of antigen (Kang *et al.*, 1999). Furthermore, suppression lasted longer when antigen was fused to IgG1 (Kang *et al.*, 1999). In this system, suppression was independent of IL-10 but antigen presentation was crucial as B cells lacking MHC-II or B7.2 co-stimulatory molecule were not suppressive (El-Amine *et al.*, 2000; Litzinger *et al.*, 2005). Suppression was also abolished by blocking anti-CTLA-4 antibodies, indicating a key role for B7.2-CTLA-4 interaction (El-Amine *et al.*, 2000).

Scott and colleagues have tested their approach in several autoimmune models, including experimental autoimmune uveitis (EAU) (Agarwal *et al.*, 2000), EAE (Melo *et al.*, 2002), type I diabetes (Melo *et al.*, 2002), and adjuvant-induced arthritis (Satpute *et al.*, 2007). EAU is a T cell-mediated autoimmune disease of the neural retina that can be induced either by immunization with retinal antigen, or by adoptive transfer of activated antigen-reactive T cells. Remarkably, a single administration of LPS-activated B cells expressing the retinal antigen fused to IgG1 is sufficient to protect mice against EAU induction for more than 2 months (Agarwal *et al.*, 2000). Protection results from a suppression of both T and B cell immunity towards the retinal antigen. Such vaccination protocol can also inhibit disease in animals primed by immunization or by transfer of activated T cells as long as the B cells are injected before manifestation of clinical signs. Similarly, LPS-activated B cells expressing a chimeric myelin antigen-IgG1 fusion protein can prevent EAE induction, and even suppress EAE progression when administered shortly after appearance of the first clinical signs (Melo *et al.*, 2002). Notably, suppression was in that model restricted to the antigen presented by the B cell blasts. Nonetheless, the approach of Scott and colleagues was effective at preventing spontaneous diabetes in NOD mice. NOD mice spontaneously develop an autoimmune response to glutamic acid decarboxylase (GAD) 65 and insulin by 3–7 weeks of age, and destructive islet infiltration is usually observable at 10–12 weeks of age. Injection of LPS-activated B cell blasts retrovirally transduced to express GAD-IgG1 or insulin-IgG1 fusion proteins into 7–12 weeks NOD mice decreased incidence of diabetes, delayed its onset, and prolonged survival (Melo *et al.*, 2002; Song *et al.*, 2004). Thus, TLR-activated B cells expressing key antigens can suppress a complex and spontaneous autoimmune disease. Protection was in that case associated with a prolonged increase of $CD4^+$ $CD25^+$ $CD62L^+$ T cell numbers (Song *et al.*, 2004; Soukhareva *et al.*, 2006). These studies certainly provide an interesting ground for further explorations of the therapeutic potential of TLR-activated transduced B cells.

4.7. Conclusions and perspectives

TLR-activated B cells provide a new option as tolerogenic APC. B cells present the advantage of being easily accessible in large numbers in peripheral blood. During the last years, many studies have assessed the possibility of suppressing autoimmune diseases with DC, based on the concept that immature DC are tolerogenic (Hawiger et al., 2001). However, this approach does not seem straightforward because injection of DC can suffice to trigger autoimmune symptoms, although it can protect from autoimmune disease in some cases (Huang et al., 2000; Weir et al., 2002). Stabilizing DC in a tolerogenic phenotype may be difficult due to their high sensitivity to microbial products and inflammatory mediators, a property crucial for their role as sentinels for stressed tissues and pathogens. Furthermore, isolation and ex vivo manipulation can mature DC (Steinman et al., 2003). In contrast, B cells are less sensitive to stress signals. Altogether, TLR-activated B cells constitute an interesting alternative to immature DC for suppressing autoimmune diseases, even if using an activated APC to induce tolerance appears counter-intuitive at first glance.

The participation of TLR-activated B cells to resolution of disease in EAE suggests that B cells link microbial recognition to suppression of autoimmune diseases. Indeed, components of Mycobacterium tuberculosis present in the complete Freund's adjuvant used to induce disease provide the agonists of TLR-2 and/or TLR-4 triggering a suppressive function in B cells in that model (Lampropoulou et al., 2008). It is notable that TLR-2 and TLR-4 control remission from EAE by B cells but are not necessary for disease induction. In contrast, TLR-9 contributes to disease severity but does not participate to the recovery process (Lampropoulou et al., 2008; Prinz et al., 2006). Thus, Mycobacterium tuberculosis contains distinct types of TLR-agonists exerting opposite effects on clinical outcome. Some preferentially trigger a regulatory function in B cells, while others stimulate the innate immune system and the pathogenic process. It is conceivable that microbes enriched in agonists stimulating a suppressive function in B cells can protect from autoimmune diseases. Mycobacterium tuberculosis can indeed suppress autoimmune diseases. An inverse correlation has been observed between exposure to Mycobacterium tuberculosis and MS incidence (Andersen et al., 1981). Furthermore, MS patients treated with a Mycobacterium bovis strain bacillus Calmette-Guérin (BCG) vaccine showed a 51% reduction in CNS lesions measured by magnetic resonance imaging (Ristori et al., 1999). Helminths can also stimulate IL-10 production by B cells and thereby limit immunopathology. For instance, infection with Schistosoma mansoni induces IL-10 production by B cells, which protects mice from anaphylaxis (Mangan et al., 2004; Velupillai and Harn, 1994). These observations collectively integrate B cells within the framework of the "hygiene hypothesis", which proposes that microbial exposure can prevent immune-mediated pathologies (Bach, 2002). Thus, reduced exposure to microbes such as mycobacteria and helminths will cause a loss of B cell-mediated regulation, and thereby facilitate the onset of immunopathology.

ACKNOWLEDGMENTS

We thank David Gray for critical reading of the manuscript. SF is supported by the Deutsche Forschungsgemeinschaft grant no. SFB650, Association pour la Recherche sur la Sclerose en Plaques (ARSEP), and the Hertie Stiftung.

REFERENCES

(1969) Prevention of primary Rh immunization: First report of the Western Canadian trial, 1966–1968. *Can. Med. Assoc. J.* **100**, 1021–1024.

(1971) Prevention of Rh-haemolytic disease: Final results of the "high-risk" clinical trial. A combined study from centres in England and Baltimore. *Br. Med. J.* **2**, 607–609.

(1999) Primary immunodeficiency diseases. Report of an IUIS Scientific Committee. International Union of Immunological Societies. *Clin. Exp. Immunol.* **118**(Suppl 1), 1–28.

(2005) MS drug withdrawn from market. *FDA. Consum.* **39**, 3.

Adib, M., Ragimbeau, J., Avrameas, S., and Ternynck, T. (1990). IgG autoantibody activity in normal mouse serum is controlled by IgM. *J. Immunol.* **145**, 3807–3813.

Agarwal, R. K., Kang, Y., Zambidis, E., Scott, D. W., Chan, C. C., and Caspi, R. R. (2000). Retroviral gene therapy with an immunoglobulin-antigen fusion construct protects from experimental autoimmune uveitis. *J. Clin. Invest.* **106**, 245–252.

Agematsu, K., Futatani, T., Hokibara, S., Kobayashi, N., Takamoto, M., Tsukada, S., Suzuki, H., Koyasu, S., Miyawaki, T., Sugane, K., Komiyama, A., and Ochs, H. D. (2002). Absence of memory B cells in patients with common variable immunodeficiency. *Clin. Immunol.* **103**, 34–42.

Alonzi, T., Fattori, E., Lazzaro, D., Costa, P., Probert, L., Kollias, G., De, B. F., Poli, V., and Ciliberto, G. (1998). Interleukin 6 is required for the development of collagen-induced arthritis. *J. Exp. Med.* **187**, 461–468.

Andersen, E., Isager, H., and Hyllested, K. (1981). Risk factors in multiple sclerosis: Tuberculin reactivity, age at measles infection, tonsillectomy and appendectomy. *Acta Neurol. Scand.* **63**, 131–135.

Anderton, S. M. (2001). Peptide-based immunotherapy of autoimmunity: A path of puzzles, paradoxes and possibilities. *Immunology* **104**, 367–376.

Arnold, J. N., Wormald, M. R., Sim, R. B., Rudd, P. M., and Dwek, R. A. (2007). The impact of glycosylation on the biological function and structure of human immunoglobulins. *Annu. Rev. Immunol.* **25**, 21–50.

Asakura, K., Miller, D. J., Pease, L. R., and Rodriguez, M. (1998). Targeting of IgMkappa antibodies to oligodendrocytes promotes CNS remyelination. *J. Neurosci.* **18**, 7700–7708.

Augener, W., Friedmann, B., and Brittinger, G. (1985). Are aggregates of IgG the effective part of high-dose immunoglobulin therapy in adult idiopathic thrombocytopenic purpura (ITP)? *Blut* **50**, 249–252.

Bach, J. F. (2002). The effect of infections on susceptibility to autoimmune and allergic diseases. *N. Engl. J. Med.* **347**, 911–920.

Banchereau, J., and Steinman, R. M. (1998). Dendritic cells and the control of immunity. *Nature* **392**, 245–252.

Barr, T. A., Brown, S., Ryan, G., Zhao, J., and Gray, D. (2007). TLR-mediated stimulation of APC: Distinct cytokine responses of B cells and dendritic cells. *Eur. J. Immunol.* **37**, 3040–3053.

Bayry, J., Lacroix-Desmazes, S., Carbonneil, C., Misra, N., Donkova, V., Pashov, A., Chevailler, A., Mouthon, L., Weill, B., Bruneval, P., Kazatchkine, M. D., and Kaveri, S. V. (2003a). Inhibition of maturation and function of dendritic cells by intravenous immunoglobulin. *Blood* **101**, 758–765.

Bayry, J., Lacroix-Desmazes, S., Delignat, S., Mouthon, L., Weill, B., Kazatchkine, M. D., and Kaveri, S. V. (2003b). Intravenous immunoglobulin abrogates dendritic cell differentiation induced by interferon-alpha present in serum from patients with systemic lupus erythematosus. *Arthritis Rheum.* **48**, 3497–3502.

Bayry, J., Kazatchkine, M. D., and Kaveri, S. V. (2007). Shortage of human intravenous immunoglobulin--reasons and possible solutions. *Nat. Clin. Pract. Neurol.* **3**, 120–121.

Bazin, R., Lemieux, R., Tremblay, T., and St-Amour, I. (2004). Tetramolecular immune complexes are more efficient than IVIg to prevent antibody-dependent *in vitro* and *in vivo* phagocytosis of blood cells. *Br. J. Haematol.* **127**, 90–96.

Bennett, C. M., Rogers, Z. R., Kinnamon, D. D., Bussel, J. B., Mahoney, D. H., Abshire, T. C., Sawaf, H., Moore, T. B., Loh, M. L., Glader, B. E., McCarthy, M. C., Mueller, B. U., *et al.* (2006). Prospective phase 1/2 study of rituximab in childhood and adolescent chronic immune thrombocytopenic purpura. *Blood* **107**, 2639–2642.

Bettelli, E., Das, M. P., Howard, E. D., Weiner, H. L., Sobel, R. A., and Kuchroo, V. K. (1998). IL-10 is critical in the regulation of autoimmune encephalomyelitis as demonstrated by studies of IL-10- and IL-4-deficient and transgenic mice. *J. Immunol.* **161**, 3299–3306.

Bieber, A. J., Warrington, A., Pease, L. R., and Rodriguez, M. (2001). Humoral autoimmunity as a mediator of CNS repair. *Trends Neurosci.* **24**, S39–S44.

Boes, M., Schmidt, T., Linkemann, K., Beaudette, B. C., Marshak-Rothstein, A., and Chen, J. (2000). Accelerated development of IgG autoantibodies and autoimmune disease in the absence of secreted IgM. *Proc. Natl. Acad. Sci. USA* **97**, 1184–1189.

Bolland, S., and Ravetch, J. V. (2000). Spontaneous autoimmune disease in Fc(gamma)RIIB-deficient mice results from strain-specific epistasis. *Immunity* **13**, 277–285.

Borel, Y. (1980). Haptens bound to self IgG induce immunologic tolerance, while when coupled to syngeneic spleen cells they induce immune suppression. *Immunol. Rev.* **50**, 71–104.

Botto, M., Dell'Agnola, C., Bygrave, A. E., Thompson, E. M., Cook, H. T., Petry, F., Loos, M., Pandolfi, P. P., and Walport, M. J. (1998). Homozygous C1q deficiency causes glomerulonephritis associated with multiple apoptotic bodies. *Nat. Genet.* **19**, 56–59.

Bouaziz, J. D., Yanaba, K., Venturi, G. M., Wang, Y., Tisch, R. M., Poe, J. C., and Tedder, T. F. (2007). Therapeutic B cell depletion impairs adaptive and autoreactive CD4+ T cell activation in mice. *Proc. Natl. Acad. Sci. USA* **104**, 20878–20883.

Bouchard, C., Fridman, W. H., and Sautes, C. (1994). Mechanism of inhibition of lipopolysaccharide-stimulated mouse B-cell responses by transforming growth factor-beta 1. *Immunol. Lett.* **40**, 105–110.

Bouchard, C., Galinha, A., Tartour, E., Fridman, W. H., and Sautes, C. (1995). A transforming growth factor beta-like immunosuppressive factor in immunoglobulin G-binding factor. *J. Exp. Med.* **182**, 1717–1726.

Bourquin, C., Schubart, A., Tobollik, S., Mather, I., Ogg, S., Liblau, R., and Linington, C. (2003). Selective unresponsiveness to conformational B cell epitopes of the myelin oligodendrocyte glycoprotein in H-2b mice. *J. Immunol.* **171**, 455–461.

Bowman, J. M. (1988). The prevention of Rh immunization. *Transfus. Med. Rev.* **2**, 129–150.

Bowman, J. M. (1998). RhD hemolytic disease of the newborn. *N. Engl. J. Med.* **339**, 1775–1777.

Bowman, J. M., and Pollock, J. M. (1987). Failures of intravenous Rh immune globulin prophylaxis: An analysis of the reasons for such failures. *Transfus. Med. Rev.* **1**, 101–112.

Brighton, T. A., Evans, S., Castaldi, P. A., Chesterman, C. N., and Chong, B. H. (1996). Prospective evaluation of the clinical usefulness of an antigen-specific assay (MAIPA) in idiopathic thrombocytopenic purpura and other immune thrombocytopenias. *Blood* **88**, 194–201.

Brinc, D., Le-Tien, H., Crow, A. R., Freedman, J., and Lazarus, A. H. (2007). IgG-mediated immunosuppression is not dependent on erythrocyte clearance or immunological evasion: Implications for the mechanism of action of anti-D in the prevention of haemolytic disease of the newborn? *Br. J. Haematol.* **139**, 275–279.

Bruhns, P., Samuelsson, A., Pollard, J. W., and Ravetch, J. V. (2003). Colony-stimulating factor-1-dependent macrophages are responsible for IVIG protection in antibody-induced autoimmune disease. *Immunity* **18**, 573–581.

Burdach, S. E., Evers, K. G., and Geursen, R. G. (1986). Treatment of acute idiopathic thrombocytopenic purpura of childhood with intravenous immunoglobulin G: Comparative efficacy of 7S and 5S preparations. *J. Pediatr.* **109**, 770–775.

Bussel, J. (2006). Treatment of immune thrombocytopenic purpura in adults. *Semin. Hematol.* **43**, S3–10.

Bussel, J. B., and Pham, L. C. (1987). Intravenous treatment with gammaglobulin in adults with immune thrombocytopenic purpura: Review of the literature. *Vox Sang.* **52**, 206–211.

Byrne, S. N., and Halliday, G. M. (2005). B cells activated in lymph nodes in response to ultraviolet irradiation or by interleukin-10 inhibit dendritic cell induction of immunity. *J. Invest. Dermatol.* **124**, 570–578.

Carlsen, H. S., Baekkevold, E. S., Morton, H. C., Haraldsen, G., and Brandtzaeg, P. (2004). Monocyte-like and mature macrophages produce CXCL13 (B cell-attracting chemokine 1) in inflammatory lesions with lymphoid neogenesis. *Blood* **104**, 3021–3027.

Cassell, D. J., and Schwartz, R. H. (1994). A quantitative analysis of antigen-presenting cell function: Activated B cells stimulate naive CD4 T cells but are inferior to dendritic cells in providing costimulation. *J. Exp. Med.* **180**, 1829–1840.

Catani, L., Fagioli, M. E., Tazzari, P. L., Ricci, F., Curti, A., Rovito, M., Preda, P., Chirumbolo, G., Amabile, M., Lemoli, R. M., Tura, S., Conte, R., *et al.* (2006). Dendritic cells of immune thrombocytopenic purpura (ITP) show increased capacity to present apoptotic platelets to T lymphocytes. *Exp. Hematol.* **34**, 879–887.

Cella, M., Jarrossay, D., Facchetti, F., Alebardi, O., Nakajima, H., Lanzavecchia, A., and Colonna, M. (1999). Plasmacytoid monocytes migrate to inflamed lymph nodes and produce large amounts of type I interferon. *Nat. Med.* **5**, 919–923.

Chan, O. T., Hannum, L. G., Haberman, A. M., Madaio, M. P., and Shlomchik, M. J. (1999). A novel mouse with B cells but lacking serum antibody reveals an antibody-independent role for B cells in murine lupus. *J. Exp. Med.* **189**, 1639–1648.

Chen, C., Rivera, A., Ron, N., Dougherty, J. P., and Ron, Y. (2001). A gene therapy approach for treating T-cell-mediated autoimmune diseases. *Blood* **97**, 886–894.

Chen, C. C., Rivera, A., Dougherty, J. P., and Ron, Y. (2004). Complete protection from relapsing experimental autoimmune encephalomyelitis induced by syngeneic B cells expressing the autoantigen. *Blood* **103**, 4616–4618.

Choi, C. W., Kim, B. S., Seo, J. H., Shin, S. W., Kim, Y. H., Kim, J. S., Sohn, S. K., Kim, J. S., Shin, D. G., Ryoo, H. M., Lee, K. H., Lee, J. J., *et al.* (2001). Response to high-dose intravenous immune globulin as a valuable factor predicting the effect of splenectomy in chronic idiopathic thrombocytopenic purpura patients. *Am. J. Hematol.* **66**, 197–202.

Cines, D. B., and Blanchette, V. S. (2002). Immune thrombocytopenic purpura. *N. Engl. J. Med.* **346**, 995–1008.

Cines, D. B., and Bussel, J. B. (2005). How I treat idiopathic thrombocytopenic purpura (ITP). *Blood* **106**, 2244–2251.

Clarke, C. A., Donohoe, W. T., McConnell, R. B., Woodrow, J. C., Finn, R., Krevans, J. R., Kulke, W., Lehane, D., and Sheppard, P. M. (1963). Further experimental studies on the prevention of Rh haemolytic disease. *Br. Med. J.* **1,** 979–984.

Clarkson, S. B., Bussel, J. B., Kimberly, R. P., Valinsky, J. E., Nachman, R. L., and Unkeless, J. C. (1986). Treatment of refractory immune thrombocytopenic purpura with an anti-Fc gamma-receptor antibody. *N. Engl. J. Med.* **314,** 1236–1239.

Cohen, Y. C., Djulbegovic, B., Shamai-Lubovitz, O., and Mozes, B. (2000). The bleeding risk and natural history of idiopathic thrombocytopenic purpura in patients with persistent low platelet counts. *Arch. Intern. Med.* **160,** 1630–1638.

Cooper, N., Heddle, N. M., Haas, M., Reid, M. E., Lesser, M. L., Fleit, H. B., Woloski, B. M., and Bussel, J. B. (2004). Intravenous (IV) anti-D and IV immunoglobulin achieve acute platelet increases by different mechanisms: Modulation of cytokine and platelet responses to IV anti-D by FcgammaRIIa and FcgammaRIIIa polymorphisms. *Br. J. Haematol.* **124,** 511–518.

Crocker, P. R. (2002). Siglecs: Sialic-acid-binding immunoglobulin-like lectins in cell-cell interactions and signalling. *Curr. Opin. Struct. Biol.* **12,** 609–615.

Crocker, P. R., Paulson, J. C., and Varki, A. (2007). Siglecs and their roles in the immune system. *Nat. Rev. Immunol.* **7,** 255–266.

Croft, M., Joseph, S. B., and Miner, K. T. (1997). Partial activation of naive CD4 T cells and tolerance induction in response to peptide presented by resting B cells. *J. Immunol.* **159,** 3257–3265.

Cross, A. H., Stark, J. L., Lauber, J., Ramsbottom, M. J., and Lyons, J. A. (2006). Rituximab reduces B cells and T cells in cerebrospinal fluid of multiple sclerosis patients. *J. Neuroimmunol.* **180,** 63–70.

Crow, A. R., Song, S., Freedman, J., Helgason, C. D., Humphries, R. K., Siminovitch, K. A., and Lazarus, A. H. (2003). IVIg-mediated amelioration of murine ITP via FcgammaRIIB is independent of SHIP1, SHP-1, and Btk activity. *Blood* **102,** 558–560.

Croxford, J. L., Miyake, S., Huang, Y. Y., Shimamura, M., and Yamamura, T. (2006). Invariant V(alpha)19i T cells regulate autoimmune inflammation. *Nat. Immunol.* **7,** 987–994.

Cunningham-Rundles, C., and Bodian, C. (1999). Common variable immunodeficiency: Clinical and immunological features of 248 patients. *Clin. Immunol.* **92,** 34–48.

Cunningham-Rundles, C., and Ponda, P. P. (2005). Molecular defects in T- and B-cell primary immunodeficiency diseases. *Nat. Rev. Immunol.* **5,** 880–892.

Day, M. J., Tse, A. G., Puklavec, M., Simmonds, S. J., and Mason, D. W. (1992). Targeting autoantigen to B cells prevents the induction of a cell-mediated autoimmune disease in rats. *J. Exp. Med.* **175,** 655–659.

De Magistris, M. T., Alexander, J., Coggeshall, M., Altman, A., Gaeta, F. C., Grey, H. M., and Sette, A. (1992). Antigen analog-major histocompatibility complexes act as antagonists of the T cell receptor. *Cell* **68,** 625–634.

Deng, R., and Balthasar, J. P. (2005). Investigation of antibody-coated liposomes as a new treatment for immune thrombocytopenia. *Int. J. Pharm.* **304,** 51–62.

Dietrich, G., Algiman, M., Sultan, Y., Nydegger, U. E., and Kazatchkine, M. D. (1992). Origin of anti-idiotypic activity against anti-factor VIII autoantibodies in pools of normal human immunoglobulin G (IVIg). *Blood* **79,** 2946–2951.

Dilillo, D. J., Hamaguchi, Y., Ueda, Y., Yang, K., Uchida, J., Haas, K. M., Kelsoe, G., and Tedder, T. F. (2008). Maintenance of Long-Lived Plasma Cells and Serological Memory Despite Mature and Memory B Cell Depletion during CD20 Immunotherapy in Mice. *J. Immunol.* **180,** 361–371.

Duddy, M. E., Alter, A., and Bar-Or, A. (2004). Distinct profiles of human B cell effector cytokines: A role in immune regulation? *J. Immunol.* **172,** 3422–3427.

Duddy, M., Niino, M., Adatia, F., Hebert, S., Freedman, M., Atkins, H., Kim, H. J., and Bar-Or, A. (2007). Distinct effector cytokine profiles of memory and naive human B cell subsets and implication in multiple sclerosis. *J. Immunol.* **178**, 6092–6099.

Ehrenstein, M. R., Cook, H. T., and Neuberger, M. S. (2000). Deficiency in serum immunoglobulin (Ig)M predisposes to development of IgG autoantibodies. *J. Exp. Med.* **191**, 1253–1258.

El-Amine, M., Melo, M., Kang, Y., Nguyen, H., Qian, J., and Scott, D. W. (2000). Mechanisms of tolerance induction by a gene-transferred peptide-IgG fusion protein expressed in B lineage cells. *J. Immunol.* **165**, 5631–5636.

Ephrem, A., Misra, N., Hassan, G., Dasgupta, S., Delignat, S., Van Huyen, J. P., Chamat, S., Prost, F., Lacroix-Desmazes, S., Kavery, S. V., and Kazatchkine, M. D. (2005). Immunomodulation of autoimmune and inflammatory diseases with intravenous immunoglobulin. *Clin. Exp. Med.* **5**, 135–140.

Ephrem, A., Chamat, S., Miquel, C., Fisson, S., Mouthon, L., Caligiuri, G., Delignat, S., Elluru, S., Bayry, J., Lacroix-Desmazes, S., Cohen, J. L., Salomon, B. L., *et al.* (2007). Expansion of CD4+ CD25+ regulatory T cells by intravenous immunoglobulin: A critical factor in controlling experimental autoimmune encephalomyelitis. *Blood* **111**(2), 715–722.

Eynon, E. E., and Parker, D. C. (1992). Small B cells as antigen-presenting cells in the induction of tolerance to soluble protein antigens. *J. Exp. Med.* **175**, 131–138.

Fehr, J., Hofmann, V., and Kappeler, U. (1982). Transient reversal of thrombocytopenia in idiopathic thrombocytopenic purpura by high-dose intravenous gamma globulin. *N. Engl. J. Med.* **306**, 1254–1258.

Fillatreau, S., and Anderton, S. M. (2007). B-cell function in CNS inflammatory demyelinating disease: A complexity of roles and a wealth of possibilities. *Expert Rev. Clin. Immunol.* **3**, 565–578.

Fillatreau, S., Sweenie, C. H., McGeachy, M. J., Gray, D., and Anderton, S. M. (2002). B cells regulate autoimmunity by provision of IL-10. *Nat. Immunol.* **3**, 944–950.

Finnegan, A., Kaplan, C. D., Cao, Y., Eibel, H., Glant, T. T., and Zhang, J. (2003). Collagen-induced arthritis is exacerbated in IL-10-deficient mice. *Arthritis Res. Ther.* **5**, R18–R24.

Fletcher, A., and Thomson, A. (1995). The introduction of human monoclonal anti-D for therapeutic use. *Transfus. Med. Rev.* **9**, 314–326.

Franco, A., Southwood, S., Arrhenius, T., Kuchroo, V. K., Grey, H. M., Sette, A., and Ishioka, G. Y. (1994). T cell receptor antagonist peptides are highly effective inhibitors of experimental allergic encephalomyelitis. *Eur. J. Immunol.* **24**, 940–946.

Freda, V. J., Gorman, J. G., and Pollack, W. (1964). Successful prevention of experimental Rh sensitization in man with an anti-Rh gamma2-globulin antibody preparation: A preliminary report. *Transfusion* **4**, 26–32.

Godeau, B., Oksenhendler, E., Brossard, Y., Bartholeyns, J., Leaute, J. B., Duedari, N., Schaeffer, A., and Bierling, P. (1996). Treatment of chronic autoimmune thrombocytopenic purpura with monoclonal anti-D. *Transfusion* **36**, 328–330.

Gray, M., Miles, K., Salter, D., Gray, D., and Savill, J. (2007). Apoptotic cells protect mice from autoimmune inflammation by the induction of regulatory B cells. *Proc. Natl. Acad. Sci. USA* **104**, 14080–14085.

Hamerman, J. A., and Lanier, L. L. (2006). Inhibition of immune responses by ITAM-bearing receptors. *Sci. STKE.* 2006, re1.

Hansen, R. J., and Balthasar, J. P. (2002a). Effects of intravenous immunoglobulin on platelet count and antiplatelet antibody disposition in a rat model of immune thrombocytopenia. *Blood* **100**, 2087–2093.

Hansen, R. J., and Balthasar, J. P. (2002b). Intravenous immunoglobulin mediates an increase in anti-platelet antibody clearance via the FcRn receptor. *Thromb. Haemost.* **88**, 898–899.

Harrington, W. J., Minnich, V., Hollingsworth, J. W., and Moore, C. V. (1951). Demonstration of a thrombocytopenic factor in the blood of patients with thrombocytopenic purpura. *J. Lab Clin. Med.* **38,** 1–10.

Harris, D. P., Haynes, L., Sayles, P. C., Duso, D. K., Eaton, S. M., Lepak, N. M., Johnson, L. L., Swain, S. L., and Lund, F. E. (2000). Reciprocal regulation of polarized cytokine production by effector B and T cells. *Nat. Immunol.* **1,** 475–482.

Hawiger, D., Inaba, K., Dorsett, Y., Guo, M., Mahnke, K., Rivera, M., Ravetch, J. V., Steinman, R. M., and Nussenzweig, M. C. (2001). Dendritic cells induce peripheral T cell unresponsiveness under steady state conditions *in vivo. J. Exp. Med.* **194,** 769–779.

Healy, J. I., and Goodnow, C. C. (1998). Positive versus negative signaling by lymphocyte antigen receptors. *Annu. Rev. Immunol.* **16,** 645–670.

Hoyer, B. F., Moser, K., Hauser, A. E., Peddinghaus, A., Voigt, C., Eilat, D., Radbruch, A., Hiepe, F., and Manz, R. A. (2004). Short-lived plasmablasts and long-lived plasma cells contribute to chronic humoral autoimmunity in NZB/W mice. *J. Exp. Med.* **199,** 1577–1584.

Huang, Y. M., Yang, J. S., Xu, L. Y., Link, H., and Xiao, B. G. (2000). Autoantigen-pulsed dendritic cells induce tolerance to experimental allergic encephalomyelitis (EAE) in Lewis rats. *Clin. Exp. Immunol.* **122,** 437–444.

Hurez, V., Kaveri, S. V., and Kazatchkine, M. D. (1993). Expression and control of the natural autoreactive IgG repertoire in normal human serum. *Eur. J. Immunol.* **23,** 783–789.

Iijima, H., Takahashi, I., Kishi, D., Kim, J. K., Kawano, S., Hori, M., and Kiyono, H. (1999). Alteration of interleukin 4 production results in the inhibition of T helper type 2 cell-dominated inflammatory bowel disease in T cell receptor alpha chain-deficient mice. *J. Exp. Med.* **190,** 607–615.

Imbach, P., Barandun, S., d'Apuzzo, V., Baumgartner, C., Hirt, A., Morell, A., Rossi, E., Schoni, M., Vest, M., and Wagner, H. P. (1981). High-dose intravenous gammaglobulin for idiopathic thrombocytopenic purpura in childhood. *Lancet* **1,** 1228–1231.

Imbach, P., Tani, P., Berchtold, W., Blanchette, V., Burek-Kozlowska, A., Gerber, H., Jacobs, P., Newland, A., Turner, C., Wood, L., and McMillan, R. (1991). Different forms of chronic childhood thrombocytopenic purpura defined by antiplatelet autoantibodies. *J. Pediatr.* **118,** 535–539.

Ivan, E., and Colovai, A. I. (2006). Human Fc receptors: Critical targets in the treatment of autoimmune diseases and transplant rejections. *Hum. Immunol.* **67,** 479–491.

Jego, G., Palucka, A. K., Blanck, J. P., Chalouni, C., Pascual, V., and Banchereau, J. (2003). Plasmacytoid dendritic cells induce plasma cell differentiation through type I interferon and interleukin 6. *Immunity* **19,** 225–234.

Jones, N. C., Mollison, P. L., and Veall, N. (1957). Removal of incompatible red cells by the spleen. *Br. J. Haematol.* **3,** 125–133.

Jude, B. A., Pobezinskaya, Y., Bishop, J., Parke, S., Medzhitov, R. M., Chervonsky, A. V., and Golovkina, T. V. (2003). Subversion of the innate immune system by a retrovirus. *Nat. Immunol.* **4,** 573–578.

Kaneko, Y., Nimmerjahn, F., and Ravetch, J. V. (2006). Anti-inflammatory activity of immunoglobulin G resulting from Fc sialylation. *Science* **313,** 670–673.

Kang, Y., Melo, M., Deng, E., Tisch, R., El-Amine, M., and Scott, D. W. (1999). Induction of hyporesponsiveness to intact foreign protein via retroviral-mediated gene expression: The IgG scaffold is important for induction and maintenance of immune hyporesponsiveness. *Proc. Natl. Acad. Sci. USA* **96,** 8609–8614.

Kappos, L., Comi, G., Panitch, H., Oger, J., Antel, J., Conlon, P., and Steinman, L. (2000). Induction of a non-encephalitogenic type 2 T helper-cell autoimmune response in multiple sclerosis after administration of an altered peptide ligand in a placebo-controlled, randomized phase II trial. *The Altered Peptide Ligand in Relapsing MS Study Group. Nat. Med.* **6,** 1176–1182.

Karlsson, M. C., Wernersson, S., az de, S. T., Gustavsson, S., and Heyman, B. (1999). Efficient IgG-mediated suppression of primary antibody responses in Fcgamma receptor-deficient mice. *Proc. Natl. Acad. Sci. USA* **96,** 2244–2249.

Karlsson, M. C., Getahun, A., and Heyman, B. (2001). FcgammaRIIB in IgG-mediated suppression of antibody responses: Different impact *in vivo* and *in vitro*. *J. Immunol.* **167,** 5558–5564.

Katz, U., Achiron, A., Sherer, Y., and Shoenfeld, Y. (2007). Safety of intravenous immunoglobulin (IVIG) therapy. *Autoimmun. Rev.* **6,** 257–259.

Kazatchkine, M. D., and Kaveri, S. V. (2001). Immunomodulation of autoimmune and inflammatory diseases with intravenous immune globulin. *N. Engl. J. Med.* **345,** 747–755.

Kennedy, M. K., Mohler, K. M., Shanebeck, K. D., Baum, P. R., Picha, K. S., Otten-Evans, C. A., Janeway, C. A., Jr, and Grabstein, K. H. (1994). Induction of B cell costimulatory function by recombinant murine CD40 ligand. *Eur. J. Immunol.* **24,** 116–123.

Kersh, E. N., Kersh, G. J., and Allen, P. M. (1999). Partially phosphorylated T cell receptor zeta molecules can inhibit T cell activation. *J. Exp. Med.* **190,** 1627–1636.

Kiefel, V., Freitag, E., Kroll, H., Santoso, S., and Mueller-Eckhardt, C. (1996). Platelet autoantibodies (IgG, IgM, IgA) against glycoproteins IIb/IIIa and Ib/IX in patients with thrombocytopenia. *Ann. Hematol.* **72,** 280–285.

Kim, S. J., Gershov, D., Ma, X., Brot, N., and Elkon, K. B. (2002). I-PLA(2) activation during apoptosis promotes the exposure of membrane lysophosphatidylcholine leading to binding by natural immunoglobulin M antibodies and complement activation. *J. Exp. Med.* **196,** 655–665.

Kimberly, R. P., Salmon, J. E., Bussel, J. B., Crow, M. K., and Hilgartner, M. W. (1984). Modulation of mononuclear phagocyte function by intravenous gamma-globulin. *J. Immunol.* **132,** 745–750.

Kjaersgaard, M., Aslam, R., Kim, M., Speck, E. R., Freedman, J., Stewart, D. I., Wiersma, E. J., and Semple, J. W. (2007). Epitope specificity and isotype of monoclonal anti-D antibodies dictate their ability to inhibit phagocytosis of opsonized platelets. *Blood* **110,** 1359–1361.

Korganow, A. S., Ji, H., Mangialaio, S., Duchatelle, V., Pelanda, R., Martin, T., Degott, C., Kikutani, H., Rajewsky, K., Pasquali, J. L., Benoist, C., and Mathis, D. (1999). From systemic T cell self-reactivity to organ-specific autoimmune disease via immunoglobulins. *Immunity* **10,** 451–461.

Kuchroo, V. K., Martin, C. A., Greer, J. M., Ju, S. T., Sobel, R. A., and Dorf, M. E. (1993). Cytokines and adhesion molecules contribute to the ability of myelin proteolipid protein-specific T cell clones to mediate experimental allergic encephalomyelitis. *J. Immunol.* **151,** 4371–4382.

Kuhn, R., Lohler, J., Rennick, D., Rajewsky, K., and Muller, W. (1993). Interleukin-10-deficient mice develop chronic enterocolitis. *Cell* **75,** 263–274.

Kumpel, B. M. (2002). Monoclonal anti-D development programme. *Transpl. Immunol.* **10,** 199–204.

Kumpel, B. M., and Elson, C. J. (2001). Mechanism of anti-D-mediated immune suppression–a paradox awaiting resolution? *Trends Immunol.* **22,** 26–31.

Kumpel, B. M., Goodrick, M. J., Pamphilon, D. H., Fraser, I. D., Poole, G. D., Morse, C., Standen, G. R., Chapman, G. E., Thomas, D. P., and Anstee, D. J. (1995). Human Rh D monoclonal antibodies (BRAD-3 and BRAD-5) cause accelerated clearance of Rh D+ red blood cells and suppression of Rh D immunization in Rh D– volunteers. *Blood* **86,** 1701–1709.

Lamoureux, J., Aubin, E., and Lemieux, R. (2003). Autoimmune complexes in human serum in presence of therapeutic amounts of intravenous immunoglobulins. *Blood* **101,** 1660–1662.

Lamoureux, J., Aubin, E., and Lemieux, R. (2004). Autoantibodies purified from therapeutic preparations of intravenous immunoglobulins (IVIg) induce the formation of autoimmune complexes in normal human serum: A role in the *in vivo* mechanisms of action of IVIg? *Int. Immunol.* **16**, 929–936.

Lampropoulou, V., Hoehlig, K., Roch, T., Neves, P., Calderon-Gomez, E., Sweenie, C. H., Hao, Y., Freitas, A. A., Steinhoff, U., Anderton, S. M., and Fillatreau, S. (2008). Toll-like receptor-activated B cells suppress T cell-mediated autoimmunity. *J. Immunol.* **180**, 4763–4773.

Langer-Gould, A., Atlas, S. W., Green, A. J., Bollen, A. W., and Pelletier, D. (2005). Progressive multifocal leukoencephalopathy in a patient treated with natalizumab. *N. Engl. J. Med.* **353**, 375–381.

Lassila, O., Vainio, O., and Matzinger, P. (1988). Can B cells turn on virgin T cells? *Nature* **334**, 253–255.

Lehuen, A., Lantz, O., Beaudoin, L., Laloux, V., Carnaud, C., Bendelac, A., Bach, J. F., and Monteiro, R. C. (1998). Overexpression of natural killer T cells protects Valpha14-Jalpha281 transgenic nonobese diabetic mice against diabetes. *J. Exp. Med.* **188**, 1831–1839.

Lei, T. C., Su, Y., and Scott, D. W. (2005). Tolerance induction via a B-cell delivered gene therapy-based protocol: Optimization and role of the Ig scaffold. *Cell Immunol.* **235**, 12–20.

Litzinger, M. T., Su, Y., Lei, T. C., Soukhareva, N., and Scott, D. W. (2005). Mechanisms of gene therapy for tolerance: B7 signaling is required for peptide-IgG gene-transferred tolerance induction. *J. Immunol.* **175**, 780–787.

Liu, B., Zhao, H., Poon, M. C., Han, Z., Gu, D., Xu, M., Jia, H., Yang, R., and Han, Z. C. (2007). Abnormality of CD4(+)CD25(+) regulatory T cells in idiopathic thrombocytopenic purpura. *Eur. J. Haematol.* **78**, 139–143.

Lyons, D. S., Lieberman, S. A., Hampl, J., Boniface, J. J., Chien, Y., Berg, L. J., and Davis, M. M. (1996). A TCR binds to antagonist ligands with lower affinities and faster dissociation rates than to agonists. *Immunity* **5**, 53–61.

MacDonald, K. P., Rowe, V., Bofinger, H. M., Thomas, R., Sasmono, T., Hume, D. A., and Hill, G. R. (2005). The colony-stimulating factor 1 receptor is expressed on dendritic cells during differentiation and regulates their expansion. *J. Immunol.* **175**, 1399–1405.

Madrenas, J., Wange, R. L., Wang, J. L., Isakov, N., Samelson, L. E., and Germain, R. N. (1995). Zeta phosphorylation without ZAP-70 activation induced by TCR antagonists or partial agonists. *Science* **267**, 515–518.

Malynn, B. A., Romeo, D. T., and Wortis, H. H. (1985). Antigen-specific B cells efficiently present low doses of antigen for induction of T cell proliferation. *J. Immunol.* **135**, 980–988.

Mangan, N. E., Fallon, R. E., Smith, P., van, R. N., McKenzie, A. N., and Fallon, P. G. (2004). Helminth infection protects mice from anaphylaxis via IL-10-producing B cells. *J. Immunol.* **173**, 6346–6356.

Martin, F., and Chan, A. C. (2006). B cell immunobiology in disease: Evolving concepts from the clinic. *Annu. Rev. Immunol.* **24**, 467–496.

Matejtschuk, P., Chidwick, K., Prince, A., More, J. E., and Goldblatt, D. (2002). A direct comparison of the antigen-specific antibody profiles of intravenous immunoglobulins derived from US and UK donor plasma. *Vox Sang.* **83**, 17–22.

Matsumura, Y., Byrne, S. N., Nghiem, D. X., Miyahara, Y., and Ullrich, S. E. (2006). A role for inflammatory mediators in the induction of immunoregulatory B cells. *J. Immunol.* **177**, 4810–4817.

Matsushita, T., Fujimoto, M., Hasegawa, M., Komura, K., Takehara, K., Tedder, T. F., and Sato, S. (2006). Inhibitory role of CD19 in the progression of experimental autoimmune encephalomyelitis by regulating cytokine response. *Am. J. Pathol.* **168**, 812–821.

Mauri, C., Gray, D., Mushtaq, N., and Londei, M. (2003). Prevention of arthritis by interleukin 10-producing B cells. *J. Exp. Med.* **197**, 489–501.

McMillan, R. (1997). Therapy for adults with refractory chronic immune thrombocytopenic purpura. *Ann. Intern. Med.* **126**, 307–314.

Melo, M. E., Qian, J., El-Amine, M., Agarwal, R. K., Soukhareva, N., Kang, Y., and Scott, D. W. (2002). Gene transfer of Ig-fusion proteins into B cells prevents and treats autoimmune diseases. *J. Immunol.* **168,** 4788–4795.

Mimura, Y., Church, S., Ghirlando, R., Ashton, P. R., Dong, S., Goodall, M., Lund, J., and Jefferis, R. (2000). The influence of glycosylation on the thermal stability and effector function expression of human IgG1-Fc: Properties of a series of truncated glycoforms. *Mol. Immunol.* **37,** 697–706.

Mimura, Y., Sondermann, P., Ghirlando, R., Lund, J., Young, S. P., Goodall, M., and Jefferis, R. (2001). Role of oligosaccharide residues of IgG1-Fc in Fc gamma RIIb binding. *J. Biol. Chem.* **276,** 45539–45547.

Miyamoto, K., Miyake, S., and Yamamura, T. (2001). A synthetic glycolipid prevents auto-immune encephalomyelitis by inducing TH2 bias of natural killer T cells. *Nature* **413,** 531–534.

Mizoguchi, A., and Bhan, A. K. (2006). A case for regulatory B cells. *J. Immunol.* **176,** 705–710.

Mizoguchi, A., Mizoguchi, E., Smith, R. N., Preffer, F. I., and Bhan, A. K. (1997). Suppressive role of B cells in chronic colitis of T cell receptor alpha mutant mice. *J. Exp. Med.* **186,** 1749–1756.

Mizoguchi, A., Mizoguchi, E., Takedatsu, H., Blumberg, R. S., and Bhan, A. K. (2002). Chronic intestinal inflammatory condition generates IL-10-producing regulatory B cell subset characterized by CD1d upregulation. *Immunity* **16,** 219–230.

Moore, K. W., de Waal, M. R., Coffman, R. L., and O'Garra, A. (2001). Interleukin-10 and the interleukin-10 receptor. *Annu. Rev. Immunol.* **19,** 683–765.

Moulin, V., Andris, F., Thielemans, K., Maliszewski, C., Urbain, J., and Moser, M. (2000). B lymphocytes regulate dendritic cell (DC) function *in vivo*: Increased interleukin 12 production by DCs from B cell-deficient mice results in T helper cell type 1 deviation. *J. Exp. Med.* **192,** 475–482.

Mouzaki, A., Theodoropoulou, M., Gianakopoulos, I., Vlaha, V., Kyrtsonis, M. C., and Maniatis, A. (2002). Expression patterns of Th1 and Th2 cytokine genes in childhood idiopathic thrombocytopenic purpura (ITP) at presentation and their modulation by intravenous immunoglobulin G (IVIg) treatment: Their role in prognosis. *Blood* **100,** 1774–1779.

Musaji, A., Cormont, F., Thirion, G., Cambiaso, C. L., and Coutelier, J. P. (2004). Exacerbation of autoantibody-mediated thrombocytopenic purpura by infection with mouse viruses. *Blood* **104,** 2102–2106.

Musaji, A., Meite, M., Detalle, L., Franquin, S., Cormont, F., Preat, V., Izui, S., and Coutelier, J. P. (2005). Enhancement of autoantibody pathogenicity by viral infections in mouse models of anemia and thrombocytopenia. *Autoimmun. Rev.* **4,** 247–252.

Nakahara, J., Tan-Takeuchi, K., Seiwa, C., Gotoh, M., Kaifu, T., Ujike, A., Inui, M., Yagi, T., Ogawa, M., Aiso, S., Takai, T., and Asou, H. (2003). Signaling via immunoglobulin Fc receptors induces oligodendrocyte precursor cell differentiation. *Dev. Cell* **4,** 841–852.

Ogawara, H., Handa, H., Morita, K., Hayakawa, M., Kojima, J., Amagai, H., Tsumita, Y., Kaneko, Y., Tsukamoto, N., Nojima, Y., and Murakami, H. (2003). High Th1/Th2 ratio in patients with chronic idiopathic thrombocytopenic purpura. *Eur. J. Haematol.* **71,** 283–288.

Olsson, B., Andersson, P. O., Jernas, M., Jacobsson, S., Carlsson, B., Carlsson, L. M., and Wadenvik, H. (2003). T-cell-mediated cytotoxicity toward platelets in chronic idiopathic thrombocytopenic purpura. *Nat. Med.* **9,** 1123–1124.

Ozsoylu, S., Karabent, A., Irken, G., and Tuncer, M. (1991). Antiplatelet antibodies in childhood idiopathic thrombocytopenic purpura. *Am. J. Hematol.* **36,** 82–85.

Parekh, R. B., Dwek, R. A., Sutton, B. J., Fernandes, D. L., Leung, A., Stanworth, D., Rademacher, T. W., Mizuochi, T., Taniguchi, T., Matsuta, K., *et al.* (1985). Association of rheumatoid arthritis and primary osteoarthritis with changes in the glycosylation pattern of total serum IgG. *Nature* **316,** 452–457.

Park, H., Li, Z., Yang, X. O., Chang, S. H., Nurieva, R., Wang, Y. H., Wang, Y., Hood, L., Zhu, Z., Tian, Q., and Dong, C. (2005). A distinct lineage of CD4 T cells regulates tissue inflammation by producing interleukin 17. *Nat. Immunol.* **6,** 1133–1141.

Park-Min, K. H., Serbina, N. V., Yang, W., Ma, X., Krystal, G., Neel, B. G., Nutt, S. L., Hu, X., and Ivashkiv, L. B. (2007). FcgammaRIII-dependent inhibition of interferon-gamma responses mediates suppressive effects of intravenous immune globulin. *Immunity* **26,** 67–78.

Pasare, C., and Medzhitov, R. (2003). Toll pathway-dependent blockade of CD4+ CD25+ T cell-mediated suppression by dendritic cells. *Science* **299,** 1033–1036.

Pasquier, B., Launay, P., Kanamaru, Y., Moura, I. C., Pfirsch, S., Ruffie, C., Henin, D., Benhamou, M., Pretolani, M., Blank, U., and Monteiro, R. C. (2005). Identification of FcalphaRI as an inhibitory receptor that controls inflammation: Dual role of FcRgamma ITAM. *Immunity* **22,** 31–42.

Paul, S. P., Taylor, L. S., Stansbury, E. K., and McVicar, D. W. (2000). Myeloid specific human CD33 is an inhibitory receptor with differential ITIM function in recruiting the phosphatases SHP-1 and SHP-2. *Blood* **96,** 483–490.

Pearse, R. N., Kawabe, T., Bolland, S., Guinamard, R., Kurosaki, T., and Ravetch, J. V. (1999). SHIP recruitment attenuates Fc gamma RIIB-induced B cell apoptosis. *Immunity* **10,** 753–760.

Pedotti, R., Mitchell, D., Wedemeyer, J., Karpuj, M., Chabas, D., Hattab, E. M., Tsai, M., Galli, S. J., and Steinman, L. (2001). An unexpected version of horror autotoxicus: Anaphylactic shock to a self-peptide. *Nat. Immunol.* **2,** 216–222.

Pedotti, R., Sanna, M., Tsai, M., DeVoss, J., Steinman, L., McDevitt, H., and Galli, S. J. (2003). Severe anaphylactic reactions to glutamic acid decarboxylase (GAD) self peptides in NOD mice that spontaneously develop autoimmune type 1 diabetes mellitus. *BMC. Immunol.* **4,** 2.

Pendergrast, J. M., Sher, G. D., and Callum, J. L. (2005). Changes in intravenous immunoglobulin prescribing patterns during a period of severe product shortages, 1995–2000. *Vox Sang.* **89,** 150–160.

Perrier, P., Martinez, F. O., Locati, M., Bianchi, G., Nebuloni, M., Vago, G., Bazzoni, F., Sozzani, S., Allavena, P., and Mantovani, A. (2004). Distinct transcriptional programs activated by interleukin-10 with or without lipopolysaccharide in dendritic cells: Induction of the B cell-activating chemokine, CXC chemokine ligand 13. *J. Immunol.* **172,** 7031–7042.

Phillips, N. E., and Parker, D. C. (1984). Cross-linking of B lymphocyte Fc gamma receptors and membrane immunoglobulin inhibits anti-immunoglobulin-induced blastogenesis. *J. Immunol.* **132,** 627–632.

Pike, B. E., Battye, F. L., and Nossal, G. J. (1981). Effect of hapten valency and carrier composition on the tolerogenic potential of hapten-protein conjugates. *J. Immunol.* **126,** 89–94.

Prinz, M., Garbe, F., Schmidt, H., Mildner, A., Gutcher, I., Wolter, K., Piesche, M., Schroers, R., Weiss, E., Kirschning, C. J., Rochford, C. D., Bruck, W., *et al.* (2006). Innate immunity mediated by TLR9 modulates pathogenicity in an animal model of multiple sclerosis. *J. Clin. Invest.* **116,** 456–464.

Rand, M. L., and Wright, J. F. (1998). Virus-associated idiopathic thrombocytopenic purpura. *Transfus. Sci.* **19,** 253–259.

Ravetch, J. V., and Lanier, L. L. (2000). Immune inhibitory receptors. *Science* **290,** 84–89.

Reff, M. E., Carner, K., Chambers, K. S., Chinn, P. C., Leonard, J. E., Raab, R., Newman, R. A., Hanna, N., and Anderson, D. R. (1994). Depletion of B cells *in vivo* by a chimeric mouse human monoclonal antibody to CD20. *Blood* **83,** 435–445.

Ristori, G., Buzzi, M. G., Sabatini, U., Giugni, E., Bastianello, S., Viselli, F., Buttinelli, C., Ruggieri, S., Colonnese, C., Pozzilli, C., and Salvetti, M. (1999). Use of Bacille Calmette-Guerin (BCG) in multiple sclerosis. *Neurology* **53,** 1588–1589.

Rochman, I., Paul, W. E., and Ben-Sasson, S. Z. (2005). IL-6 increases primed cell expansion and survival. *J. Immunol.* **174,** 4761–4767.

Ronchese, F., and Hausmann, B. (1993). B lymphocytes *in vivo* fail to prime naive T cells but can stimulate antigen-experienced T lymphocytes. *J. Exp. Med.* **177,** 679–690.

Rook, G. A., Steele, J., Brealey, R., Whyte, A., Isenberg, D., Sumar, N., Nelson, J. L., Bodman, K. B., Young, A., Roitt, I. M., *et al.* (1991). Changes in IgG glycoform levels are associated with remission of arthritis during pregnancy. *J. Autoimmun.* **4,** 779–794.

Roopenian, D. C., Christianson, G. J., Sproule, T. J., Brown, A. C., Akilesh, S., Jung, N., Petkova, S., Avanessian, L., Choi, E. Y., Shaffer, D. J., Eden, P. A., and Anderson, C. L. (2003). The MHC class I-like IgG receptor controls perinatal IgG transport, IgG homeostasis, and fate of IgG-Fc-coupled drugs. *J. Immunol.* **170,** 3528–3533.

Rowley, D. A., and Stach, R. M. (1998). B lymphocytes secreting IgG linked to latent transforming growth factor-beta prevent primary cytolytic T lymphocyte responses. *Int. Immunol.* **10,** 355–363.

Salama, A., Mueller-Eckhardt, C., and Kiefel, V. (1983). Effect of intravenous immunoglobulin in immune thrombocytopenia. *Lancet* **2,** 193–195.

Salama, A., Kiefel, V., Amberg, R., and Mueller-Eckhardt, C. (1984). Treatment of autoimmune thrombocytopenic purpura with rhesus antibodies [anti-Rh0(D)]. *Blut* **49,** 29–35.

Salama, A., Kiefel, V., and Mueller-Eckhardt, C. (1986). Effect of IgG anti-Rho(D) in adult patients with chronic autoimmune thrombocytopenia. *Am. J. Hematol.* **22,** 241–250.

Samoilova, E. B., Horton, J. L., Hilliard, B., Liu, T. S., and Chen, Y. (1998). IL-6-deficient mice are resistant to experimental autoimmune encephalomyelitis: Roles of IL-6 in the activation and differentiation of autoreactive T cells. *J. Immunol.* **161,** 6480–6486.

Samuelsson, A., Towers, T. L., and Ravetch, J. V. (2001). Anti-inflammatory activity of IVIG mediated through the inhibitory Fc receptor. *Science* **291,** 484–486.

Saoudi, A., Simmonds, S., Huitinga, I., and Mason, D. (1995). Prevention of experimental allergic encephalomyelitis in rats by targeting autoantigen to B cells: Evidence that the protective mechanism depends on changes in the cytokine response and migratory properties of the autoantigen-specific T cells. *J. Exp. Med.* **182,** 335–344.

Satpute, S. R., Soukhareva, N., Scott, D. W., and Moudgil, K. D. (2007). Mycobacterial Hsp65-IgG-expressing tolerogenic B cells confer protection against adjuvant-induced arthritis in Lewis rats. *Arthritis Rheum.* **56,** 1490–1496.

Scaradavou, A., Woo, B., Woloski, B. M., Cunningham-Rundles, S., Ettinger, L. J., Aledort, L. M., and Bussel, J. B. (1997). Intravenous anti-D treatment of immune thrombocytopenic purpura: Experience in 272 patients. *Blood* **89,** 2689–2700.

Scott, D. W., Venkataraman, M., and Jandinski, J. J. (1979). Multiple pathways of B lymphocyte tolerance. *Immunol. Rev.* **43,** 241–280.

Scott, R. S., McMahon, E. J., Pop, S. M., Reap, E. A., Caricchio, R., Cohen, P. L., Earp, H. S., and Matsushima, G. K. (2001). Phagocytosis and clearance of apoptotic cells is mediated by MER. *Nature* **411,** 207–211.

Semple, J. W. (2002). Immune pathophysiology of autoimmune thrombocytopenic purpura. *Blood Rev.* **16,** 9–12.

Semple, J. W., and Freedman, J. (1991). Increased antiplatelet T helper lymphocyte reactivity in patients with autoimmune thrombocytopenia. *Blood* **78,** 2619–2625.

Shimomura, T., Fujimura, K., Takafuta, T., Fujii, T., Katsutani, S., Noda, M., Fujimoto, T., and Kuramoto, A. (1996). Oligoclonal accumulation of T cells in peripheral blood from patients with idiopathic thrombocytopenic purpura. *Br. J. Haematol.* **95,** 732–737.

Siegal, F. P., Kadowaki, N., Shodell, M., Fitzgerald-Bocarsly, P. A., Shah, K., Ho, S., Antonenko, S., and Liu, Y. J. (1999). The nature of the principal type 1 interferon-producing cells in human blood. *Science* **284,** 1835–1837.

Siragam, V., Brinc, D., Crow, A. R., Song, S., Freedman, J., and Lazarus, A. H. (2005). Can antibodies with specificity for soluble antigens mimic the therapeutic effects of intravenous IgG in the treatment of autoimmune disease? *J. Clin. Invest* **115,** 155–160.

Siragam, V., Crow, A. R., Brinc, D., Song, S., Freedman, J., and Lazarus, A. H. (2006). Intravenous immunoglobulin ameliorates ITP via activating Fc gamma receptors on dendritic cells. *Nat. Med.* **12,** 688–692.

Sloan-Lancaster, J., Shaw, A. S., Rothbard, J. B., and Allen, P. M. (1994). Partial T cell signaling: Altered phospho-zeta and lack of zap70 recruitment in APL-induced T cell anergy. *Cell* **79,** 913–922.

Smith, C. E., Eagar, T. N., Strominger, J. L., and Miller, S. D. (2005). Differential induction of IgE-mediated anaphylaxis after soluble vs. *cell-bound tolerogenic peptide therapy of autoimmune encephalomyelitis. Proc. Natl. Acad. Sci. USA* **102,** 9595–9600.

Solal-Celigny, P., Bernard, J. F., Herrera, A., and Boivin, P. (1983). Treatment of adult autoimmune thrombocytopenic purpura with high-dose intravenous plasmin-cleaved gammaglobulins. *Scand. J. Haematol.* **31,** 39–44.

Song, S., Crow, A. R., Freedman, J., and Lazarus, A. H. (2003). Monoclonal IgG can ameliorate immune thrombocytopenia in a murine model of ITP: An alternative to IVIG. *Blood* **101,** 3708–3713.

Song, L., Wang, J., Wang, R., Yu, M., Sun, Y., Han, G., Li, Y., Qian, J., Scott, D. W., Kang, Y., Soukhareva, N., and Shen, B. (2004). Retroviral delivery of GAD-IgG fusion construct induces tolerance and modulates diabetes: A role for CD4+ regulatory T cells and TGF-beta? *Gene Ther.* **11,** 1487–1496.

Song, S., Crow, A. R., Siragam, V., Freedman, J., and Lazarus, A. H. (2005). Monoclonal antibodies that mimic the action of anti-D in the amelioration of murine ITP act by a mechanism distinct from that of IVIg. *Blood* **105,** 1546–1548.

Sood, R., Wong, W., Gotlib, J., Jeng, M., and Zehnder, J. L. (2008). Gene expression and pathway analysis of immune thrombocytopenic purpura. *Br. J. Haematol.* **140,** 99–103.

Soukhareva, N., Jiang, Y., and Scott, D. W. (2006). Treatment of diabetes in NOD mice by gene transfer of Ig-fusion proteins into B cells: Role of T regulatory cells. *Cell Immunol.* **240,** 41–46.

Stahl, D., Lacroix-Desmazes, S., Heudes, D., Mouthon, L., Kaveri, S. V., and Kazatchkine, M. D. (2000). Altered control of self-reactive IgG by autologous IgM in patients with warm autoimmune hemolytic anemia. *Blood* **95,** 328–335.

Stasi, R., Pagano, A., Stipa, E., and Amadori, S. (2001). Rituximab chimeric anti-CD20 monoclonal antibody treatment for adults with chronic idiopathic thrombocytopenic purpura. *Blood* **98,** 952–957.

Stasi, R., Del, P. G., Stipa, E., Evangelista, M. L., Trawinska, M. M., Cooper, N., and Amadori, S. (2007). Response to B-cell depleting therapy with rituximab reverts the abnormalities of T-cell subsets in patients with idiopathic thrombocytopenic purpura. *Blood* **110,** 2924–2930.

Steinman, R. M., Hawiger, D., and Nussenzweig, M. C. (2003). Tolerogenic dendritic cells. *Annu. Rev. Immunol.* **21,** 685–711.

Svensson, L., Jirholt, J., Holmdahl, R., and Jansson, L. (1998). B cell-deficient mice do not develop type II collagen-induced arthritis (CIA). *Clin. Exp. Immunol.* **111,** 521–526.

Tian, J., Zekzer, D., Hanssen, L., Lu, Y., Olcott, A., and Kaufman, D. L. (2001). Lipopolysaccharide-activated B cells down-regulate Th1 immunity and prevent autoimmune diabetes in nonobese diabetic mice. *J. Immunol.* **167,** 1081–1089.

Tiller, T., Tsuji, M., Yurasov, S., Velinzon, K., Nussenzweig, M. C., and Wardemann, H. (2007). Autoreactivity in human IgG+ memory B cells. *Immunity* **26,** 205–213.

Urbaniak, S. J., and Greiss, M. A. (2000). RhD haemolytic disease of the fetus and the newborn. *Blood Rev.* **14,** 44–61.

Velupillai, P., and Harn, D. A. (1994). Oligosaccharide-specific induction of interleukin 10 production by B220+ cells from schistosome-infected mice: A mechanism for regulation of CD4+ T-cell subsets. *Proc. Natl. Acad. Sci. USA* **91,** 18–22.

Velupillai, P., Garcea, R. L., and Benjamin, T. L. (2006). Polyoma virus-like particles elicit polarized cytokine responses in APCs from tumor-susceptible and -resistant mice. *J. Immunol.* **176,** 1148–1153.

Wang, J., Wiley, J. M., Luddy, R., Greenberg, J., Feuerstein, M. A., and Bussel, J. B. (2005). Chronic immune thrombocytopenic purpura in children: Assessment of rituximab treatment. *J. Pediatr.* **146,** 217–221.

Ward, E. S., Zhou, J., Ghetie, V., and Ober, R. J. (2003). Evidence to support the cellular mechanism involved in serum IgG homeostasis in humans. *Int. Immunol.* **15,** 187–195.

Ware, R. E., and Zimmerman, S. A. (1998). Anti-D: Mechanisms of action. *Semin. Hematol.* **35,** 14–22.

Warner, M. N., Moore, J. C., Warkentin, T. E., Santos, A. V., and Kelton, J. G. (1999). A prospective study of protein-specific assays used to investigate idiopathic thrombocytopenic purpura. *Br. J. Haematol.* **104,** 442–447.

Weir, C. R., Nicolson, K., and Backstrom, B. T. (2002). Experimental autoimmune encephalomyelitis induction in naive mice by dendritic cells presenting a self-peptide. *Immunol. Cell Biol.* **80,** 14–20.

Wolf, S. D., Dittel, B. N., Hardardottir, F., and Janeway, C. A., Jr. (1996). Experimental autoimmune encephalomyelitis induction in genetically B cell-deficient mice. *J. Exp. Med.* **184,** 2271–2278.

Woodrow, J. C., Clarke, C. A., Donohow, W. T., Finn, R., McConnell, R. B., Sheppard, P. M., Lehane, D., Roberts, F. M., and Gimlette, T. M. (1975). Mechanism of Rh prophylaxis: An experimental study on specificity of immunosuppression. *Br. Med. J.* **2,** 57–59.

Yuasa, T., Kubo, S., Yoshino, T., Ujike, A., Matsumura, K., Ono, M., Ravetch, J. V., and Takai, T. (1999). Deletion of fcgamma receptor IIB renders H-2(b) mice susceptible to collagen-induced arthritis. *J. Exp. Med.* **189,** 187–194.

Zambidis, E. T., and Scott, D. W. (1996). Epitope-specific tolerance induction with an engineered immunoglobulin. *Proc. Natl. Acad. Sci. USA* **93,** 5019–5024.

Zambidis, E. T., Barth, R. K., and Scott, D. W. (1997). Both resting and activated B lymphocytes expressing engineered peptide-Ig molecules serve as highly efficient tolerogenic vehicles in immunocompetent adult recipients. *J. Immunol.* **158,** 2174–2182.

Zhang, F., Chu, X., Wang, L., Zhu, Y., Li, L., Ma, D., Peng, J., and Hou, M. (2006). Cell-mediated lysis of autologous platelets in chronic idiopathic thrombocytopenic purpura. *Eur. J. Haematol.* **76,** 427–431.

CHAPTER 2

Cumulative Environmental Changes, Skewed Antigen Exposure, and the Increase of Allergy

Tse Wen Chang* and **Ariel Y. Pan***

Contents

* Genomics Research Center, Academia Sinica, Nankang, Taipei 11529, Taiwan

Advances in Immunology, Volume 98
ISSN 0065-2776, DOI: 10.1016/S0065-2776(08)00402-1

Abstract

The human immune system evolved over many hundreds of million of years in the ancestors of vertebrates and mammals to defend them against infectious and parasitic organisms in their natural habitats. By the time the Primates and Rodentia orders diverged about 88 million years ago, the human immune system was largely configured. From about 125,000 years ago, marked by the use of fire, *Homo sapiens* began to make substantial changes in their living environment and lifestyle. Here, we examine those changes in two phases, before and after the Industrial Revolution, and analyze their impact on the exposure of our immune system to infectious organisms and to harmless environmental antigens. Our analyses show that the cumulative changes in environment and lifestyle in many regions of the world have drastically altered the pattern by which humans are exposed to infectious organisms and harmless environmental antigens and that these changes have profoundly impacted the function of the immune system and enhanced the development of allergy.

Our analyses expand the hygiene hypothesis by taking into consideration the immunological impact of a broader range of antigen exposure changes than simply decreased microbial infection during childhood. We subsequently examine the proposed mechanisms of TH1 to TH2 shift and Treg downregulation with regard to the hygiene hypothesis and present an immunological basis for the increased activity of the IgE-mediated pathway in allergic patients.

In our "skewed antigen exposure" theory, we propose that, for many individuals living in modern societies: (i) reduced exposure to a large variety of infectious organisms and environmental antigens and (ii) increased exposure to a small variety of environmental antigens, resulting from the cumulative changes in individuals' living environment and lifestyle, together alter the balance of the immune system, and increase production of IgE and the sensitization of mast cells toward a limited variety environmental antigens unique to affected individuals, resulting in an overall increase in allergy.

1. INTRODUCTION

Antigens are generally foreign substances to which our immune system is exposed. These substances, either living or nonliving, usually come into contact with our bodies' surface, penetrate through our mucosal epithelium or skin, and interact with the cells and molecules of our immune system. The immune system has evolved over more than 500 million years (Laird *et al.*, 2000; Shintani *et al.*, 2000) to recognize and clear invading living organisms and their products. Allergic reactions, however, are responses of our immune system toward generally harmless, noninfectious, living or nonliving environmental substances.

In this chapter, an animal's "antigen exposure" refers to the composition of the antigens – the amounts and relative proportions of various antigens that an animal is exposed to – at a particular location and time. Such antigens include infectious organisms such as microorganisms and parasitic worms, environmental protein-containing substances such as the debris and shed body parts of small animals and plants, airborne mites, tree and grass pollens, and potentially antigenic proteins contained in food.

Allergic diseases emerged in the early 19th century among people of the wealthy classes and have gradually become more prevalent over the past two centuries (Emanuel, 1988; Jackson, 2001). Allergic asthma, allergic rhinitis, atopic dermatitis, food allergy, and other allergic diseases affect large portions of the population in modern societies. In the first half of this chapter, we focus on how the cumulative effect of the environmental and lifestyle changes humans have adopted has drastically changed their antigen exposure. We focus particularly on the environmental and lifestyle changes people have made since the beginning of the 19th century, which have significantly impacted humans' antigen exposure. In addition, we take an evolutionary view by examining how humans have changed their living environment and lifestyle since evolutionarily primitive times. We discuss how such cumulative changes have altered the exposure of our immune system to the two types of antigens: infectious organisms and environmental antigens.

Our analyses broaden the hygiene hypothesis. As originally proposed by Strachan (1989) the hygiene hypothesis postulates that the absence of frequent microbial infection during infancy and childhood because of the adoption of small families in recent decades has led to the increase of allergic diseases. In our analyses, exposure to infectious organisms involves more than just microbial infection and longer periods than just infancy and early childhood. We discuss more of the many changes in lifestyle and environment since *Homo sapiens* started to use fire than only the improvement in hygiene seen in recent decades.

In the second half of this chapter, we review and attempt to integrate the various theories proposed by earlier investigators to explain the immunological mechanisms underlying the hygiene hypothesis. We stress that a sound immunological explanation is needed to delineate how the IgE-mediated pathway and sensitization of mast cells are enhanced. We propose a "skewed antigen exposure" theory that suggests that in modern times, decreased exposure to a large variety of infectious organisms and environmental antigens and increased exposure to a limited variety of environmental antigens unique to the individual has altered the balance of the individual's immune system and increased the human tendency to develop allergies.

2. ANTIGEN EXPOSURE DURING TIMES WHEN THE IMMUNE SYSTEM EVOLVED

Lower forms of life, including bacteria, fungi, protozoa, worms of many phyla, especially Nematoda and Platyhelminthes, and arthropods, existed long before the vertebrates appeared. Evidence indicates that while there were periods of mass extinction and explosive generation of new species, the numbers of genera of marine species (a measure indicative of biodiversity) varied within a 2–3-fold range throughout most of the Phanerozoic Eon of the last 540 million years (Alroy *et al.*, 2001).

As the immune system evolved in the vertebrates over more than half a billion years (Cannon *et al.*, 2004; Flajnik *et al.*, 1999; Laird *et al.*, 2000; Shintani *et al.*, 2000), its prime purpose was to prevent diverse microorganisms and small animals from using a vertebrate's body as a rich source of nutrients and a bed for propagation. A large variety of infectious and parasitic organisms shared the waters and soils where the ancestors of vertebrates and mammals rested, moved about, searched for and ate food, mated, excreted, and shed body debris. The mucosal surfaces of the mouth and parts of the digestive tract, the nose and parts of the respiratory tract, the genital tract, the eyes, ears and skin as well as the numerous abrasions and wounds of a vertebrate were in constant contact with these opportunistic infectious and parasitic organisms. Blood-feeding arthropods and worms adhered to the skin and body hairs (e.g., fleas, ticks, and many species of mites and leeches) and resided in the digestive, respiratory, and genitourinary tracts (e.g., many species of nematodes) and in the surroundings (e.g., mosquitoes, flies) of a vertebrate. They bit, burrowed, or penetrated through a vertebrate's skin or mucosal epithelium, transferring microorganisms or small parasites or creating openings to allow the entrance of microorganisms, small worms, and their larvae or eggs. A vertebrate drank liquid and ate food that contained a large variety of living organisms (e.g., microorganisms and worms in the digestive tracts of animal prey) or had a large variety of living organisms stuck to it. Evidence also shows that the viruses evolved within the same time frames as their host species (Rybicki, 2003). In this chapter, "microorganisms" refers to bacteria, protozoa, fungi, and other unicellular organisms, and viruses. These various infectious organisms developed a myriad of ways to gain foothold in an animal's body, some propagating in the extracellular milieu and some inside cells (Janeway *et al.*, 2005).

Various branches of the vertebrates, including mammals, birds, fish, and others developed similarly effective immune mechanisms of a common origin to combat the various types of intrusions. The immune systems of the various classes of jawed vertebrates appear to work comparably well, as they can support some mammals (e.g., elephants), birds (e.g., swans and Amazon parrots), fish (e.g., carps), and species of other classes (e.g., box turtles) to live 70 years or longer.

3. DID ALLERGY PLAY A ROLE IN THE EVOLUTION OF THE HUMAN IMMUNE SYSTEM?

3.1. The human immune system was largely developed by the time the Primates and the Rodentia orders diverged

Innate immunity, which had been in evolution long before the first vertebrates appeared during the Cambrian explosion about 542 million years ago (Kumar and Hedges, 1998; Shu *et al.*, 2001; Zarkadis *et al.*, 2001), is essential, especially during the initial periods of an encounter with a foreign intruder. However, in the context of antigen exposure, antigen recognition, and immune stimulation and regulation – adaptive immunity – is the core of the immune system in the vertebrates. The clonal selection-based adaptive immunity is crucial for the recognition of self and nonself, the extreme diversity and the specificity of immune responses, and for immune memory.

The primary elements of adaptive immunity, namely, the VDJ genes, recombinase activation genes, immunoglobulins, B and T cell receptors, and MHC molecules, had already appeared by the time the jawed vertebrates began to evolve about 500–550 million years ago (Cannon *et al.*, 2004; Eason *et al.*, 2004; Flajnik *et al.*, 1999; Kumar and Hedges, 1998; Laird *et al.*, 2000; Shintani *et al.*, 2000). By the time the Mammalia and Aves classes split about 310 million years ago (Fig. 2.1A) (Kumar and Hedges, 1998), most of the main molecules and cells that were characteristic of adaptive immunity had already been developed (Table 2.1) (Igyarto *et al.*, 2006; Six *et al.*, 1996; Zekarias *et al.*, 2002).

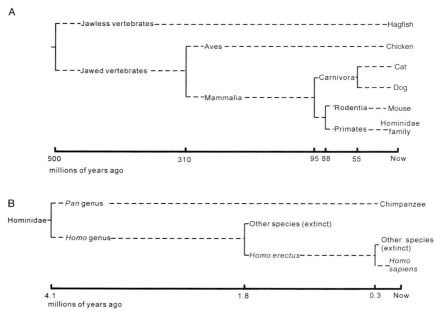

FIGURE 2.1 The evolution of the species referred to in this paper in the context of the evolution of the adaptive immunity, the IgE pathway, and the development of allergy. (A) The lineages and times of divergence for the referred species since the appearance of vertebrates. (B) The lineages and divergence of the referred species from the *Homo* genus.

During the 200 million years from the divergence of the Mammalia class out of the vertebrate lineage to the speciation of the Primates and Rodentia orders about 88 million years ago (Fig. 2.1A) (Springer *et al.*, 2003), the immune system became more developed and refined. Among various new components of the immune system, IgE (Vernersson *et al.*, 2004) and its associated molecular components, such as the α chain of type I IgE.Fc receptor (Hughes, 1996), evolved (Table 2.2). While there are numerous differences between the human and murine immune systems (Mestas and Hughes, 2004), their overall architecture, recognition and processing of antigens, induction and regulation of various immune functions, and mechanisms for clearing various foreign intruders are strikingly similar (Janeway *et al.*, 2005).

Among these similarities, the IgE-mediated mast cell-activation pathway is particularly pertinent to the subject matter of this chapter, because it has been shown to play major roles in the pathogenesis of the modern maladies – allergic diseases, including allergic asthma, allergic rhinitis, food allergy, and other allergic diseases and conditions (Chang *et al.*, 2007; Janeway *et al.*, 2005). IgE, which is also present in species of other orders of the Mammalia class (Vernersson *et al.*, 2004), such as dogs (Carnivora), goats (Artiodactyla), and horses (Perrissodactyla), is believed to play

TABLE 2.1 The evolution of components relating to adaptive immunity in human, mouse, and chicken

Species	VDJ	T/B cells[a]	IgM	IgG	IgA	MHC[b]	Dendritic cells	CD4/ CD8 cells	IL-2/ IL-4	ITAM	ITIM
Human	+	+	+	+	+	+	+	+	+	+	+
Mouse	+	+	+	+	+	+	+	+	+	+	+
Chicken	+	+	+	+	+	+	+	+	+	+	+

[a] The T and B cell receptors evolved as T and B lymphocytes evolved.
[b] Abbreviation and symbols used: MHC, major histocompatibility complex; IL, interleukin; ITAM, immunoreceptor tyrosine-based activation motif; ITIM, immunoreceptor tyrosine-based inhibition motif. "+" indicates the presence of the element in the species.

TABLE 2.2 The evolution of immune-related components involved in the IgE-mediated pathways in human, mouse, and chicken

Species	Mast cells	Basophils	IgE	FcεRI[a] α chain
Human	+	+	+	+
Mouse	+	+	+	+
Chicken	+	+	−	−

[a] FcεRI, type I IgE.Fc receptor.

beneficial roles in the defense against parasitic worms (Gould *et al.*, 2003; Janeway *et al.*, 2005; King *et al.*, 1997).

3.2. Allergic reactions probably did not play a role in the evolution of the immune system

Allergic reactions, which cause constriction of the smooth muscle of the airway, inflammation in mucosal tissues and skin, and sometimes anaphylaxis (Janeway *et al.*, 2005), do not appear to serve a defensive or sentinel purpose. On the contrary, they make breathing difficult and often cause wheezing and sneezing, which is noisy and would reveal a reactive animal's presence and location to predators. Inflamed tissues have a weakened epithelial barrier and are prone to infection (Blair *et al.*, 2001; Nathan, 2007; Talbot *et al.*, 2005). Thus, it appears compelling that since allergic reactions probably did not occur during several hundred million years of evolution, they probably did not play a role in the evolution of the immune system; in other words, because the immune system did not experience allergic reactions, it did not eliminate such a disadvantageous trait.

In addition to literature that indicates that allergic diseases are post-Industrial Revolution, modern ailments (Emanuel, 1988; Jackson, 2001) and other observations indicate that allergies are a phenomenon apparently entirely related to modern human societies. People living in primitive villages do not develop allergies (Ogunbiyi *et al.*, 2005). Chimpanzees (*Pan* genus), which diverged from the *Homo* genus out of the Hominidae family about 4.1 million years ago and are most closely related to humans genetically (Hobolth *et al.*, 2007) (Fig. 2.1B), do not develop allergies.

In many economically advanced countries, as many as 20% of domestic dogs (Chesney, 2001) and a lesser percentage of domestic cats develop allergies. House dogs and cats, which were domesticated by humans about 12,000–14,000 (Morey, 2006) and 9,000–10,000 years ago (Driscoll *et al.*, 2007), respectively, and became humans' best friends, are two species of the Carnivora order, which were separated from the human lineage about 95 million years ago (Springer *et al.*, 2003). The Canidae and Felidae families themselves diverged about 55 million years ago (Fig. 2.1A) (Springer *et al.*, 2003). Some dogs and cats develop allergies, not because they and humans carry certain allergy-prone genes, but simply because they share modern human living environments and lifestyles. They live in enclosed spaces with their human masters, eat canned or bagged sterile, processed food and drink disinfected water, and many also get bathed and shampooed (see later sections). Like allergic humans, they are sensitive to certain airborne antigenic substances and ingredients in food (Arlian *et al.*, 2003; Verlinden *et al.*, 2006) and are treated with the same medicine as human patients; some even receive allergen-based immunotherapy (Schnabl *et al.*, 2006). Laboratory mice kept in air-conditioned rooms can be induced to develop asthma-like airway reactivity, if given an antigen, such as ovalbumin or house dust mite extract, in aerosolized form or intranasally for a few weeks (Johnson *et al.*, 2004; Temelkovski *et al.*, 1998). It is conceivable that most animals, if kept in a relatively confined, clean space from birth and exposed to a specially crafted antigenic make-up, will develop allergic reactions.

3.3. Environmental, not genetic, factors often determine an individual's development of allergy

The above analysis indicates that from an evolutionary perspective, the environment, not genes, is likely the dominant factor in causing allergy in an individual. Modern *H. sapiens* were evolved only about 200,000 years ago; this time period is too short for a subpopulation of the human species to evolve genetic changes that render them susceptible to allergic diseases. People of all races and ethnic groups develop allergies. Thus, if genetic changes are not the likely cause of the increase in allergy in modern times, we should look to changes in the environment.

Additional observations that support this conclusion are statistics concerning allergy in young twins (Miller *et al.*, 2005; Rasanen *et al.*, 1998; Thomsen *et al.*, 2006). While the numbers of identical twins who both suffer from allergies (concordant) are greater than the numbers of twins of whom only one twin suffers from allergies (discordant), the existence of discordant sets suggests that environmental factors play a role in the development of allergy. This view does not negate the important roles that genes play in the susceptibility to allergy or the lack of it in an individual. A contrasting observation regarding twins with allergy is that among sets of young fraternal twins being cared for in the same households, the numbers of discordant sets are larger than those of concordant sets. This suggests that either minor differences in environmental exposure or genetic differences are responsible for the varied susceptibility to allergy development.

A person's genetic background, such as the combination of alleles (haplotypes) of MHC class I and II antigens (Li *et al.*, 2007; Torio *et al.*, 2003), the germline batteries of V, D, and J elements, and the polymorphism of various molecules or gene segments (Pawlik *et al.*, 2007; Steinke *et al.*, 2008), may affect his or her ability to recognize sequences of peptides or conformation of antigens, mount a T cell response or synthesize different repertoires of IgG, IgE, or other antibodies toward the antigens. The collective results of these immunological variations will affect an individual's susceptibility not only to certain infections but also to developing allergic reactions to certain environmental antigens.

4. ENVIRONMENTAL AND LIFESTYLE CHANGES ALTER HUMANS' ANTIGEN EXPOSURE

Because allergy was nonexistent before the beginning of the 19th century, to find the cause of allergy, here we compare humans' living environments and lifestyles before the 19th century and in contemporary times. In addition, in an attempt to delineate the underlying immunological association, we adopt an evolutionary rationale to examine the cumulative lifestyle and environmental changes humans have made and analyze how such changes have impacted the exposure of the immune system to antigens. Our rationale in taking an evolutionary approach to examine the environmental and lifestyle changes modern humans have made and the impact of such changes on humans' antigen exposure is twofold. (i) To examine how changes in antigen exposure impact the balance and function of the human immune system, we must compare the characteristics of antigen exposure in modern times with those in times when the immune system was evolved. (ii) In taking an evolutionary approach, we recognize that the changes in environment and lifestyle and the effects of environmental and lifestyle changes on antigenic make-up are both cumulative, as will be discussed in the next two sections.

The hygiene hypothesis proposed by Strachan (1989) reasoned that the lack of frequent microbial infection in infancy and early life due to small family units and improved hygiene renders the immune system prone to allergic responses. From our perspective, this hypothesis recognizes only one part of the changes in humans' lifestyle and living environment that have affected the exposure of our immune system to infectious microorganisms and consequently cause immune imbalance. In Sections 9–13, the immunological basis underlying the hygiene hypothesis will be analyzed in depth. This section and the following few (Sections 5–7) intend to substantiate the concept of antigen exposure and broaden the analysis of the impact of various environmental and lifestyle changes on antigen exposure.

One way to appreciate how substantially the living environment and lifestyle of humans have changed from the times when the immune system evolved in their vertebrate and mammal ancestors is to compare how humans live differently from the lobe-finned fish, coelacanth (a 400 million-year-old "living fossil" from the jawed vertebrate lineage (Betz *et al.*, 1994)) or the small insectivorous shrew (that resembles the first mammals that appeared in the early Jurassic period (about 195 million years ago) (Luo *et al.*, 2001)). The most significant changes actually started very recently, as the *Homo* genus developed (i) abilities to move on only two hind limbs and to use the front two limbs to manipulate objects and perform other activities, and (ii) larger brains for abstract reasoning and language.

On an evolutionary timescale, the bulk of the changes in *H. sapiens'* living environment and lifestyle have occurred in a minuscule speck of time. The supreme abilities of the human species have enabled them to expand out of proportion relative to other animal species and to vastly change the terrestrial surface of earth in macroscopic and microscopic ways. These changes have profoundly impacted both quantitatively and qualitatively the ways in which humans contact other organisms, in particular, infectious microorganisms and parasitic worms, and also environmental biological substances.

Humans nowadays are living in an environment drastically different from those of the ancestors of *H. sapiens*, primates, mammals, and vertebrates. The sheer size of the human population and the things humans do have affected the distribution of habitats among species and distorted Nature's food cycle chains and the relative proportions of species. Humans occupy large portions of the terrestrial space on earth and also execute many activities in aqua space (in both fresh and sea waters). Humans essentially control the existence or extinction, expansion or reduction, of nearly all other species. The materials humans have produced and spread over the earth and the various other changes, such as the climate change contributed to in part by humans' extremely rapid consumption of fossil fuels, have affected other living species to a massive degree at an extremely rapid speed (evolution biologists claim that we are in the Holocene mass extinction epoch (Pimm *et al.*, 1995)). Most of these human abilities and activities are the results of progress in civilization and in science and technology. While such progress is desirable and has brought human lives multitudes of benefits, including benefits to health, it has profoundly impacted the makeup of biological and antigenic substances that humans come into physical contact with. These drastic changes in antigen exposure may have resulted in an imbalance of human immune function and hence the development of allergic problems in some people living in economically advanced societies.

5. SIGNIFICANT CHANGES IN LIFESTYLE AND ENVIRONMENT SINCE *H. SAPIENS* STATED TO USE FIRE

Humans have not only reduced the variety and amounts of organisms in or near the areas where they reside but have also developed many means and products which shield them purposely or unintentionally from contact with other organisms and biological substances. But, while humans' contact with a large variety of harmless environmental biological substances has generally decreased, many individuals are exposed to certain environmental antigens unique to them in increased amounts and proportions. In our opinion, the first significant introduction that altered humans' lifestyle and environment was the use of fire. Pieces of evidence of varying credibility suggest that the first control of fire was achieved by the extinct *Homo erectus* about 790,000 years ago (Goren-Inbar *et al.*, 2004) or earlier; however, widespread use of fire was implemented by *H. sapiens* around 125,000 years ago (James, 1989). Fire transformed food preparation and facilitated the making of tools, materials, and utensils, which reduced humans' contact with infectious organisms. Various other changes were gradual and became frequent as humans advanced in knowledge and skills. Some inventions and discoveries, such as the written language, printing, various means of transportation, coal, petroleum, electricity, and telecommunications, which sped up the propagation of culture, science, and technology development, have obviously had a huge indirect impact on environmental and lifestyle changes. Here, we focus on those major changes that have had direct effects on the exposure of the human immune system to antigens.

Considering that allergic diseases started to emerge in the early 19th century and were probably nonexistent or very rare before then, we have grouped human lifestyle and living environment changes into two phases, Phase I and II, before and after the Industrial Revolution, respectively. Because the Industrial Revolution lasted from the late 18th to the early 19th century, a practical division for Phase I and II is the year 1800. Tables 2.3 and 2.4 list changes in Phases I and II, respectively. Clearly, the changes in the environment and lifestyle that have occurred since primitive, evolutionary times have been cumulative and have influenced changes in exposure to antigens. It is conceivable that the human immune system, which has evolved through various environments over several hundreds of millions of years, can adjust to a wide spectrum of changes in antigen exposure. Thus, while the environment and lifestyle changes in Phase I did not affect antigen exposure to an extent large enough to cause allergy, they and the additional changes in Phase II have accumulated to such an extent that antigen exposure has been affected to extreme levels and now causes allergy in some people in certain situations.

In our analysis of exposure to infectious organisms, contact with natural or outdoor water and soil is considered very important. For millions of years, the ancestors of humans established habitats near available water supplies. Until several decades to a century ago, most people used natural water, taken from rivers, streams, lakes, ponds, and less frequently, from springs or wells, without purification to drink and to prepare and cook food. The water contained large numbers of microorganisms and worms, which lived in the water or in the animals living in the water (e.g., snails, which are intermediate hosts for many species of parasitic worms), or which were washed into the water from the surrounding soil by rain. The soil harbored a large number of varieties of microorganisms and small animals, including parasitic worms and their

TABLE 2.3 Changes in the environment and lifestyle introduced in Phase I (before the year 1800) that have probably had major impacts on human exposure to (i) infectious organisms (IO) and (ii) environmental antigens with certain unique features (EAu[a])

| Event of changes | Impact on exposure to | | Remarks |
	IO	EAu	
Use of fire	↓[b]		Widespread use ~125,000 years ago
Cooking foods	↓		
Protective shelters	↓		
Domesticating dogs, cats, cattle, pigs, sheep, chicken, ducks, etc.	↑		Infections from animal to men ↑
Body and foot covering	↓		Contact with soil ↓
Agriculture	↓		Started 10,000–5,000 years ago; habitats for many species ↓
Containers for food, water, made with clay, metal, etc.	↓		Contamination by soil and natural water ↓
Forming villages	↓		Started 15,000–10,000 years ago
Cement, bricks, tiles, concrete, metal, etc. for construction	↓		Blocking contacts with soil
Plasters, paints	↓		Sealed off wall surfaces
Spoons, forks, chopsticks	↓		Food contamination by dirty hands↓
Preserving foods with salt	↓		
Washing body, feet, and hands	↓		
Wells for water	↓		Contamination by human and animal wastes ↓
Building towns, cities	↑↓		Human contact ↑; Separating humans from other species
Establishing markets, schools, armies, churches, etc.	↑		Congregation and contact among people ↑
Planting trees, grasses of selected varieties		↑	Pollens of few species in specific areas ↑
Improved toiletry habits, facilities	↓		

[a] EAu refers to those environmental antigens, such as dust mites, cat dander, tree and grass pollens, and certain food antigens, of limited varieties that are unique to individuals and that can potentially increase in total amounts and in proportions in the living spaces or food of those individuals.

[b] symbols and abbreviations: ↓: decrease, ↑: increase.

larvae and eggs, part of which came from fecal excretions of humans and various other large and small animals.

The changes introduced in Phase I, such as protective shelters, processes, and utensils for preparing food and holding water, body and foot coverings, and human waste disposal (Table 2.3), reduced human contact with outdoor water and soil and the organisms contained therein. The use of various construction materials, such as bricks, tiles, concrete, metal, plaster, paints, and others, helped build shelter walls, floors, and roofs, whose inside surfaces were largely uninhabitable by small animals and microorganisms and which prevented organisms from getting inside. Towards the end of Phase I, although the general exposure to outdoor water and soil had been reduced, even in advanced societies people probably still consumed water and food that contained substantial amounts of live, infectious organisms. Contact with those organisms might not have caused frequent productive infections and illnesses, but would constantly tune the immune system. Well into the 20th century, before helminthic medicine became available, even in economically progressive societies many people were probably chronically infected with helminths, especially intestinal geohelminths, from early childhood into adult lives (also Section 12.2).

Notably, while human exposure to infectious organisms generally decreased, the expansion of the human population, the creation villages and cities and various other social changes such as the creation of markets, armies, schools, churches, theaters, prostitution, and wars, increased the congregation and contact of people and hence the spread of infectious organisms among people. The transmission of infectious organisms among people was gradually reduced in the later part of Phase II as health care and medicines (including vaccines) advanced. The domestication of animals, including dogs, cats, cattle, pigs, sheep, goats, chickens, ducks, and others, increased people's contact with those animals and hence the transmission of infectious organisms, such as influenza viruses (Liu, 2006) and parasitic worms (Craig and Larrieu, 2006), from them to human hosts. The industrialization of livestock farming and meat production in the past century has probably gradually decreased the direct contact of large numbers of people with livestock.

6. SIGNIFICANT LIFESTYLE AND ENVIRONMENTAL CHANGES SINCE THE INDUSTRIAL REVOLUTION

The first recorded hay fever-like illness in 1819 (Jackson, 2001) occurred when the Industrial Revolution had already produced initial advances in materials, products, and utilities and the corresponding environment and lifestyle changes. The changes that occurred in Phase II over the past two centuries (Table 2.4), in addition to those in Phase I (Table 2.3), have now effectively shielded some people from nearly all contact with primitive, natural water and soil and the infectious and parasitic organisms in them. Polluted industrial waste, fertilizers, weed and insect control agents, and

TABLE 2.4 Changes in the environment and lifestyle introduced in Phase II (after the year 1800) that have probably had major impacts on human exposure to (i) infectious organisms (IO) and (ii) environmental antigens with certain unique features (EAu[a])

| Event of changes | Impact on exposure to | | Remarks |
	IO	EAu	
Industrial wastes polluting waters, soils	↓		Killing water and soil species
Large-scale deforestation	↓		Reducing lives and species
Chemical fertilizers	↓		Killing many species
Soaps, toothpastes, shampoos	↓		Microorganisms, small animals in and on human bodies ↓
Detergents, cleaning agents, disinfectants			Microorganisms, small animals in households ↓
Insecticides for mosquitoes, flies, cockroaches	↓		Reducing microbial transmitting vehicles
Poisons for mice, rats	↓		Transmitting vehicles ↓
Filtered, bleached drinking water; city water through pipes	↓		
Refrigerators, freezers	↓		
Antibiotics for helminths, bacteria, fungi, and protozoa	↓		Reducing these species in human bodies; other effects
Vaccination programs	↓		Productive infection ↓
Industrial enzymes and proteins		↑	Ag[b] in factories ↑
Paints (latex or resin based), tar	↓		Sealing off wall surfaces
Urbanization, migration to cities	↓	↑	
Homes high above ground (in multi-story buildings)	↓		Separation from soil and outdoor water

(*continued*)

Table 2.4 (*continued*)

Event of changes	Impact on exposure to		Remarks
	IO	EAu	
Long-distance migration		↑	New environmental Ag ↑
Family members living apart	↓		Human contact ↓
Small families, few siblings	↓		Sibling contact ↓
Processed food, e.g. infant formulas	↓	↑	Sterile, enriched protein Ag ↑
Nonindigenous foods		↑	New Ag ↑
New products, e.g. rubber latex		↑	
City sewage treatment systems	↓		Processed, covered
Modern toiletry facilities	↓		
Poorly ventilated or closed homes	↓↑	↑	Shielding Ag both ways
Use of couches, carpets, curtains, etc.		↑	Dander deposit ↑, bed for mites
Keeping cats and dogs indoors		↑	Dander Ag ↑

[a] EAu and arrows are defined as in Table 2.3.
[b] Abbreviation: Ag, antigens.

other chemicals have killed many species in the water and soil in and near cities. The sewage systems are ducted and buried underground. Water for household use is filtered, sterilized by bleaching or other means, and piped in. In recent years, bottled purified mineral water and pure water have become the main source of drinking water for many people living in economically advanced societies. The use of detergents, a large variety of cleaning chemicals, insecticides, rodent poisons and other disinfecting agents is widespread in modern homes. Some people have aversion to or even paranoia about any germ contamination in their homes. The unnatural, biodeprived environment in such homes is further aggravated by the fact that the space in such homes is often insulated by tight enclosure and air-filtration. Many of these homes are high above ground (in high-rise buildings) and are thus farther separated from natural soil and water. It is not an overstatement to say that in some modern homes, a near germ-free condition is maintained.

7. THE INCREASE IN PROPORTIONS OF ENVIRONMENTAL ANTIGENS OF LIMITED VARIETIES

7.1. Definition of "environmental antigens"

For the context of the present study, "environmental antigens" can be defined as those otherwise innocuous antigens that come in contact with the mucosal surfaces, particularly those of the airway, gastrointestinal tract and eyes of people and animals, and that can potentially induce immunological responses. In contrast to infectious agents that can multiply in infected hosts, environmental antigens, either living or nonliving organisms or substances, do not multiply in the hosts who take in those antigens. Live organisms, such as mites, molds, pollens, do not penetrate the mucosal epithelial barrier and are broken apart in the mucosal fluid, releasing their constituents, including proteins. Only some of the proteins penetrate through the junctions between mucosal epithelial cells to reach the basal side of the mucosal layer.

Conceptually, protein-containing substances contained in airborne particles, which we breathe in, or in the food that we eat, can be considered environmental antigens. However, not all proteins in the air and probably a very small proportion of proteins in food are immunogenic or antigenic. Environmental antigens can potentially be allergenic in some people, but not all allergens are environmental antigens. For example, some drugs, either small molecules or proteins, which are allergenic in some people, are not environmental antigens. Also, certain proteins in the venoms of some species of bees and ants, which are potentially very powerful allergens in some people, should not be considered to be environmental allergens.

A striking difference exists between the immune response to an infection and the allergic response to an allergen. For example, in the immune response against a bacterial infection, a host makes IgM, IgG, and IgA to hundreds of proteins and carbohydrate moieties, and all of these antibodies contribute to the removal of the intruding bacteria. In a typical allergic response (type I hypersensitivity reaction) to allergens, only IgE specific to a few small proteins is involved, and such IgE is mainly responsible for causing mast cell activation and allergic symptoms. As will be discussed in Sections 10–11, the total concentration of allergen-specific IgE and its proportion to total IgE are crucial in activating the IgE-mediated mast cell sensitization pathway.

7.2. Environmental antigens in evolutionary times

Adopting the evolutionary view of this chapter, one could imagine that during the several hundred million years when the ancestors of vertebrates and mammals evolved, the environmental antigens would have been mostly protein-containing substances in the water and in the soil beds where those animals lived. Those primitive environments would resemble the species-rich, bio-rich shallow waters of marine habitats and the soil beds of terrestrial habitats in tropical forests. After the ancestors of humans adopted terrestrial lives, they also derived a rich source of environmental antigens from the air in their ecologically balanced natural habitats, where diverse plants, molds, mites, and other organisms coexisted.

Harmless, noninfectious environmental antigens could include proteins contained in large varieties of substances: (i) nonliving substances in feces, saliva, shed body parts and debris of living and dead microorganisms, worms, arthropods and other small and large animals, (ii) live microorganisms, molds, algae, seeds, pollens, large varieties of small animals of the phyla Nematoda, Platyhelminthes, and Arthropoda phyla (various worms, mites, insects, etc.), other small organisms, and their eggs (oocysts, embryos), and (iii) body parts of various plants. All such substances were present in the water and soil, and some of them, such as pollens, molds, mites, were also in the air. These large varieties of protein substances came in contact with the mucosal surfaces of the respiratory and alimentary tracts of resident vertebrates and mammals. Presumably some proteins contained in these environmental substances in the marine and terrestrial habitats mentioned above could induce mucosal immune responses and IgE production in host mammals. It is conceivable that the IgE produced in mammals in those evolutionary times was composed of IgE specific to very large varieties of proteins that stimulated the animal via the mucosal surfaces.

7.3. How did the ancestors of humans receive environmental antigens?

It is perhaps difficult for us to picture precisely how the ancestors of humans were exposed to environmental antigens and how these antigens made contact with their mucosa. This analysis depends on which period of human evolution we intend to focus on. Logically, the most relevant period in this context is the time period just before *H. sapiens* started to change their natural environment and primitive lifestyle in substantial ways, about 125,000 years ago when they started using fire. A rough idea of how the ancestors of humans contacted environmental antigens may be obtained by observing how chimpanzees, the closest relatives of humans, live in their natural habitats.

In addition to breathing a diverse mix of protein-containing substances in the air, chimpanzees are in constant contact with the environmental antigens contained in the soil beds in their habitats. Their mouths and noses frequently contact the bio-rich soil. They walk on four limbs most of the time, and handle food directly with their mouths or with their front limbs (hands). Chimpanzees are omnivorous and eat fruits, leaves, nuts, and insects, and hunt small animals for meat. They contact environmental substances in food or stuck to the food they eat, or stuck to their hands. Chimpanzees sleep and roll on their soil beds, catching residue on their hair and skin. They do not wash their bodies, and instead groom their own or group members' hairy bodies with their hands and mouths. When they get physically close to play or show social bonds, they touch, embrace, lick, and kiss their group members. We can imagine that the proteins contained in the environmental substances that get into the mouth and nose of a chimpanzee contact the mucosal surfaces in the mouth and nose. The proteins get deeper in the gastrointestinal tract and in the airway, as the chimpanzee swallows saliva and breathes.

As will be discussed in Section 9.2, even in modern times, humans during their "crawling" stage, when placed on the ground, will explore the ground actively, very much like other animals at young ages. They frequently lick the ground, put fingers into their mouths to suck, and grab small objects to put into their mouths. They secrete a lot of saliva and constantly swallow it. This behavior might help them get exposed to large varieties of not only common "household" infectious organisms but also common environmental antigens in the surroundings.

7.4. Environmental antigens in modern times

Along with the environmental and lifestyle changes in Phase I and II, humans' exposure to the large variety of environmental antigens contained in natural water and soil and their primitive shelters also gradually decreased. However, beginning in the later part of Phase II, several types of environmental antigens began to increase in total amount and in relative proportions in the living space of human individuals, especially in their surrounding air and in food (Table 2.4). While a large variety of environmental antigens are known to cause allergy, the variety present at high concentrations and in high proportions in the living space of a particular individual is usually limited. For example, diverse species of mites and pollens from hundreds of species of grasses and trees can sensitize people and cause allergy in different regions of the world. However, during any specific time period, in the living space of an individual residing in a specific location, the range of environmental antigens present at high concentrations and in high proportions is relatively limited.

The shielding effects of modern homes, especially those with poor ventilation, prevent not only outdoor substances coming into the house but also indoor substances getting out. House dust mites and their fecal particles, molds, shed skin, dried saliva, and the hair of cats, and other minute particles, which are small and easily become airborne, are kept indoors in closed homes. The use of mattresses, couches, carpets, and curtains, which help deposit and accumulate these particles, further elevates the density of such indoor protein-containing particles.

The increase in the total amount and relative proportions of certain environmental antigens in modern homes can be illustrated by the problem of house dust mites, which can be considered the most serious allergen source worldwide. House dust mites, which are microscopic, small 8-legged animals of the Arachnida class (Arthropoda phylum), feed on flakes of shed human skin in human dwellings (Arlian and Morgan, 2003; Arlian and Platts-Mills, 2001). The mites secrete a number of enzymes, including proteases, onto the flakes of dead skin cells to digest or breakdown the constituents, they then eat the flakes, absorb some nutrients, and excrete the remaining bulk. They secrete more enzymes onto the partially digested, excreted particles and then eat them again. They repeat this process several times, until all the nutrients are absorbed. In a closed human dwelling in a modern society, which may otherwise be very "clean" and contain relatively few other biological substances, the mites, their partially or fully digested food particles, and the enzymes contained therein, which are allergenic (see Section 12.1), exist at high concentrations and account for a high proportion of the total environmental antigens in that living space. Furthermore, in the closed living space, the residents are immersed in mite-enriched air for long periods, including when they sleep. In the natural habitat of early humans, in a primitive shelter, or even in an older-style home (without tight enclosure), this phenomenon would not occur.

In modern times, agriculture, the lumber industry, and planned forestry have utilized large proportions of the land on earth. On this land, only selected crops and trees are planted. In modern communities, a few selected varieties of trees, grasses, or other plants of mostly ornamental value, such as birch, oak, cedar trees, Timothy grass, and Bermuda grass of indigenous or nonindigenous varieties, are planted. In their pollinating seasons, these plants produce large amounts of a few varieties of pollen unique to the individual communities. Furthermore, in modern times, many people migrate across long distances for jobs, education, refuge, or other reasons, and hence are exposed to a new mix of flora, to which their immune systems are not adapted (Rosenberg *et al.*, 1999; Rottem *et al.*, 2005). The pollens from those plants sensitize some migrants to develop allergic states.

While the airborne environmental antigens discussed above contact mucosal areas in the airway, many environmental antigens contained in foods contact the mucosal surface of the gastrointestinal tract. Before *H. sapiens* routinely cooked food, some of the proteins contained in raw food or stuck to the food were probably antigenic and/or immunogenic at the mucosa of the gastrointestinal tract. Cooking generally denatures proteins, including those that have protease activity and can potentially destroy the mucosal epithelial junctions and penetrate across mucosal epithelia. Since most modern humans cook most of the food they eat, the environmental antigens coming from food or carried by food would seem to have decreased. However, people living modern lifestyles also derive many environmental antigens from food, which are unique to the individuals' lifestyle, likings, and affordability.

Cow and goat's milk was probably the first environmental antigen humans created for themselves, as they domesticated these animals about 10,000–12,000 years ago (Table 2.3). Only a few decades ago, most babies were breast fed by their mothers; today many infants are fed formula food, which contains a variety of antigens derived from wheat, soy, eggs, milk, and so on. People in modern societies eat food, including grains, vegetables, fruits, nuts, fish, shell fish, meats, and a large variety of other foods from regions of the world other than their own. These people are exposed to new proteins which they have not yet experienced by the time their immune system has matured at a young age.

It appears puzzling that certain proteins contained in some common foods, such as milk, wheat, eggs, and nuts are allergenic (Maleki, 2004). It is probable that although the large majority of proteins are denatured and destroyed swiftly by the acid and various enzymes in gastric fluid, some probably survive and exhibit antigenic and immunogenic properties. Furthermore, certain proteins in food after being denatured or altered due to preparation procedures, such as heating or other physical or chemical processes, become more antigenic and allergenic (Beyer *et al.*, 2001). For example, ingesting small amounts of peanuts or peanut products can cause severe allergy, often in the form of anaphylaxis, among many affected people (mostly children) living in economically advanced countries (Sicherer, 2002). It is now understood that the major peanut allergens, Ara h1 and h2, undergo structural alteration upon roasting but not boiling, and become more potent in inhibiting trypsin and in binding to IgE isolated from allergic patients (Beyer *et al.*, 2001; Maleki *et al.*, 2003).

8. PREVALENCE OF ALLERGY IS RELATED TO CHANGES IN ANTIGEN EXPOSURE

8.1. Increase in allergy keeps pace with modernization

Generally, the proportion of people with allergies in total populations in different regions of the world correlates with the region's economic advancement. Specifically, the timing and the rate of the increase in allergy incidences in a region keeps pace with the "modernization" of people's living environment and lifestyle. Based on surveys such as the

European Community Respiratory Health Survey in Adults (Janson *et al.*, 2001) and the International Study of Asthma and Allergy in Childhood (Beasley *et al.*, 2000; Weiland *et al.*, 2004), it is clear that the prevalence of asthma, allergic rhinitis, and atopic dermatitis has increased several fold in the past few decades in most regions worldwide. The prevalence is generally highest in developed countries, intermediate in developing countries, and lowest in underdeveloped countries (Beasley *et al.*, 2000; Gupta *et al.*, 2007).

The correlation between the increase in allergy prevalence and the pace of economic development becomes evident when one compares allergy prevalence among people with the same ethnic background who live in areas with different degrees of economic advancement. In developing countries, such as Taiwan (Chang *et al.*, 2006), Turkey (Cingi *et al.*, 2005), South Africa (Steinman *et al.*, 2003), and Mongolia (Viinanen *et al.*, 2005), the allergy prevalence is higher in the cities than in adjacent rural areas. In a series of surveys from 1990 to 2001 comparing the prevalence of atopy in Hamburg in West Germany and Erhurt in East Germany after Unification, allergy prevalence was initially found to be higher in Hamburg than in Erfurt, but as the "westernization" of Erfurt increased, its allergy prevalence tended towards the levels in Hamburg (Heinrich *et al.*, 1998; Richter *et al.*, 2000).

8.2. Modernization speeds up antigen exposure changes

So how does modernization and urbanization bring about an increase in allergy occurrence in a community? Through the process of modernization, amenities of "improved" living are installed in a community and its individual households. These include housing developments, governmental office buildings, industrial zones, paved roads, electricity power lines, city gas lines, piped sterile water, sewage treatment systems, supermarkets, schools, organized planting of trees, and many others. The homes, government offices, and industrial facilities have continually "improved" interior air control, insulation, weather-proofing, and separation from exterior influences.

The overall result of economic advances is that they bring about urban development and move larger proportions of populations out of agricultural and into industrial occupations. Increasing populations of people spend increasing proportions of their time indoors in artificially controlled environments. Even if the environmental and lifestyle changes listed in Tables 2.3 and 2.4 have not already been introduced to a relatively rural or old-styled community, the modernization process brings about some of the additional changes listed. Thus, modernization leads to large environmental and lifestyle changes that affect the antigen exposure of people residing in the newly modernized community.

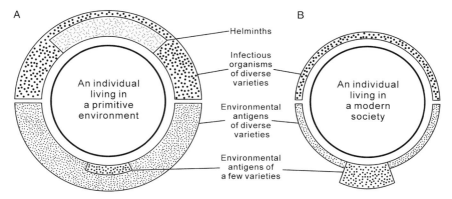

FIGURE 2.2   A schematic illustration of the "skewed antigen exposure" of people living in a modern society. (A) An individual living in a primitive environment is exposed to diverse varieties of infectious organisms and environmental antigens. (B) An individual living a modern lifestyle in a modern environment is exposed to greatly reduced amounts of infectious organisms and environmental antigens of diverse varieties, and almost no helminths, but is exposed to environmental antigens of a limited variety, unique to the individual's situation, at increased amounts and as an increased proportion of total environmental antigens.

Section 3 reasons that allergy probably did not play a role in the evolution of our immune system and that the environment, not genes, is the dominant factor in the development of allergy in most cases. Compared to the times when our immune system evolved in the ancestors of primates, mammals, and vertebrates, humans are changing their living environment and lifestyle on a massive scale at an extremely rapid speed. In the next sections, we present analyses showing that increased exposure to environmental antigens, which are generally of limited varieties and specific to individuals, and decreased exposure to infectious microorganisms and environmental antigens of broad antigenic varieties (Fig. 2.2), are together altering the balance of the human immune system leading to increased activation of IgE-mediated pathways in allergy affected persons.

9. THE HYGIENE HYPOTHESIS AND ITS IMMUNOLOGICAL BASIS

9.1. The essence of the hygiene hypothesis

The hygiene hypothesis (Strachan, 1989) proposes that the lack of frequent microbial infections in infancy and early childhood due to the reduced size of family units and improved hygiene increases the likelihood of developing allergy. Over the past two decades, many investigators have researched this hypothesis from several epidemiological and immunological perspectives. Among many findings supporting the hypothesis is that young children cared for at home are more likely to develop allergies than those cared for in nurseries, because cross-infections are more frequent among children in the nurseries than among siblings at home (Haby *et al.*, 2000; Kramer *et al.*, 1999).

9.2. "Mouthing" and geophagy behaviors of young children

The hygiene hypothesis focuses mainly on the reduced microbial infection during infancy and young childhood. Here, we further present an interesting aspect of human behavior in the first 2–3 years of life and propose that it has great implications in the exposure of the human immune system to common surrounding antigens, including not only broad "household" infectious organisms but also common environmental antigens in our surroundings.

New-born humans, like many other mammals, begin to explore, experience, and make contact with their surroundings as soon as they develop some mobility. This behavior lasts for 2–3 years, during the period when they can roll, crawl, and toddle around. A central part of their exploration of the surroundings is "mouthing" (Fessler and Abrams, 2004; Juberg *et al.*, 2001). During this period, they secrete a lot of saliva and constantly swallow the saliva. When they crawl on the ground, they lick, taste, and swallow their saliva, as though they were sampling the ground. They frequently put their fingers into their mouths, "wash" their fingers with their saliva, and swallow the saliva. They either reach and grab small objects and put them into their mouths or move their mouths to contact large objects. The substances, soluble or in particular form, contained in the saliva contact the mucosal surfaces in a young child's mouth, nasal area, and gastrointestinal tract. It is possible that in such a way, the young human child exposes his or her immune system to common "household" infectious organisms and environmental antigens, both living and non-living, and achieves immunological adjustments to the environment, a process phrased "immunocalibration" by anthropologists Fessler and Abrams (2004). As part of this adjustment process, the young child establishes a "home-grown" flora of commensal microorganisms and helminths in the intestine.

As recently as 10 years ago, children and adults living in some under-developed regions habitually ate soil, a behavior termed "geophagy" by researchers who study animal behavior (Geissler *et al.*, 1997; Wong *et al.*, 1988). Several hypotheses have suggested that humans, nonhuman primates, and various other animals eat soil, for nutrition and detoxification-related bodily requirements (Krishnamani and Mahaney, 2000). A few reports have pointed out that soil consumed by children living in those regions contained a variety of helminths and their eggs, and that geophagy promoted infection by those helminths (Geissler *et al.*, 1998; Saathoff *et al.*, 2002). In keeping with the notion of immune adjustment, it is possible that if the intestine of a child was to be occupied by helminths anyway, those transmitted from the soil near the child's shelter should be the least "foreign" and risky.

In economically advanced modern societies, the measurement of the amount of soil ingested by young children has been performed in many environmental toxicological studies to assess the risk of exposure of children to toxic or poisonous substances present in certain neighborhoods (Calabrese *et al.*, 1989; Davis *et al.*, 1990). While these studies were not specifically immunology-related, the methodology used and the data obtained indicate that that if young children are allowed outdoor activities, they ingest substantial amounts of soil from direct mouth-soil and indirect soil-hand-mouth contacts. The biological substances, living or nonliving, contained in soil are naturally taken in by young children ingesting the soil.

Today, in primitive shelters in the remote villages of tribal people, human crawlers and toddlers are still exposed to rich sources of indigenous, domestic infectious organisms and environmental antigens in the earthy ground and bio-rich surroundings. Exposure to these biological substances, both living and nonliving, helped establish a balanced immune system in the ancestors of humans during their millions of years of evolution. However, in the hygienic, bio-deprived modern homes of economically advanced modern societies, crawlers (who are often confined to cribs) and toddlers are exposed to plastic toys, disinfected floors or vacuumed carpets, and very little biological substance, except possibly dust mites and their fecal particles, cat dander and other indoor antigens. These few environmental antigens account for a high proportion of the doses of antigens their immune systems receive.

9.3. Reduced microbial infections shift TH1 to TH2

After the hygiene hypothesis was introduced, the explanation that frequent microbial infections in infancy and early childhood are required to tune the immune system for adequate TH1 functions drew much research, discussion and debate with generally acquiescent opinions emerging (Matricardi *et al.*, 2002; Prioult and Nagler-Anderson, 2005; Romagnani, 2004; Schaub *et al.*, 2006; von Hertzen, 2000). Primitive conditions of hygiene support frequent infections, as typified by hepatitis A virus, *Helicobacter pylori* bacterium, and *Toxoplasma gondii* fungus infections (Linneberg *et al.*, 2003), allowing TH1-associated immune mechanisms to become robust. Under improved hygiene conditions, these infections are reduced, resulting in under-stimulation of TH1, hence unleashing TH2-skewed pathways.

The main immune effectors for defending against microbial infections are (i) IgM, IgG, and IgA, which mediate mechanisms to clear cell-free bacteria, fungi, protozoa, viruses, and their components, (ii) CD4 Th type-I cells, which activate infected macrophages to clear bacteria and protozoa in vesicles, and (iii) CD8 cytotoxic T cells, which lyse host cells that harbor viruses in their cytosol. These mechanisms are mainly generated in TH1 pathways. Invading parasitic worms, on the other hand, are mainly dealt with by IgE and activated eosinophils generated in TH2 pathways (Janeway et al., 2005). The TH1 and TH2 domains are self-enhancing and mutually inhibitory. Among such inhibitory effects, IFN-γ of TH1 inhibits TH2 and IL-4 of TH2 inhibits TH1 by effectively blocking the differentiation of the counterpart subsets from naïve CD4 precursors (Coffman, 2006). Thus, frequent infections by microorganisms elicit TH1 responses and thereby deflate TH2, whereas the lack of such infections mutes TH1 mode and releases the inhibitory pressure on TH2 expansion.

While the TH1/TH2 explanation offers an attractive rationale, it has been challenged. Firstly, helminthic infection, a hallmark of poor hygiene, drives a TH2-accented response, producing high levels of IgE in infected persons, yet helminthic infection has an inverse relationship with the occurrence of allergy (Kitagaki et al., 2006; Yazdanbakhsh et al., 2002). Second, people living in modern societies also suffer from increased TH1-promoted autoimmune diseases, including those caused by auto-reactive IgG and IgM, type-I CD4 helper cells, and CD8 cytotoxic T cells. There is now preliminary evidence that the TH1-related autoimmune diseases, such as rheumatoid arthritis, type-1 diabetes, and psoriasis, occur at higher frequencies in patients who also have asthma, allergic rhinitis, and eczema (Kero et al., 2001; Simpson et al., 2002). Thus, if TH1 is under-stimulated from a lack of infections, how does autoimmunity increase?

9.4. The loss of regulation of Treg cells and IL-10

In recent years, the "Treg downregulation" theory appears to offer a unified rationale for explaining the increases in both allergy and autoim-munity. The foundation of this theory is that Foxp3$^+$, CD4$^+$, CD25$^+$ Treg cells and their cytokine IL-10 play important roles in maintaining homeo-static TH1 and TH2 functions, exerting suppressive activities on both (Curotto de Lafaille and Lafaille, 2002; Ostroukhova and Ray, 2005). In some people, Treg cells are downregulated, thus allowing both TH1 and TH2 to expand, leading to increased unwanted immune activities towards self antigens and environmental antigens (Robinson et al., 2004; Yazdanbakhsh et al., 2002).

The "Treg downregulation" theory appears to be in conflict with the "TH1 to TH2 shift" theory in that the former suggests that TH1 activities are generally enhanced, while the latter suggests that the overall TH1 activities are reduced as a result of infrequent microbial infection. Thus, a proposition claiming that the entire TH1 or TH2 domain is under- or over-activated does not provide an adequate explanation for the increase in allergy. The recent finding that IL-10 suppresses IgE production and enhances IgG4 and IgA production (Taylor *et al.*, 2006; Verhagen *et al.*, 2006) seems to be one step closer to explaining the increase in allergy occurrence. Furthermore, recently, TH17 cells, a CD4 T cell-derived subset distinct from TH1 and TH2 and countering TH1 in functions, and their cytokine, IL-17, have been implicated in allergic asthma, leukocyte movement, inflammation, and tissue damage (Schnyder-Candrian *et al.*, 2006; Steinman, 2007; Umetsu and DeKruyff, 2006). These findings suggest that all TH T cell subsets (TH1, TH2, and TH17) and Treg and a horde of their cytokines play intricate, concerted or counter-acting roles in the pathological processes of autoimmune and allergic diseases. It remains to be determined whether frequent microbial infection affects TH17 activities.

9.5. The hygiene hypothesis neglects the impact of changes in environmental antigens

The hygiene hypothesis focuses on the immune imbalance caused by the lack of frequent microbial infections during infancy and young childhood. The various immunological mechanisms proposed for the hygiene hypothesis stress that the lack of frequent microbial infections somehow elevates the TH2 pathway. We would suggest that the various immunological explanations for the hygiene hypothesis fall short of explaining why allergic responses occur as the result of the shift to TH2 and that the hygiene hypothesis, as it is, provides only half of the answer to the question of why allergy has become a modern day malady. The other half of the answer is that the composition of the environmental antigens in the living space of many individuals living in modern societies has drastically changed.

IgE has been used most commonly as an indicator of TH2 pathway activity. The fact that some patients with allergy, even very severe allergy, have serum IgE levels lower than 30 IU/ml (Lanier, 2005; Spector and Tan, 2007), which is lower than the average of serum IgE levels in people without allergy, argues against the suggestion that elevated TH2 activity is the cause of allergy in those patients. Indeed, it is fair to say that if house dust mites and cat dander were not present in many modern homes at high concentrations and did not account for high proportions of total airborne antigens, most of the asthma cases in the world would not have occurred. Similarly, if grass and tree pollens of selected varieties were not present in the air at high concentrations and did not account for high proportions of total airborne antigens, a significant portion of the allergic rhinitis cases in the world would probably not have occurred. Exposure to environmental antigens of a limited variety at high concentrations and in high proportions, which are unique to individuals in their living spaces, definitely seems to be a critical factor in the prevalence of allergy in modern societies.

A primary feature of allergic diseases is that they are episodic. Patients with allergic asthma, allergic rhinitis, and food allergy develop asthma attacks, sneezing attacks, or anaphylactic reactions, respectively, in bouts when they are exposed to allergens (or triggers). Most patients with seasonal allergy suffer in the spring and fall, when tree and grass pollen counts climb. Therefore, when we study how environmental and lifestyle changes, including improvements in hygiene, have brought about the increase in allergy occurrence, we should examine how those changes affect our exposure to environmental antigens.

For many people living in a highly hygienic environment following the hygiene-conscious lifestyle of modern societies, the only remaining major protein antigens that induce mucosal IgE production appear to be the potentially allergic substances in the air or food. In the next few sections, we will analyze how changes in antigen exposure to both infectious organisms and to environmental antigens work in concert to cause the increase in IgE-mediated mast cell sensitization pathway and allergic diseases.

10. KEY FACTORS AFFECTING THE IgE-MEDIATED PATHWAY IN ALLERGIC RESPONSES

10.1. Allergic state should be expressed in terms of the sensitivity of the IgE pathway

To delineate the molecular basis underlying the elevated sensitivity toward environmental antigens in allergic individuals, it is important to understand the etiology and the manifestation of allergy. Studies of a new therapeutic humanized IgG anti-IgE antibody (abbreviated anti-IgE) in patients with different allergic diseases have helped us to understand the central roles of IgE in allergy (Chang, 2000; Chang *et al.*, 2007). Anti-IgE can neutralize free IgE and, unlike ordinary anti-IgE antibodies, does not sensitize mast cells and basophils. Among its multiple immunoregulatory effects, anti-IgE indirectly causes the downregulation of FcεRI on mast cells and basophils, rendering these cells insensitive to allergen stimulation (Chang and Shiung, 2006; MacGlashan *et al.*, 1997). Anti-IgE has been shown in more than 30 Phase II and III human clinical trials to be efficacious in treating allergic asthma, allergic rhinitis, and peanut allergy (Holgate *et al.*, 2005; Lanier *et al.*, 2003; Leung *et al.*, 2003). It may also ameliorate other allergic diseases, such as atopic dermatitis (Chang *et al.*, 2007). Anti-IgE reduces nearly all symptoms of allergy, such as the inflammation in the affected mucosa, overall inflammatory status in the entire immune system, skin rashes, eosinophil infiltration to the lung, bronchial constriction, asthma exacerbations, sneezing attacks, anaphylactic reactions, and others (Holgate *et al.*, 2005; Lanier *et al.*, 2003; Leung *et al.*, 2003).

These clinical studies of anti-IgE indicate that the hygiene hypothesis, TH1/TH2/TH17 imbalance, skewed expression of IL-2, IL-4, Il-10, IL-17, and TGF-β, and other theories proposed to explain the increase of allergy in modern times must be reduced to the level of how the IgE-mediated pathway becomes more sensitive or elevated. In another words, in the establishment of an allergic state and in response to an allergen, the TH pathways and various cytokines can be regarded as regulators, while IgE-mediated mast cell and basophil sensitization is an effector. However the regulation process has taken place, the manifestation must be elucidated at the effector level.

10.2. Key factors dictating mast cell sensitivity

In the human body, mast cells residing underneath the mucosal epithelium and the skin, and basophils in the blood are the primary cells that release the pharmacological mediators that manifest various allergic symptoms. Many factors, including cytokines, chemokines, neurological products, environmental factors, and others both immunological and nonimmunological, can influence the sensitivity and activity of these inflammatory cells (Chang and Shiung, 2006). We propose that among these factors, two dictate whether a mast cell or basophil can be activated upon its exposure to an allergen: (i) the density of FcεRIs that are occupied by allergen-specific IgE, and (ii) the concentration of the allergen in the fluid surrounding the cell (Fig. 2.3A). This underlying state of immunological sensitization can also be influenced by non-IgE and non-allergen-related "triggers", such as sudden temperature change, automobile exhaust, and exercise.

The sensitivity of a mast cell or basophil toward an allergen is mainly determined by the density of FcεRIs (number of FcεRIs/cell) that are bound by IgE specific to the allergen. This density is further determined by two parameters: (i) the baseline level of total IgE, and (ii) the proportion of allergen-specific IgE in total IgE (Fig. 2.3B). The density of total FcεRI is related to the concentration of total IgE: the higher the IgE concentration, the denser the FcεRI. It was found that in a range of total IgE concentration from 3 to 10,000 ng/ml, the numbers of FcεRIs per

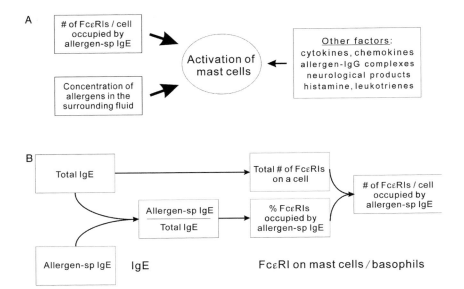

FIGURE 2.3   The key factors that determine the sensitivity and activity of mast cells and basophils. (A) The two factors that dictate the activation of mast cells and basophils. (B) The density of FcεRI on a mast cell or basophil that is occupied by allergen-specific IgE is a function of total IgE and allergen-specific IgE in the fluid surrounding the cell. Such a density corresponds to the sensitivity of the cell toward the allergen.

basophil are in the range of 30,000–600,000 (Malveaux *et al.*, 1978). In other words, while the concentration of total IgE varies over a 3,000-fold range, the density of FcεRI varies 20-fold.

Unlike other antibody classes, IgE by itself, without prior binding to antigens, binds to FcεRI with very high affinity ($K_d \sim 1 \times 10^{-10}$ M). Thus, in a steady state, the proportion of total IgE-occupied FcεRI on a mast cell or basophil that is occupied by allergen-specific IgE is the same as the ratio of allergen-specific IgE to total IgE in the surrounding fluids. In addition, because of the high affinity of IgE binding to FcεRI, most of the FcεRIs are occupied by IgE in the range of IgE concentrations found in most people (Chang *et al.*, 2007). Non-allergen-specific IgE, such as IgE specific for antigens of parasitic worms, has two opposing effects: (i) it increases total IgE concentration and hence the density of FcεRI on cells, and (ii) it decreases the proportion of allergen-specific IgE in total IgE and hence the proportion of FcεRIs occupied by allergen-specific IgE, with the second effect more pronounced than the first.

The other major determinant in the IgE-mediated mast cell activation process is the concentration of allergens in the surrounding fluids. This should be related to the concentration of the allergen and, to some extent, to the relative proportion of this allergen in the total allergens in the air surrounding the affected person or in the food the person is ingesting. Because allergen-containing particles may be retained to different extents in different areas of the mucosal surfaces, the mast cells in different mucosal locations are exposed to different concentrations of allergens.

11. TOTAL IgE *VS.* ALLERGEN-SPECIFIC IgE

Based on the above analysis, a key question pertaining to how antigen exposure changes lead to the increase of allergy is whether the proportions of IgE specific to certain environmental antigens in total IgE are increased significantly in people with allergy, hence preventing the non-allergen-specific IgE from competing sufficiently with allergen-specific IgE to occupy FcεRI. If the analysis in Section 7 is correct, in the modern day environment where there is much reduced diversity of environmental antigens and much increased concentrations and proportions of certain environmental antigens unique to affected individuals, IgE induced by such environmental antigens and specific to such allergens should account for increased proportions of total IgE in patients and the allergen-specific IgE should occupy increased proportions of FcεRI on mast cells and basophils.

11.1. Is the "IgE-blocking mechanism" valid?

The "IgE-blocking mechanism" was first suggested to explain why helminth infection has an inverse relationship with atopic disease occurrence (Godfrey and Gradidge, 1976). It was reasoned that the large amounts of IgE produced as a result of a helminth infection out-compete allergen-specific IgE in binding to FcεRI. The loss of such a protective mechanism due to the lack of helminth infection has not been well received as part of the explanation for the hygiene hypothesis, however, because experimental results addressing this possibility were controversial. For example, in *in vitro* experiments mixing allergen-specific IgE with IgE from helminth-infected individuals, ratios as high as 1:388 could not prevent allergens from sensitizing basophils that had been stripped of their FcεRI-bound IgE by treating with lactic acid (Mitre *et al.*, 2005).

MacGlashan *et al.* (1997) estimated that a basophil with only 200 mono-specific IgE-occupied FcεRIs can be sensitized by the antigen to which the IgE is specific. Thus, if the basophils used in the above IgE-mixing experiments had 220,000 FcεRIs per cell (an average density) (Malveaux *et al.*, 1978), an IgE preparation with antigen-specific IgE at a 1:388 ratio would still allow 566 FcεRIs to be bound by antigen-specific IgE. In fact, the researchers performing the IgE-mixing experiments, who did not determine the density of FcεRI on the basophils used, commented that the ratios of helminth IgE to allergen-specific IgE were probably not high enough (Mitre *et al.*, 2005).

In our opinion, the "IgE-blocking mechanism" is sensible and the loss of this mechanism should be incorporated as part of the explanation for the hygiene hypothesis, in view of the major role of IgE in allergic diseases. However, as discussed above, the concentration and composition of IgE are involved in delicate relationship which affects the sensitivity of mast cells and basophils. In many cases, the concentration of either total IgE or allergen-specific IgE alone cannot determine the level of mast cell sensitization and the allergic tendency of patients (Erwin *et al.*, 2007). The ratios of allergen-specific IgE to total IgE would be a more reliable predictor (Lynch *et al.*, 1998).

The data in the literature concerning the effects of helminth infections in prohibiting or protecting against allergy are compelling. Relevant studies include numerous epidemiological analyses comparing the rates of incidences of allergy in people chronically infected and not infected with helminths (Cooper *et al.*, 2003; Flohr *et al.*, 2006; Lynch *et al.*, 1998). A number of papers have also demonstrated that in mouse models, infection with helminths protects against allergy-like reactions (Kitagaki *et al.*, 2006; Mangan *et al.*, 2004). While various immunological mechanisms, including the changes in certain cytokine levels, such IL-4, IL-10, and IL-13, and the secretion of immunomodulatory substances by helminths, have been suggested, invariably the helminth-infected people and mice have higher total IgE levels than those not infected with helminths.

11.2. Allergic patients with very low IgE

The claims that the IgE levels in patients with allergy are higher than those in people without allergy, as concluded in many surveys, are a statistical matter. The range of IgE levels in allergic patients and that in normal people both cover several orders of magnitude and overlap broadly. Some people without allergies have very high IgE levels. Thus, it would be very crude to conclude that an elevated TH2 pathway based on high total IgE levels will lead to an allergic state. Similarly, some patients with allergic asthma, allergic rhinitis, or other allergic conditions have very low levels of serum IgE, say, below 30 IU/ml (Lanier, 2005; Spector and Tan, 2007), which is much lower than the average serum IgE levels in normal people (Carosso *et al.*, 2007; Sharma *et al.*, 2006). These facts suggest that a seemingly depressed TH2 pathway based on very low total IgE levels cannot block the occurrence of allergy. To better assess the likelihood of the occurrence of allergic response towards a set of antigens, it is important to know whether the IgE specific for those antigens account for a significant proportion in total IgE.

Some patients develop extreme sensitivity, even anaphylactic reactions, to single allergens, such as bee venom proteins or peanut allergens. A minute amount of an allergen, e.g., one wasp sting (Steen *et al.*, 2005) or one half a peanut (Leung *et al.*, 2004) will cause anaphylactic reactions in such patients. It is interesting to investigate how allergen-specific IgE occupies the FcεRI on mast cells and basophils in those patients. Allergic patients with serum IgE below 30 IU/ml have been excluded from the organized clinical trials of omalizumab (an anti-IgE drug, see Section 10.1), but sporadic studies have shown that anti-IgE works for those patients (Lanier, 2005; Leung *et al.*, 2003), indicating that although IgE is very low in those patients, it plays a pathogenic role. Based on the analysis in Section 10.2, one would surmise that although the total IgE is very low and hence the density of FcεRI on mast cells and basophils will also be low, the proportions of allergen-specific IgE is probably high, and therefore, the density of FcεRI occupied by allergen-specific IgE is sufficiently high for the mast cells and basophils to be sensitized. A recent study showed that among 220 patients who developed allergic responses to Hymenoptera venom, most of those with the most severe allergic reactions had lowest levels of total IgE (Sturm *et al.*, 2007).

12. ALLERGENIC SUBSTANCES BECOME DOMINANT MUCOSAL IgE STIMULATORS

12.1. Certain protein antigens are powerful mucosal IgE stimulators

Microorganisms that successfully break through the mucosal epithelium of a mouse or human and establish some degree of proliferation locally induce TH1-accented immune mechanisms, including the production of IgG and IgA. On the contrary, nonproliferating protein substances induce TH2-skewed responses in the mucosa (Lambrecht *et al.*, 2000). Besides being abundant in the air surrounding an allergic individual or present in the food the person is ingesting, allergenic proteins are usually highly soluble, small in size, structurally stable, and possess protease activity (Janeway *et al.*, 2005). The proteolytic activity helps destroy the transmembrane protein occludin, which holds the tight junction between mucosal epithelial cells (Wan *et al.*, 2001). Additional properties of allergens, as revealed by the structurally altered Ara h1 and h2 allergens in roasted peanuts, are their resistance to proteases in mucosal fluids and their enhanced antigenicity with IgE (Maleki *et al.*, 2003).

Based on the comparable amounts of IgE produced in some people (not all allergic) living in highly hygienic, helminth-free conditions, the stimulation at mucosal areas of the airway by minute amounts of airborne protein antigens is remarkably effective. The amplitude of mucosal stimulation by allergenic proteins contained in food, as shown by the extremely high IgE levels in some patients with atopic dermatitis (Laske *et al.*, 2003; Scaglia *et al.*, 1979), is comparable to that by the heaviest loads of helminth infection (see below).

12.2. Parasitic worms and the immune system strike a balance

Certain helminths, e.g., the soil-transmitted geohelminths, infest and reside only in the intestine of humans and do not migrate to different organs or tissues (Murray *et al.*, 2005). These parasitic worms induce a range of immune responses, including the production of large amounts of IgE, at least some of which is specific for antigens on the parasites. It is not clear how helminth infection induces the production of IgE. A possibility is that some antigens shed by mature worms, their eggs, or larvae induce mucosal IgE production in a manner similar to the way environmental antigens induce mucosal IgE production.

Parasitic worms are called "parasitic", because we think that they take nutrients from our bodies without doing any good (Murray *et al.*, 2005). However, those worms have been an invariant and integral part of the lives of vertebrates and mammals during the hundred million years of evolution, as evidenced by the fact that all free-ranging animals (El-Shehabi *et al.*, 1999), chimpanzees in their natural habitats (Muehlenbein, 2005), and tribal people in their primitive villages (Sugunan *et al.*, 1996) are infected by helminths. Since the immune system has not been able to expel the helminths effectively, a forced commensal relationship has evolved between the worms and their hosts. The fact that most people chronically infected by parasitic worms are asymptomatic and that the clinical studies with "helminthic therapy" yield positive results on a number of autoimmune (Croese *et al.*, 2006; Summers *et al.*, 2005) and allergic diseases (Falcone and Pritchard, 2005; Moncayo and Cooper, 2006) suggest that a commensal relationship does exist, at least in an immunological sense.

12.3. Allergenic substances are the major remaining protein antigens at the mucosa in modern life

As discussed above, before medicine became widespread in the last century, a variety of parasitic worms were regular residents in the small and large intestines of most people (Murray *et al.*, 2005). The antigens shed by those organisms are presumably powerful stimulators of IgE production in the intestinal mucosa. The IgE produced by these antigens compete for the FcεRI on mast cells and basophils and can help mute the allergic reactions driven by potential allergens. With the advent of

medicine and the greatly improved hygiene of modern times, most people do not carry parasitic worms and hence lose this source of powerful
mucosal IgE stimulators (Bart *et al.*, 2006; Tanaka and Tsuji, 1997).

As discussed in Sections 7.2 and 7.3, over the hundreds of millions of
years until humans started to change their environment and lifestyle, the
ancestors of humans came into contact with rich sources of environmental
antigens. Because of the very hygienic environment and lifestyle of modern times, we are no longer exposed to bio-rich and protein-rich soil and
natural water untouched by human influence (Fig. 2.2). The remaining
major protein antigens that our mucosa is exposed to are the airborne
indoor and outdoor antigens, which are harder for us to avoid, and
antigens contained in certain foods. The particular makeup of those antigens is not natural, but highly artificial and unique to each individual's
situation and lifestyle. Thus, it is likely that these proteins have become
the major stimulators of IgE production in the mucosa in many people
living a modern lifestyle in a modern environment.

13. MOLECULAR MECHANISMS DELINEATING THE EFFECTS OF THE SKEWED ANTIGEN EXPOSURE ON THE SENSITIZATION OF THE IgE PATHWAY

Based on the above analysis, the immunological mechanisms linking
antigen exposure changes to the increase of allergy in modern societies
are summarized in Figure 2.4. This summary delineates the effects of
(i) decreased exposure to a variety of infectious microorganisms and
parasitic worms, (ii) decreased exposure to a large variety of environmental antigens, and (iii) exposure to certain environmental antigens unique
to affected individuals at high concentrations and in high proportions
(Fig. 2.2), on the production of allergen-specific IgE and total IgE. The
combined effects of these antigen exposure changes bring about elevated
proportions of allergen-specific IgE in total IgE, and hence the proportions of allergen-specific IgE-occupied FcεRI and the density of such
receptors on mast cells and basophils, rendering these cells very sensitive
to allergen sensitization.

It should be reiterated that while the environmental antigens that can
cause allergy among people living different regions of the world are very
diverse, generally only a limited variety of allergens are present at high
concentrations in the surroundings of an allergic person, relating to the
person's residential location, living environment, lifestyle, diet and so on.
These antigens may therefore induce antigen-specific IgEs that account
for a high proportion of total IgE. The relative abundance of the specific
antigens in the air or in the food at particular times, in combination with

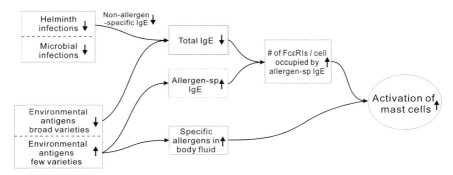

FIGURE 2.4   The molecular basis delineating how "skewed antigen exposure" enhances the IgE-mediated sensitization of mast cells and basophils. The lack of helminth infection and the decreased exposure to environmental antigens of diverse varieties lowers total IgE synthesis in the mucosa. The exposure to environmental antigens of a few varieties unique to the affected individuals at higher proportions increases allergen-specific IgE production. As a result, allergen-specific IgE accounts for higher proportions of total IgE, thus increasing the density of allergen-IgE-occupied FcεRI on mast cells and basophils, and the sensitivity of those cells toward the allergens (Figure 2.3).

the relatively high occupancy of FcεRI by IgE-specific to the allergens, renders the already sensitized host very sensitive to the antigens.

The under-stimulation of the TH1 domain by microorganisms turns up TH2 differentiation and activities. The downregulation of Treg and IL-10 also raises TH2 activity. The stimulation by environmental antigens at the mucosa is biased toward TH2; the resultant IL-4 drives effective IgE production and inhibits TH1 differentiation. Thus, the decreased microbial infection and exposure to environmental antigens at increased concentrations act in concert to increase the proportion of allergen-specific IgE in total IgE and hence the density of FcεRI that is occupied by allergen-specific IgE.

Many adults, who previously lived in conditions with poor hygiene and had frequent infections in childhood and were thus free of allergy, develop allergies a few years after moving to a modern environment and living a modern lifestyle (e.g., many immigrants living in economically advanced countries who came from underdeveloped countries (Rosenberg *et al.*, 1999; Rottem *et al.*, 2005)). Additionally, many healthy people who move to new equally modernized locations and live the same lifestyle develop allergy. These facts indicate that the exposure to environmental antigens is a powerful and sometimes dominant factor in causing allergy. In these few scenarios, the shift of TH1 to TH2 due to the lack of frequent microbial infection during young childhood is not as relevant as the new encounter with environmental antigens not previously experienced. Furthermore, as discussed in Section 11.2, some allergic patients have very low IgE (below 30 IU/ml). To suggest that the TH2 pathway is stimulated by any immunological imbalance in these patients is not persuasive.

14. CONCLUDING REMARKS

Here we have proposed a "skewed antigen exposure" theory to explain the appearance and rise of allergy in modern societies. The fundamental basis of this theory is that in many people living in modern societies, a small variety of environmental antigens have become the major driver of mucosal IgE stimulation, and as a result, IgE specific for such environmental antigens accounts for a high proportion of total IgE, rendering the IgE-mediated allergic pathway very sensitive to such antigens.

Our theory adopts the concept and claims of the hygiene hypothesis and expands on several major aspects of it. First, our theory incorporates not only the impact of the reduced exposure to microorganisms but also the reduced exposure to parasitic worms that drive parasite antigen-specific IgE and probably enhance overall IgE production. Second, our theory considers that the reduced mucosal exposure to a diverse variety of environmental antigens and the increased mucosal exposure to a limited variety of environmental antigens unique to individuals are crucial in the development of allergy. Third, the hygiene hypothesis examines the improvement in hygiene brought about by changes in environment and lifestyle over the past decades; our theory on the other hand takes an evolutionary view and examines how the cumulative changes in humans' living environment and lifestyle, not limited to those that improve hygiene, have affected the exposure of our immune system to antigens, especially the exposure of our mucosa to nonpenetrating, nonreplicating protein antigens.

Our skewed antigen exposure theory should be critically tested in future research. A central issue is whether allergen-specific IgE accounts for a higher proportion of total IgE in patients with allergy than in people without allergy who live in the same environment and have a similar lifestyle. Investigating the profiles of antigenic specificities in allergic patients with very low IgE (some patients with allergic rhinitis or allergic asthma) or very high IgE (many patients with atopic dermatitis), why some patients develop extreme sensitivity to peanuts, and why some infants are very susceptible to food allergy, will also help attest the concepts discussed in this chapter.

We believe that further research on the subject discussed in this paper will help our understanding of allergy and our goal to develop better care for patients with allergies. If our theory is largely valid, it bears important implications in the development of new prophylactic treatments and therapies for IgE-mediated allergic diseases. Administering allergic patients with live intestinal geohelminths, their larvae, or eggs (Falcone and Pritchard, 2005; Moncayo and Cooper, 2006) sounds repulsive as a treatment, but it has a rational basis and provides leads for future pharmaceutical development. It would be interesting to study whether oral administration of a mixture of helminth-derived proteins can drive safe and active mucosal IgE production. Therapeutic approaches, such as anti-IgE (Chang *et al.*, 2007), that intervene in the IgE-mediated pathway would be attractive, if they can be made broadly available to the large populations of patients suffering from allergy.

ACKNOWLEDGMENTS

The authors thank Dr. Carmay Lim and Ms. Miranda Loney for reading the manuscript and giving valuable suggestions and Mr. Donic Lu for preparing the figures. This work is supported by a grant, #96–2320-B001–014-MY3, from the National Science Council, Taiwan.

REFERENCES

Alroy, J., Marshall, C. R., Bambach, R. K., Bezusko, K., Foote, M., Fursich, F. T., Hansen, T. A., Holland, S. M., Ivany, L. C., Jablonski, D., Jacobs, D. K., Jones, D. C., *et al.* (2001). Effects of sampling standardization on estimates of Phanerozoic marine diversification. *Proc. Natl. Acad. Sci. USA* **98,** 6261–6266.

Arlian, L. G., and Morgan, M. S. (2003). Biology, ecology, and prevalence of dust mites. *Immunol. Allergy Clin. North Am.* **23,** 443–468.

Arlian, L. G., and Platts-Mills, T. A. (2001). The biology of dust mites and the remediation of mite allergens in allergic disease. *J. Allergy Clin. Immunol.* **107,** S406–S413.

Arlian, L. G., Schumann, R. J., Morgan, M. S., and Glass, R. L. (2003). Serum immunoglobulin E against storage mite allergens in dogs with atopic dermatitis. *Am. J. Vet. Res.* **64,** 32–36.

Bart, J. M., Abdukader, M., Zhang, Y. L., Lin, R. Y., Wang, Y. H., Nakao, M., Ito, A., Craig, P. S., Piarroux, R., Vuitton, D. A., and Wen, H. (2006). Genotyping of human cystic echinococcosis in Xinjiang, PR China. *Parasitology* **133,** 571–579.

Beasley, R., Crane, J., Lai, C. K., and Pearce, N. (2000). Prevalence and etiology of asthma. *J. Allergy Clin. Immunol.* **105,** S466–S472.

Betz, U. A., Mayer, W. E., and Klein, J. (1994). Major histocompatibility complex class I genes of the coelacanth Latimeria chalumnae. *Proc. Natl. Acad. Sci. USA* **91,** 11065–11069.

Beyer, K., Morrow, E., Li, X. M., Bardina, L., Bannon, G. A., Burks, A. W., and Sampson, H. A. (2001). Effects of cooking methods on peanut allergenicity. *J. Allergy Clin. Immunol.* **107,** 1077–1081.

Blair, C., Nelson, M., Thompson, K., Boonlayangoor, S., Haney, L., Gabr, U., Baroody, F. M., and Naclerio, R. M. (2001). Allergic inflammation enhances bacterial sinusitis in mice. *J. Allergy Clin. Immunol.* **108,** 424–429.

Calabrese, E. J., Barnes, R., Stanek, E. J., 3rd, Pastides, H., Gilbert, C. E., Veneman, P., Wang, X. R., Lasztity, A., and Kostecki, P. T. (1989). How much soil do young children ingest: An epidemiologic study. *Regul. Toxicol. Pharmacol.* **10,** 123–137.

Cannon, J. P., Haire, R. N., Rast, J. P., and Litman, G. W. (2004). The phylogenetic origins of the antigen-binding receptors and somatic diversification mechanisms. *Immunol. Rev.* **200,** 12–22.

Carosso, A., Bugiani, M., Migliore, E., Anto, J. M., and DeMarco, R. (2007). Reference values of total serum IgE and their significance in the diagnosis of allergy in young European adults. *Int. Arch. Allergy Immunol.* **142,** 230–238.

Chang, J. W., Lin, C. Y., Chen, W. L., and Chen, C. T. (2006). Higher incidence of Dermatophagoides pteronyssinus allergy in children of Taipei city than in children of rural areas. *J. Microbiol. Immunol. Infect.* **39,** 316–320.

Chang, T. W. (2000). The pharmacological basis of anti-IgE therapy. *Nat. Biotechnol.* **18,** 157–162.

Chang, T. W., and Shiung, Y. Y. (2006). Anti-IgE as a mast cell-stabilizing therapeutic agent. *J. Allergy Clin. Immunol.* **117,** 1203–1212; quiz 1213.

Chang, T. W., Wu, P. C., Hsu, C. L., and Hung, A. F. (2007). Anti-IgE Antibodies for the Treatment of IgE-Mediated Allergic Diseases. *Adv. Immunol.* **93,** 63–119.

Chesney, C. (2001). Systematic review of evidence for the prevalence of food sensitivity in dogs. *Vet. Rec.* **148,** 445–448.

Cingi, C., Cakli, H., Us, T., Akgun, Y., Kezban, M., Ozudogru, E., Cingi, E., and Ozdamar, K. (2005). The prevalence of allergic rhinitis in urban and rural areas of Eskisehir-Turkey. *Allergol. Immunopathol. (Madr).* **33,** 151–156.

Coffman, R. L. (2006). Origins of the T(H)1-T(H)2 model: A personal perspective. *Nat. Immunol.* **7,** 539–541.

Cooper, P. J., Chico, M. E., Rodrigues, L. C., Ordonez, M., Strachan, D., Griffin, G. E., and Nutman, T. B. (2003). Reduced risk of atopy among school-age children infected with geohelminth parasites in a rural area of the tropics. *J. Allergy Clin. Immunol.* **111,** 995–1000.

Craig, P. S., and Larrieu, E. (2006). Control of cystic echinococcosis/hydatidosis: 1863-2002. *Adv. Parasitol.* **61,** 443–508.

Croese, J., O'Neil, J., Masson, J., Cooke, S., Melrose, W., Pritchard, D., and Speare, R. (2006). A proof of concept study establishing Necator americanus in Crohn's patients and reservoir donors. *Gut.* **55,** 136–137.

Curotto de Lafaille, M. A., and Lafaille, J. J. (2002). CD4(+) regulatory T cells in autoimmunity and allergy. *Curr. Opin. Immunol.* **14,** 771–778.

Davis, S., Waller, P., Buschbom, R., Ballou, J., and White, P. (1990). Quantitative estimates of soil ingestion in normal children between the ages of 2 and 7 years: Population-based estimates using aluminum, silicon, and titanium as soil tracer elements. *Arch. Environ. Health* **45,** 112–122.

Driscoll, C. A., Menotti-Raymond, M., Roca, A. L., Hupe, K., Johnson, W. E., Geffen, E., Harley, E. H., Delibes, M., Pontier, D., Kitchener, A. C., Yamaguchi, N., O'Brien S., J., *et al.* (2007). The Near Eastern origin of cat domestication. *Science* **317,** 519–523.

Eason, D. D., Cannon, J. P., Haire, R. N., Rast, J. P., Ostrov, D. A., and Litman, G. W. (2004). Mechanisms of antigen receptor evolution. *Semin. Immunol.* **16,** 215–226.

El-Shehabi, F. S., Abdel-Hafez, S. K., and Kamhawi, S. A. (1999). Prevalence of intestinal helminths of dogs and foxes from Jordan. *Parasitol. Res.* **85,** 928–934.

Emanuel, M. B. (1988). Hay fever, a post industrial revolution epidemic: A history of its growth during the 19th century. *Clin. Allergy* **18,** 295–304.

Erwin, E. A., Ronmark, E., Wickens, K., Perzanowski, M. S., Barry, D., Lundback, B., Crane, J., and Platts-Mills, T. A. (2007). Contribution of dust mite and cat specific IgE to total IgE: Relevance to asthma prevalence. *J. Allergy Clin. Immunol.* **119,** 359–365.

Falcone, F. H., and Pritchard, D. I. (2005). Parasite role reversal: Worms on trial. *Trends. Parasitol.* **21,** 157–160.

Fessler, D. M., and Abrams, E. T. (2004). Infant mouthing behavior: The immunocalibration hypothesis. *Med. Hypotheses* **63,** 925–932.

Flajnik, M. F., Ohta, Y., Namikawa-Yamada, C., and Nonaka, M. (1999). Insight into the primordial MHC from studies in ectothermic vertebrates. *Immunol. Rev.* **167,** 59–67.

Flohr, C., Tuyen, L. N., Lewis, S., Quinnell, R., Minh, T. T., Liem, H. T., Campbell, J., Pritchard, D., Hien, T. T., Farrar, J., Williams, H., and Britton, J. (2006). Poor sanitation and helminth infection protect against skin sensitization in Vietnamese children: A cross-sectional study. *J. Allergy Clin. Immunol.* **118,** 1305–1311.

Geissler, P. W., Mwaniki, D., Thiong, F., and Friis, H. (1998). Geophagy as a risk factor for geohelminth infections: A longitudinal study of Kenyan primary schoolchildren. *Trans. R. Soc. Trop. Med. Hyg.* **92,** 7–11.

Geissler, P. W., Mwaniki, D. L., Thiong'o, F., and Friis, H. (1997). Geophagy among school children in western Kenya. *Trop. Med. Int. Health* **2,** 624–630.

Godfrey, R. C., and Gradidge, C. F. (1976). Allergic sensitisation of human lung fragments prevented by saturation of IgE binding sites. *Nature* **259,** 484–486.

Goren-Inbar, N., Alperson, N., Kislev, M. E., Simchoni, O., Melamed, Y., Ben-Nun, A., and Werker, E. (2004). Evidence of hominin control of fire at Gesher Benot Ya'aqov, Israel. *Science* **304,** 725–727.

Gould, H. J., Sutton, B. J., Beavil, A. J., Beavil, R. L., McCloskey, N., Coker, H. A., Fear, D., and Smurthwaite, L. (2003). The biology of IGE and the basis of allergic disease. *Annu. Rev. Immunol.* **21,** 579–628.

Gupta, R., Sheikh, A., Strachan, D. P., and Anderson, H. R. (2007). Time trends in allergic disorders in the UK. *Thorax* **62,** 91–96.

Haby, M. M., Marks, G. B., Peat, J. K., and Leeder, S. R. (2000). Daycare attendance before the age of two protects against atopy in preschool age children. *Pediatr. Pulmonol.* **30,** 377–384.

Heinrich, J., Richter, K., Magnussen, H., and Wichmann, H. E. (1998). Is the prevalence of atopic diseases in East and West Germany already converging? *Eur. J. Epidemiol.* **14,** 239–245.

Hobolth, A., Christensen, O. F., Mailund, T., and Schierup, M. H. (2007). Genomic relation-ships and speciation times of human, chimpanzee, and gorilla inferred from a coalescent hidden Markov model. *PLoS. Genet.* **3,** e7.

Holgate, S., Casale, T., Wenzel, S., Bousquet, J., Deniz, Y., and Reisner, C. (2005). The anti-inflammatory effects of omalizumab confirm the central role of IgE in allergic inflamma-tion. *J. Allergy Clin. Immunol.* **115,** 459–465.

Hughes, A. L. (1996). Gene duplication and recombination in the evolution of mammalian Fc receptors. *J. Mol. Evol.* **43,** 4–10.

Igyarto, B. Z., Lacko, E., Olah, I., and Magyar, A. (2006). Characterization of chicken epider-mal dendritic cells. *Immunology* **119,** 278–288.

Jackson, M. (2001). Allergy: The making of a modern plague. *Clin. Exp. Allergy* **31,** 1665–1671.

James, S. (1989). Hominid use of fire in the lower and middle Pleistocene. *Curr. Anthropol.* **30,** 1–26.

Janeway, C. A., Jr., Travers, P., Walport, M., and Shlomchik, M. J. (2005). "Immunobiology: The immune system in health and disease " 6th ed. ed.. Garland Science Publishing, New York.

Janson, C., Anto, J., Burney, P., Chinn, S., de Marco, R., Heinrich, J., Jarvis, D., Kuenzli, N., Leynaert, B., Luczynska, C., Neukirch, F., Svanes, C., *et al.* (2001). The European Commu-nity Respiratory Health Survey: What are the main results so far? European Community Respiratory Health Survey II. *Eur. Respir. J.* **18,** 598–611.

Johnson, J. R., Wiley, R. E., Fattouh, R., Swirski, F. K., Gajewska, B. U., Coyle, A. J., Gutierrez-Ramos, J. C., Ellis, R., Inman, M. D., and Jordana, M. (2004). Continuous exposure to house dust mite elicits chronic airway inflammation and structural remodeling. *Am. J. Respir. Crit. Care. Med.* **169,** 378–385.

Juberg, D. R., Alfano, K., Coughlin, R. J., and Thompson, K. M. (2001). An observational study of object mouthing behavior by young children. *Pediatrics* **107,** 135–142.

Kero, J., Gissler, M., Hemminki, E., and Isolauri, E. (2001). Could TH1 and TH2 diseases coexist? Evaluation of asthma incidence in children with coeliac disease, type 1 diabetes, or rheumatoid arthritis: A register study. *J. Allergy Clin. Immunol.* **108**, 781–783.

King, C. L., Xianli, J., Malhotra, I., Liu, S., Mahmoud, A. A., and Oettgen, H. C. (1997). Mice with a targeted deletion of the IgE gene have increased worm burdens and reduced granulomatous inflammation following primary infection with *Schistosoma mansoni*. *J. Immunol.* **158**, 294–300.

Kitagaki, K., Businga, T. R., Racila, D., Elliott, D. E., Weinstock, J. V., and Kline, J. N. (2006). Intestinal helminths protect in a murine model of asthma. *J. Immunol.* **177**, 1628–1635.

Kramer, U., Heinrich, J., Wjst, M., and Wichmann, H. E. (1999). Age of entry to day nursery and allergy in later childhood. *Lancet* **353**, 450–454.

Krishnamani, R., and Mahaney, W. C. (2000). Geophagy among primates: Adaptive significance and ecological consequences. *Anim. Behav.* **59**, 899–915.

Kumar, S., and Hedges, S. B. (1998). A molecular timescale for vertebrate evolution. *Nature* **392**, 917–920.

Laird, D. J., De Tomaso, A. W., Cooper, M. D., and Weissman, I. L. (2000). 50 million years of chordate evolution: Seeking the origins of adaptive immunity. *Proc. Natl. Acad. Sci. USA* **97**, 6924–6926.

Lambrecht, B. N., De Veerman, M., Coyle, A. J., Gutierrez-Ramos, J. C., Thielemans, K., and Pauwels, R. A. (2000). Myeloid dendritic cells induce Th2 responses to inhaled antigen, leading to eosinophilic airway inflammation. *J. Clin. Invest.* **106**, 551–559.

Lanier, B. Q. (2005). Unanswered questions and warnings involving anti-immunoglobulin E therapy based on 2-year observation of clinical experience. *Allergy Asthma Proc.* **26**, 435–439.

Lanier, B. Q., Corren, J., Lumry, W., Liu, J., Fowler-Taylor, A., and Gupta, N. (2003). Omalizumab is effective in the long-term control of severe allergic asthma. *Ann. Allergy Asthma Immunol.* **91**, 154–159.

Laske, N., Bunikowski, R., and Niggemann, B. (2003). Extraordinarily high serum IgE levels and consequences for atopic phenotypes. *Ann. Allergy Asthma Immunol.* **91**, 202–204.

Leung, D. Y., Sampson, H. A., Yunginger, J. W., Burks, A. W., Jr., Schneider, L. C., Wortel, C. H., Davis, F. M., Hyun, J. D., and Shanahan, W. R., Jr. (2003). Effect of anti-IgE therapy in patients with peanut allergy. *N Engl. J. Med.* **348**, 986–993.

Leung, D. Y., Shanahan, W. R., Jr., Li, X. M., and Sampson, H. A. (2004). New approaches for the treatment of anaphylaxis. *Novartis Found Symp.* **257**, 248–260; discussion 260-244, 276-285.

Li, S., Jiao, H., Yu, X., Strong, A. J., Shao, Y., Sun, Y., Altfeld, M., and Lu, Y. (2007). Human leukocyte antigen class I and class II allele frequencies and HIV-1 infection associations in a Chinese cohort. *J. Acquir. Immune Defic. Syndr.* **44**, 121–131.

Linneberg, A., Ostergaard, C., Tvede, M., Andersen, L. P., Nielsen, N. H., Madsen, F., Frolund, L., Dirksen, A., and Jorgensen, T. (2003). IgG antibodies against microorganisms and atopic disease in Danish adults: The Copenhagen Allergy Study. *J. Allergy Clin. Immunol.* **111**, 847–853.

Liu, A. H., and Leung, D. Y. (2006). Renaissance of the hygiene hypothesis. *J. Allergy Clin. Immunol.* **117**, 1063–1066.

Luo, Z. X., Crompton, A. W., and Sun, A. L. (2001). A new mammaliaform from the early Jurassic and evolution of mammalian characteristics. *Science* **292**, 1535–1540.

Lynch, N. R., Hagel, I. A., Palenque, M. E., Di Prisco, M. C., Escudero, J. E., Corao, L. A., Sandia, J. A., Ferreira, L. J., Botto, C., Perez, M., and Le Souef, P. N. (1998). Relationship between helminthic infection and IgE response in atopic and nonatopic children in a tropical environment. *J. Allergy Clin. Immunol.* **101**, 217–221.

MacGlashan, D. W., Jr., Bochner, B. S., Adelman, D. C., Jardieu, P. M., Togias, A., McKenzie-White, J., Sterbinsky, S. A., Hamilton, R. G., and Lichtenstein, L. M. (1997).

Down-regulation of Fc(epsilon)RI expression on human basophils during *in vivo* treatment of atopic patients with anti-IgE antibody. *J. Immunol.* **158**, 1438–1445.

Maleki, S. J. (2004). Food processing: Effects on allergenicity. *Curr. Opin. Allergy Clin. Immunol.* **4**, 241–245.

Maleki, S. J., Viquez, O., Jacks, T., Dodo, H., Champagne, E. T., Chung, S. Y., and Landry, S. J. (2003). The major peanut allergen, Ara h 2, functions as a trypsin inhibitor, and roasting enhances this function. *J. Allergy Clin. Immunol.* **112**, 190–195.

Malveaux, F. J., Conroy, M. C., Adkinson, N. F., Jr., and Lichtenstein, L. M. (1978). IgE receptors on human basophils. Relationship to serum IgE concentration. *J. Clin. Invest.* **62**, 176–181.

Mangan, N. E., Fallon, R. E., Smith, P., van Rooijen, N., McKenzie, A. N., and Fallon, P. G. (2004). Helminth infection protects mice from anaphylaxis via IL-10-producing B cells. *J. Immunol.* **173**, 6346–6356.

Matricardi, P. M., Rosmini, F., Panetta, V., Ferrigno, L., and Bonini, S. (2002). Hay fever and asthma in relation to markers of infection in the United States. *J. Allergy Clin. Immunol.* **110**, 381–387.

Mestas, J., and Hughes, C. C. (2004). Of mice and not men: Differences between mouse and human immunology. *J. Immunol.* **172**, 2731–2738.

Miller, M. E., Levin, L., and Bernstein, J. A. (2005). Characterization of a population of monozygotic twins with asthma. *J. Asthma* **42**, 325–330.

Mitre, E., Norwood, S., and Nutman, T. B. (2005). Saturation of immunoglobulin E (IgE) binding sites by polyclonal IgE does not explain the protective effect of helminth infections against atopy. *Infect. Immun.* **73**, 4106–4111.

Moncayo, A. L., and Cooper, P. J. (2006). Geohelminth infections: Impact on allergic diseases. *Int. J. Biochem. Cell. Biol.* **38**, 1031–1035.

Morey, D. F. (2006). Burying key evidence: The social bond between dogs and people. *J. Archaeol. Sci.* **33**, 158–175.

Muehlenbein, M. P. (2005). Parasitological analyses of the male chimpanzees (Pan troglodytes schweinfurthii) at Ngogo, Kibale National Park, Uganda. *Am. J. Primatol.* **65**, 167–179.

Murray, P. R., Rosenthal, K. S., and Pfaller, M. A. (2005). Medical Microbiology 5th ed. Elsevier Mosby.

Nathan, R. A. (2007). The burden of allergic rhinitis. *Allergy Asthma Proc.* **28**, 3–9.

Ogunbiyi, A. O., Owoaje, E., and Ndahi, A. (2005). Prevalence of skin disorders in school children in Ibadan, Nigeria. *Pediatr. Dermatol.* **22**, 6–10.

Ostroukhova, M., and Ray, A. (2005). CD25+ T cells and regulation of allergen-induced responses. *Curr. Allergy Asthma Rep.* **5**, 35–41.

Pawlik, A., Kaminski, M., Kusnierczyk, P., Kurzawski, M., Dziedziejko, V., Adamska, M., Safranow, K., and Gawronska-Szklarz, B. (2007). Interleukin-18 promoter polymorphism in patients with atopic asthma. *Tissue Antigens* **70**, 314–318.

Pimm, S. L., Russell, G. J., Gittleman, J. L., and Brooks, T. M. (1995). The Future of Biodiversity. *Science* **269**, 347–350.

Prioult, G., and Nagler-Anderson, C. (2005). Mucosal immunity and allergic responses: Lack of regulation and/or lack of microbial stimulation? *Immunol. Rev.* **206**, 204–218.

Rasanen, M., Laitinen, T., Kaprio, J., Koskenvuo, M., and Laitinen, L. A. (1997). Hay fever, asthma and number of older siblings--a twin study. *Clin. Exp. Allergy* **27**, 515–518.

Richter, K., Heinrich, J., Jorres, R. A., Magnussen, H., and Wichmann, H. E. (2000). Trends in bronchial hyperresponsiveness, respiratory symptoms and lung function among adults: West and East Germany. INGA Study Group. Indoor Factors and Genetics in Asthma. *Respir. Med.* **94**, 668–677.

Robinson, D. S., Larche, M., and Durham, S. R. (2004). Tregs and allergic disease. *J. Clin. Invest.* **114**, 1389–1397.

Romagnani, S. (2004). The increased prevalence of allergy and the hygiene hypothesis: Missing immune deviation, reduced immune suppression, or both? *Immunology* **112**, 352–363.

Rosenberg, R., Vinker, S., Zakut, H., Kizner, F., Nakar, S., and Kitai, E. (1999). An unusually high prevalence of asthma in Ethiopian immigrants to Israel. *Fam. Med.* **31**, 276–279.

Rottem, M., Szyper-Kravitz, M., and Shoenfeld, Y. (2005). Atopy and asthma in migrants. *Int. Arch. Allergy Immunol.* **136**, 198–204.

Rybicki, E. (2003). *In* "Introduction to Molecular Virology, internet tutorial, University of Cape Town".

Saathoff, E., Olsen, A., Kvalsvig, J. D., and Geissler, P. W. (2002). Geophagy and its association with geohelminth infection in rural schoolchildren from northern KwaZulu-Natal, South Africa. *Trans. R. Soc. Trop. Med. Hyg.* **96**, 485–490.

Scaglia, M., Tinelli, M., Revoltella, R., Peracino, A., Falagiani, P., Jayakar, S. D., Desmarais, J. C., and Siccardi, A. G. (1979). Relationship between serum IgE levels and intestinal parasite load in African populations. *Int. Arch. Allergy Appl. Immunol.* **59**, 465–468.

Schaub, B., Lauener, R., and von Mutius, E. (2006). The many faces of the hygiene hypothesis. *J. Allergy Clin. Immunol.* **117**, 969–977; quiz 978.

Schnabl, B., Bettenay, S. V., Dow, K., and Mueller, R. S. (2006). Results of allergen-specific immunotherapy in 117 dogs with atopic dermatitis. *Vet. Rec.* **158**, 81–85.

Schnyder-Candrian, S., Togbe, D., Couillin, I., Mercier, I., Brombacher, F., Quesniaux, V., Fossiez, F., Ryffel, B., and Schnyder, B. (2006). Interleukin-17 is a negative regulator of established allergic asthma. *J. Exp. Med.* **203**, 2715–2725.

Sharma, S., Kathuria, P. C., Gupta, C. K., Nordling, K., Ghosh, B., and Singh, A. B. (2006). Total serum immunoglobulin E levels in a case-control study in asthmatic/allergic patients, their family members, and healthy subjects from India. *Clin. Exp. Allergy* **36**, 1019–1027.

Shintani, S., Terzic, J., Sato, A., Saraga-Babic, M., O'HUigin, C., Tichy, H., and Klein, J. (2000). Do lampreys have lymphocytes? The Spi evidence. *Proc. Natl. Acad. Sci. USA* **97**, 7417–7422.

Shu, D. G., Chen, L., Han, J., and Zhang, X. L. (2001). An Early Cambrian tunicate from China. *Nature* **411**, 472–473.

Sicherer, S. H. (2002). Clinical update on peanut allergy. *Ann. Allergy Asthma. Immunol.* **88**, 350–361; quiz 361-352, 394.

Simpson, C. R., Anderson, W. J., Helms, P. J., Taylor, M. W., Watson, L., Prescott, G. J., Godden, D. J., and Barker, R. N. (2002). Coincidence of immune-mediated diseases driven by Th1 and Th2 subsets suggests a common aetiology. *A population-based study using computerized general practice data. Clin. Exp. Allergy* **32**, 37–42.

Six, A., Rast, J. P., McCormack, W. T., Dunon, D., Courtois, D., Li, Y., Chen, C. H., and Cooper, M. D. (1996). Characterization of avian T-cell receptor gamma genes. *Proc. Natl. Acad. Sci. USA* **93**, 15329–15334.

Spector, S. L., and Tan, R. A. (2007). Effect of omalizumab on patients with chronic urticaria. *Ann. Allergy Asthma Immunol.* **99**, 190–193.

Springer, M. S., Murphy, W. J., Eizirik, E., and O'Brien, S. J. (2003). Placental mammal diversification and the Cretaceous-Tertiary boundary. *Proc. Natl. Acad. Sci. USA* **100**, 1056–1061.

Steen, C. J., Janniger, C. K., Schutzer, S. E., and Schwartz, R. A. (2005). Insect sting reactions to bees, wasps, and ants. *Int. J. Dermatol.* **44**, 91–94.

Steinke, J. W., Rich, S. S., and Borish, L. (2008). 5. Genetics of allergic disease. *J. Allergy Clin. Immunol.* **121**, S384–387; quiz S416.

Steinman, H. A., Donson, H., Kawalski, M., Toerien, A., and Potter, P. C. (2003). Bronchial hyper-responsiveness and atopy in urban, peri-urban and rural South African children. *Pediatr. Allergy Immunol.* **14,** 383–393.

Steinman, L. (2007). A brief history of T(H)17, the first major revision in the T(H)1/T(H)2 hypothesis of T cell-mediated tissue damage. *Nat. Med.* **13,** 139–145.

Strachan, D. P. (1989). Hay fever, hygiene, and household size. *Bmj.* **299,** 1259–1260.

Sturm, G. J., Heinemann, A., Schuster, C., Wiednig, M., Groselj-Strele, A., Sturm, E. M., and Aberer, W. (2007). Influence of total IgE levels on the severity of sting reactions in Hymenoptera venom allergy. *Allergy* **62,** 884–889.

Sugunan, A. P., Murhekar, M. V., and Sehgal, S. C. (1996). Intestinal parasitic infestation among different population groups of Andaman and Nicobar islands. *J. Commun. Dis.* **28,** 253–259.

Summers, R. W., Elliott, D. E., Urban, J. F., Jr., Thompson, R., and Weinstock, J. V. (2005). Trichuris suis therapy in Crohn's disease. *Gut.* **54,** 87–90.

Talbot, T. R., Hartert, T. V., Mitchel, E., Halasa, N. B., Arbogast, P. G., Poehling, K. A., Schaffner, W., Craig, A. S., and Griffin, M. R. (2005). Asthma as a risk factor for invasive pneumococcal disease. *N. Engl. J. Med.* **352,** 2082–2090.

Tanaka, H., and Tsuji, M. (1997). From discovery to eradication of schistosomiasis in Japan: 1847-1996. *Int. J. Parasitol.* **27,** 1465–1480.

Taylor, A., Verhagen, J., Blaser, K., Akdis, M., and Akdis, C. A. (2006). Mechanisms of immune suppression by interleukin-10 and transforming growth factor-beta: The role of T regulatory cells. *Immunology* **117,** 433–442.

Temelkovski, J., Hogan, S. P., Shepherd, D. P., Foster, P. S., and Kumar, R. K. (1998). An improved murine model of asthma: Selective airway inflammation, epithelial lesions and increased methacholine responsiveness following chronic exposure to aerosolised allergen. *Thorax* **53,** 849–856.

Thomsen, S. F., Ulrik, C. S., Kyvik, K. O., von Bornemann Hjelmborg, J., Skadhauge, L. R., Steffensen, I., and Backer, V. (2006). Genetic and environmental contributions to hay fever among young adult twins. *Respir. Med.* **100,** 2177–2182.

Torio, A., Sanchez-Guerrero, I., Muro, M., Villar, L. M., Minguela, A., Marin, L., Moya-Quiles, M. R., Montes-Ares, O., Pagan, J., and Alvarez-Lopez, M. R. (2003). HLA class II genotypic frequencies in atopic asthma: Association of DRB1*01-DQB1*0501 genotype with Artemisia vulgaris allergic asthma. *Hum. Immunol.* **64,** 811–815.

Umetsu, D. T., and DeKruyff, R. H. (2006). The regulation of allergy and asthma. *Immunol. Rev.* **212,** 238–255.

Verhagen, J., Blaser, K., Akdis, C. A., and Akdis, M. (2006). Mechanisms of allergen-specific immunotherapy: T-regulatory cells and more. *Immunol. Allergy Clin. North. Am.* **26,** 207–231, vi.

Verlinden, A., Hesta, M., Millet, S., and Janssens, G. P. (2006). Food allergy in dogs and cats: A review. *Crit. Rev. Food Sci. Nutr.* **46,** 259–273.

Vernersson, M., Aveskogh, M., and Hellman, L. (2004). Cloning of IgE from the echidna (Tachyglossus aculeatus) and a comparative analysis of epsilon chains from all three extant mammalian lineages. *Dev. Comp. Immunol.* **28,** 61–75.

Viinanen, A., Munhbayarlah, S., Zevgee, T., Narantsetseg, L., Naidansuren, T., Koskenvuo, M., Helenius, H., and Terho, E. O. (2005). Prevalence of asthma, allergic rhinoconjunctivitis and allergic sensitization in Mongolia. *Allergy* **60,** 1370–1377.

von Hertzen, L. C. (2000). Puzzling associations between childhood infections and the later occurrence of asthma and atopy. *Ann. Med.* **32,** 397–400.

Wan, H., Winton, H. L., Soeller, C., Taylor, G. W., Gruenert, D. C., Thompson, P. J., Cannell, M. B., Stewart, G. A., Garrod, D. R., and Robinson, C. (2001). The transmembrane protein occludin of epithelial tight junctions is a functional target for serine peptidases from faecal pellets of Dermatophagoides pteronyssinus. *Clin. Exp. Allergy* **31,** 279–294.

Weiland, S. K., Bjorksten, B., Brunekreef, B., Cookson, W. O., von Mutius, E., and Strachan, D. P. (2004). Phase II of the International Study of Asthma and Allergies in Childhood (ISAAC II): Rationale and methods. *Eur. Respir. J.* **24,** 406–412.

Wong, M. S., Bundy, D. A., and Golden, M. H. (1988). Quantitative assessment of geophagous behaviour as a potential source of exposure to geohelminth infection. *Trans. R. Soc. Trop. Med. Hyg.* **82,** 621–625.

Yazdanbakhsh, M., Kremsner, P. G., and van Ree, R. (2002). Allergy, parasites, and the hygiene hypothesis. *Science* **296,** 490–494.

Zarkadis, I. K., Mastellos, D., and Lambris, J. D. (2001). Phylogenetic aspects of the complement system. *Dev. Comp. Immunol.* **25,** 745–762.

Zekarias, B., Ter Huurne, A. A., Landman, W. J., Rebel, J. M., Pol, J. M., and Gruys, E. (2002). Immunological basis of differences in disease resistance in the chicken. *Vet. Res.* **33,** 109–125.

Weiland, S. K., Bjorksten, B., Brunekreef, B., Cookson, W. O., von Mutius, E., and Strachan, D. P. (2004). Phase II of the International Study of Asthma and Allergies in Childhood (ISAAC II): Rationale and methods. *Eur. Respir. J.* **24,** 406–412.

Wong, M. S., Bundy, D. A., and Golden, M. H. (1988). Quantitative assessment of geophagous behaviour as a potential source of exposure to geohelminth infection. *Trans. R. Soc. Trop. Med. Hyg.* **82,** 621–625.

Yazdanbakhsh, M., Kremsner, P. G., and van Ree, R. (2002). Allergy, parasites, and the hygiene hypothesis. *Science* **296,** 490–494.

Zarkadis, I. K., Mastellos, D., and Lambris, J. D. (2001). Phylogenetic aspects of the complement system. *Dev. Comp. Immunol.* **25,** 745–762.

Zekarias, B., Ter Huurne, A. A., Landman, W. J., Rebel, J. M., Pol, J. M., and Gruys, E. (2002). Immunological basis of differences in disease resistance in the chicken. *Vet. Res.* **33,** 109–125.

CHAPTER 3

New Insights on Mast Cell Activation via the High Affinity Receptor for IgE[1]

Juan Rivera,* Nora A. Fierro,* Ana Olivera,* and Ryo Suzuki*

Contents		

* Laboratory of Immune Cell Signaling, National Institute of Arthritis and Musculoskeletal and Skin Diseases, National Institutes of Health, Bethesda, Maryland
[1] The research of the authors reported herein was supported by the Intramural Research Program of the National Institute of Arthritis and Musculoskeletal and Skin Diseases of the National Institutes of Health

Advances in Immunology, Volume 98
ISSN 0065-2776, DOI: 10.1016/S0065-2776(08)00403-3

Abstract   Mast cells are innate immune cells that function as regulatory or
effector cells and serve to amplify adaptive immunity. In adaptive
immunity these cells function primarily through cell surface Fc
receptors that bind immunoglobulin antibodies. The dysregulation
of their adaptive role makes them central players in allergy and
asthma. Upon encountering an allergen (antigen), which is recognized by immunoglobulin E (IgE) antibodies bound to the high
affinity IgE receptor (FcεRI) expressed on their cell surface, mast
cells secrete both preformed and newly synthesized mediators of
the allergic response. Blocking of these responses is an objective in
therapeutic intervention of allergic diseases. Thus, understanding
the mechanisms by which antigens elicit mast cell activation (via
FcεRI) holds promise toward identifying therapeutic targets. Here
we review the most recent advances in understanding antigen-
dependent mast cell activation. Specifically, we focus on the
requirements for FcεRI activation, the regulation of calcium
responses, co-stimulatory signals in FcεRI-mediated mast cell activation and function, and how genetics influences mast cell signaling
and responses. These recent discoveries open new avenues of
investigation with therapeutic potential.

Key Words: Calcium, FcεRI, IgE, Kinase, Mast Cell.   © 2008 Elsevier Inc.

1. INTRODUCTION

Mast cell activation requires coordinated events that are able to discriminate how the cell responds when encountering a given allergen challenge
(Blank and Rivera, 2004; Gilfillan and Tkaczyk, 2006). These events begin
with the allergen-dependent aggregation of the IgE antibody-occupied
high affinity receptor for IgE (FcεRI) and are propagated inside the cell to
assemble a highly sophisticated network of signaling molecules that
control the cells response to the particular challenge (Gilfillan and
Tkaczyk, 2006). The assembled signaling network may differ depending

on the type and strength of a stimulus (Gonzalez-Espinosa *et al.*, 2003). Regulation of network assembly exists at multiple levels and employs proteins and lipids that may have positive, negative, or dual roles in coordinating and regulating the signaling response (Rivera and Gilfillan, 2006; Rivera and Olivera, 2007). This context- or compartment-specific function underlies many of the mysteries remaining to be uncovered on how signals are transduced.

In this review we provide a very brief and general review of the FcεRI signaling to acquaint the reader with the major players. More detailed reviews on this topic are available (Gilfillan and Tkaczyk, 2006; Kraft and Kinet, 2007; Rivera and Gilfillan, 2006). The major focus of this review is on recent advances in understanding the molecular events elicited by engagement of FcεRI. In particular, we focus on new information regarding FcεRI signaling and responses, the vital discovery of key components of the calcium apparatus in mast cells and other immune cells, the identification of important co-stimulatory components of FcεRI signaling, and the influence of genetics on mast cell responsiveness. The rapid advances of the last several years reveal increasing molecular complexity in how mast cells are activated upon engagement of FcεRI. However, the findings also reveal new areas of investigation with therapeutic potential in disease.

2. A GENERAL OUTLINE OF FcεRI SIGNALING

The FcεRI exists in two forms. It can be expressed as a trimer or tetramer (see Fig. 3.1) comprised of an IgE-binding α chain, a membrane tetraspanning β chain that is absent in the trimeric receptor, and a disulfide-linked homodimer of γ chains (Nadler *et al.*, 2000). While the trimeric form can be expressed on a variety of immune cells (such as monocytes, eosinophils, Langerhan cells, etc.), the tetramer is expressed primarily on mast cells and basophils (Blank and Rivera, 2004; Nadler *et al.*, 2000). Both the β and γ chains contain immunoreceptor tyrosine-based activation motifs (ITAMs), which are essential for the signaling competence of immunoreceptors. FcεRI lacks intrinsic tyrosine kinase activity and associates with the nonreceptor Src family tyrosine kinase Lyn kinase (Fig. 3.1), whose activity is key for phosphorylation of the tyrosine residues in its ITAM motifs through transphosphorylation (Pribluda *et al.*, 1994). A small fraction of Lyn can be found to weakly interact with the FcεRIβ ITAM prior to engagement of this receptor. This interaction is greatly enhanced by FcεRI stimulation as the Lyn SH2 domain binds the phosphorylated Y219 in the FcεRIβ ITAM (Furumoto *et al.*, 2004; On *et al.*, 2004). Efficient phosphorylation of the receptor also requires plasma membrane liquid-ordered phase domains (commonly referred to as lipid rafts; see Fig. 3.1) (Field

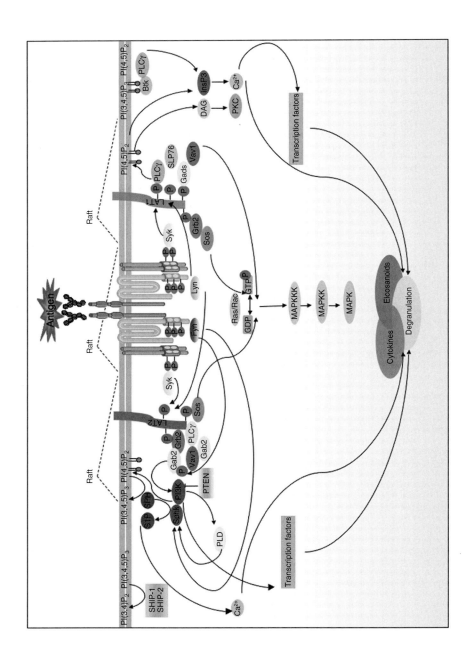

et al., 1997; Sheets *et al.*, 1999; Young *et al.*, 2003). These domains are enriched in cholesterol, sphingolipids and other saturated phospholipids, as well as with a variety of signaling proteins including Lyn kinase. Another Src family member, Fyn kinase, also appears to associate with the FcεRIβ ITAM (Fig. 3.1) and appears to be recruited similar to Lyn via SH2 domain–phosphotyrosine interaction (Fierro and Rivera, unpublished observation). However, although it does not appear to participate in the phosphorylation of FcεRI, it is required for normal mast cell degranulation and cytokine production (Gomez *et al.*, 2005). The spatial and temporal association of Fyn with the FcεRIβ ITAM remains to be unraveled.

Phosphorylated ITAMs bind a variety of proteins that are key for signal amplification (Rivera and Gilfillan, 2006). The tyrosine kinase Syk (Fig. 3.1), an essential kinase for the propagation of signals, is one of these proteins (Benhamou *et al.*, 1993; Kihara and Siraganian, 1994). Other proteins thought to interact include the tyrosine phosphatases SHP-1 and -2 (Swieter *et al.*, 1995). The temporal order of these interactions is not clear. Syk binds primarily to the phosphorylated γ chain ITAM, a step

---

FIGURE 3.1   FcεRI signaling in mast cells. Antigen-clustering of FcεRI through bound antigen-specific IgE initiates multiple events required for mast cell activation. Clustering of FcεRI drives the coalescence of cholesterol-enriched membrane microdomains (rafts) that contain signaling molecules like the Src family kinase Lyn. Lyn transphosphorylates the immunoreceptor tyrosine-based activation motifs (ITAMs) of the β and γ subunits of a neighboring FcεRI. This requires that these receptors have an appropriate distance/and or configuration within the antigen-induced receptor clusters. Lyn and Fyn can interact with the phosphorylated β subunits whereas Syk kinase interacts with the γ subunit. Fyn, Lyn, and Syk contribute to the formation of multimolecular signaling complexes that are coordinated by adaptors, like LAT1 and -2, Gab2, Grb2, Gads, among others. These signaling complexes provide docking sites for other signaling proteins including PLCγ, SLP76, Vav1, Sos, and others. The activity of the molecules in these complexes is coordinated to initiate the production of lipid second messengers that are essential for calcium mobilization and PKC activation leading to degranulation, and the *de novo* synthesis and secretion of eicosanoids and cytokines. This process is tightly regulated at multiple levels. The signaling complexes provide an environment where a balance of positive signals with negative feedback signals can occur. Lipid enzymes (like PI3K) produce and/or remove lipid messengers [by converting phosphatidylinositol 4,5-bisphosphate (PIP$_2$), the substrate of PLCγ, to phosphatidylinositol 3,4,5-trisphosphate (PIP$_3$)]. This production or loss of lipid messengers is important in the targeting, activation, and regulation of signaling protein function. Protein kinases can also phosphorylate protein and lipid enzymes modifying their activity. Protein phosphatases (like SHP-1) and lipid phosphatases (like SHIP and PTEN) will dephosphorylate phosphorylated proteins and lipids, respectively, disassembling the signaling complex or inactivating signaling proteins. This is a highly dynamic process that adjusts spatio-temporally to fine-tune mast cell responses. (See Plate 1 in Color Plate Section.)

necessary for its activation. While weak interaction of Syk with FcεRIβ phospho-ITAM peptides can be observed *in vitro*, there is no evidence of this interaction *in vivo*. Once activated, Syk is essential in amplification of mast cell signaling and in driving normal mast cell effector responses (Costello *et al.*, 1996; Zhang *et al.*, 1998). Its central role has made this kinase a prime therapeutic target in diseases where its activity is fundamental in immune responses. Syk function in mast cells and other immune cells has been recently reviewed (Berton *et al.*, 2005; Siraganian *et al.*, 2002) so it will not be covered in detail herein. However, it is important to note that Syk phosphorylates many signaling proteins, but key to its amplification function is the phosphorylation of the adaptor molecules required in the assembly of membrane localized signaling networks.

Among these adaptors, the linkers for activation of T cells (LAT) 1 and 2 (2 was formerly known as NTAL or LAB) and the Grb2-associated binder 2 (Gab2) serve as essential scaffolds in organizing, coordinating, and regulating the generated signals (Fig. 3.1) (Rivera, 2002, 2005; Tkaczyk *et al.*, 2005). Characteristic to adaptor proteins are numerous tyrosine residues that are targets for Syk and other kinases (Rivera, 2005). Once phosphorylated, these adaptors bind a variety of signaling proteins such as lipases (PLC_1 and PLC_2), phosphatases (SHIP-1 and -2), protein and lipid kinases (Lyn, Fyn, Btk, PI3K, PIP5K, etc.). Adaptors serve to coordinate and localize signals that lead to the production of a number of lipid second messenger molecules. Some of these messengers like inositol 1,4,5-trisphosphate (IP_3), phosphatidylinositol 4,5-bisphosphate (PIP_2), phosphatidylinositol 3,4,5-trisphosphate (PIP_3), and sphingosine-1-phosphate (S1P) are implicated in regulating and/or eliciting calcium mobilization in mast cells (Fig. 3.1) (Rivera and Olivera, 2007). These lipid second messengers, as well as others like diacylgycerol (DAG), play key roles in regulating the membrane recruitment and/or activity of many proteins (such as PLCs, PKCs, and others) as well as serve as ligands interacting with intra- and extracellular receptors that elicit provide signals leading to effector responses.

It is the cooperativity of proteins and lipids that in mast cells, like in many other immune cells, is key in eliciting full functional responses. For example, the cooperativity of increased cytosolic calcium with PKCs has been demonstrated as essential for mast cell degranulation (Blank and Rivera, 2004; Rivera and Beaven, 1997). However, phorbol esters alone, which mimic the function of DAG in activating PKC, do not induce mast cell degranulation but instead activate the production of some cytokines. This cooperativity is also supported by the plethora of studies where genetic deletion has led to a partial effect on mast cell signaling and responses. Moreover, redundancy in the role of signaling proteins is common and thus the presence of several isoforms of a protein or the expression of a protein with overlapping function circumvents the genetic defect (Kawakami *et al.*, 2000). For mast cells, this redundancy may be further complicated because of their well-recognized tissue heterogeneity (Galli, 1997b), reflecting differences in gene expression. Thus, it is important to keep in mind that variations of this general view of mast cell FcεRI signaling are likely to exist.

3. RECENT ADVANCES IN FcεRI PROXIMAL SIGNALING

3.1. Spatial organization in FcεRI phosphorylation

3.1.1. Spatial requirement for transphosphorylation

It has long been recognized that aggregation of FcεRI (Fig. 3.1), like for other antigen receptors, is necessary for its activation (Metzger, 1978). This requirement has been proposed to facilitate the transphosphorylation of FcεRI by receptor-associated Lyn kinase (Pribluda *et al.*, 1994). However, direct evidence of the spatial requirements for transphosphorylation was lacking until recently. Studies using trivalent ligands with rigid DNA spacers of varying lengths (16, 26, 36, or 46 bases) have revealed length-dependent stimulation of FcεRI phosphorylation (Sil *et al.*, 2007). With average distances ranging from 5 to 13 nm, the findings showed an inverse relationship between length and progressive enhancement of FcεRI phosphorylation and mast cell degranulation. These findings revealed structural constraints in FcεRI phosphorylation and demonstrate that the appropriate spatial organization of an FcεRI cluster is important for its phosphorylation. Given that the ligand with the longest length (~13 nm) is likely to prevent direct contact of aggregated receptor complexes, the findings demonstrate that the most effective ligands for FcεRI phosphorylation are those that allow proximity of aggregated FcεRI complexes thus supporting a transphosphorylation model. However, it should be noted that PLCγ phosphorylation and Ca^{2+} release from intracellular stores was not inversely correlated with ligand length suggesting that the threshold for these responses might be met under conditions that lead to minimal FcεRI phosphorylation.

3.1.2. New insights on liquid-ordered domain (lipid raft) involvement in FcεRI phosphorylation and function

Plasma membrane-localized cholesterol-enriched microdomains (lipid rafts) are well-known participants in cellular signaling and other cell functions (Simons and Toomre, 2000). Lipid rafts have been implicated as initiators of FcεRI phosphorylation (Sheets *et al.*, 1999), as domains that may facilitate exocytosis (Pombo *et al.*, 2003; Puri and Roche, 2006), and as essential for Kit-mediated survival and proliferation (Jahn *et al.*, 2007).

Considerable evidence supports the role of lipid rafts in FcεRI phosphorylation. Nonetheless, there is still substantial uncertainty as to whether lipid rafts are necessary for initiating FcεRI phosphorylation or if they are simply participants in this process, due to the concentration of Lyn kinase in these domains and the recruitment of FcεRI into these domains upon its engagement (Kovarova *et al.*, 2001; Rivera *et al.*, 2002; Sheets *et al.*, 1999). A recent study using a mouse model of Smith-Lemli-Opitz disease, a disease of cholesterol deficiency due to a mutation in 7-dehydrocholesterol reductase, revealed that partial loss of cholesterol from lipid rafts led to considerable loss (>80%) of Lyn kinase from these domains and partly diminished (40%) FcεRI phosphorylation with the most significant effect seen at late times postreceptor stimulation (Kovarova *et al.*, 2006). This suggests that Lyn in lipid rafts may participate in sustaining FcεRI phosphorylation and argues that initiation of receptor phosphorylation may not require the Lyn found in lipid raft domains. However, this may be an overly simplistic view since the methods used to biochemically define a lipid raft are unlikely to detect small microdomains.

Studies using high resolution electron microscopy (Wilson *et al.*, 2000, 2001) partly address this issue. These studies demonstrated the compartmentation of FcεRI and Lyn in small electron dense microdomains prior to FcεRI engagement. Aggregation of FcεRI enhances the size of these microdomains increasing the numbers of receptors and Lyn within. The findings also demonstrated that the adaptor molecule LAT is found in microdomains that are distinct from those that contain FcεRI. Since biochemical and genetic evidence ascribes the residence of molecules like Lyn and LAT as lipid raft microdomain residents, this would suggest that there are distinct (heterogeneous) cholesterol-enriched microdomains on the plasma membrane. This view is further supported by studies on the distribution of endogenous and transfected Thy-1 (CD90) isoforms in the plasma membrane of RBL-2H3 cells (Heneberg *et al.*, 2006). Thy-1 is a GPI-anchored membrane protein that distributes into cholesterol-enriched membranes and its aggregation can cause cellular signaling and responses. Using primarily biophysical approaches, the authors studied Thy-1 isoform distribution on the membrane and its cross-talk with FcεRI. The two Thy-1 isoforms studied were found in small autonomous clusters in resting cells whereas aggregation of one isoform led to significant co-localization of the two isoforms as detected by electron microscopic and fluorescence resonance energy transfer (FRET) analysis. The adaptor proteins LAT1 and LAT2 were also found in autonomous domains with occasional co-localization of Thy-1. This co-localization was greatly enhanced by Thy-1 aggregation. FRET also detected Thy-1 cross-talk with FcεRI upon the latters engagement. Thus, the findings revealed the presence of distinct microdomains in resting cells that are able to coalesce upon aggregation of constituent proteins or of FcεRI.

The use of two-photon fluorescence lifetime imaging microscopy and fluorescence polarization anisotropy imaging has now allowed the real-time monitoring of membrane structure and organization. Studies in which the cholesterol-rich membrane domains are labeled with the lipid analog dil-C_{18}, and FcεRI is labeled with Alexa 488-IgE, demonstrated that upon FcεRI aggregation dil-C_{18}-labeled membrane domains coalesce and redistribute to membrane patches containing FcεRI (Davey et al., 2007). This increases the fluorescence lifetime of both dil-C_{18} and Alexa 488-IgE, an increase in dil-C_{18} fluorescence lifetime was previously shown to reflect increased lipid order. In addition, upon FcεRI engagement FRET can be demonstrated to occur between the Alexa 488-labeled receptor and dil-C_{18}-labeled domains (Davey et al., 2008), suggesting the reordering of the cholesterol-enriched membrane and the aggregated FcεRI to allow FRET signals. Thus, membrane nanostructure appears to be altered and the rotational diffusion of components within these domains is decreased, suggesting increased interactions and ordered structure.

Collectively, these findings reveal important insights on the role of "lipid rafts" in FcεRI signaling. The studies begin to define these cholesterol-rich membrane microdomains as a lipid environment that is dynamic, small, heterogeneous in protein content, and that is likely to contain limited numbers of proteins prior to cell stimulation. Thus, a unifying hypothesis for the differing views on the role of lipid rafts in FcεRI phosphorylation is emerging. From these and earlier studies one might postulate that the FcεRI may be found, at any given time, within these dynamic cholesterol-rich microdomains, an environment that enhances its proximity to Lyn kinase. The aggregation of FcεRI could induce transphosphorylation with a neighboring receptor as it might bring together receptors with or without Lyn kinase. Coalescence of these dynamic cholesterol-rich domains (some of which might include Lyn) would stabilize or increase this transphosphorylation and cause assembly of a stable signaling complex (Fig. 3.1). This view of a need for stable clustering of FcεRI to form stable signaling complexes is supported by our finding that a antigen of low affinity fails to assemble a stable signaling complex even under conditions where it is able to elicit FcεRI phosphorylation to a similar extent as a high affinity antigen (Suzuki and Rivera, unpublished observation). These findings are in agreement with the view that FcεRI aggregation induces membrane changes that allow coalescence of proteins and lipids required for efficient propagation of signals.

3.2. Lyn kinase as a negative regulator

As detailed above, the positive role of Lyn kinase in mast cell activation through its control of FcεRI phosphorylation is clearly recognized. However, Lyn also plays an important role as a negative regulator of mast cell effector responses (Hernandez-Hansen *et al.*, 2004; Odom *et al.*, 2004; Xiao *et al.*, 2005). This was evident from *in vivo* studies where Lyn-deficient mice were found to develop atopic-like allergic disease and mast cells derived from these mice were hyperresponsive relative to cells from wild-type mice (Odom *et al.*, 2004). One possible cause for this hyperresponsive phenotype is that Lyn is required for phosphorylation of the lipid raft-localized Csk-binding protein (Cbp) and thus for membrane targeting of a regulatory kinase, C-terminal Src kinase (Csk). Csk negatively regulates Src family kinases by phosphorylation of a negative regulatory tyrosine in the C-terminus that causes intramolecular interaction with its own SH2 domain. We now know that this regulatory step is dependent on Lyn localization to lipid rafts and is required to downregulate Fyn kinase activity (Kovarova *et al.*, 2006). This negative role of Lyn also appears to be independent of its association with FcεRI. Another possible mechanism by which Lyn exerts negative control on mast cell effector responses is through the inositol phosphatase, SHIP. Lyn deficiency causes the loss of SHIP activity (Hernandez-Hansen *et al.*, 2004). This inositol phosphatase regulates the intracellular levels of PIP_3 upon cell activation and thus loss of its activity increases the concentration of intracellular PIP_3. As mentioned above, this key lipid second messenger is important for recruitment and formation of membrane-localized signaling networks and thus increasing its intracellular concentration would likely enhance signaling leading to enhanced cellular responses (this will be further detailed in the next section). Lyn-deficient mast cells were shown to have increased levels of intracellular PIP_3 (Odom *et al.*, 2004).

How Lyn kinase plays both a positive and negative role in mast cell activation is not completely clear. However, a recent study (Xiao *et al.*, 2005) sheds light on this apparent paradox. This study demonstrates that low or high strength stimulation of wild-type and Lyn-deficient mast cells distinguishes the positive versus negative role of Lyn, respectively. Thus, the findings showed that Lyn activity is required for mast cell degranulation and cytokine production when encountering a low strength stimulus whereas under high strength stimulation enhanced degranulation and cytokine production was seen when Lyn was absent (Xiao *et al.*, 2005). Given that the expression of FcεRIβ (but not the ITAM-tyrosine mutated FcεRIβ) in Lyn and FcεRIβ double-deficient mast cells gave a similar profile of functional responses to high strength stimulation in Lyn-deficient cells, the authors conclude that negative regulation by Lyn is mediated through its interaction with the FcεRIβ. Evidence that much of Lyn's negative role is predominantly mediated by the pool of Lyn in lipid rafts (Kovarova *et al.*, 2006) suggests that this interaction with FcεRIβ takes place upon coalescence of these domains. Thus, this fits well with the concept that a strong stimulus, which leads to extensive coalescence of microdomains

(Davey *et al.*, 2007, 2008), would serve to promote the negative role of Lyn kinase in order to control the extent of the inflammatory response.

An important caveat in defining a negative role for Lyn kinase has been the exclusive use of Lyn-deficient mice, as some of the observed effects might be attributed to the possible importance of Lyn in development. However, several new models (Hong *et al.*, 2007; Kitaura *et al.*, 2007) have recently emerged that strongly support the concept of Lyn as a negative regulator of mast cell responses. EL mice have been used as a model of epilepsy with the susceptibility locus mapped to chromosomes 2 and 9. ASK mice are spontaneous variants derived from the EL strain that are epilepsy resistant. ASK mice have been shown to be highly susceptible to anaphylaxis whereas EL mice are resistant. Studies exploring the mechanism for this susceptibility demonstrate that mast cells from ASK mice exhibited a similar hyperresponsive phenotype as Lyn-deficient mast cells. ASK mast cells showed enhanced cytokine production, although degranulation was not elevated relative to EL mice. Exploration of Lyn function in mast cells from EL and ASK mice showed reduced Lyn kinase activity in ASK mast cells that was coupled to reduced Cbp phosphorylation and increased Fyn and Src kinase activities (Kitaura *et al.*, 2007). This of course differs from Lyn deficiency in that Lyn protein is still present in ASK cells, Src kinase activity was increased in ASK mast cells but normal in Lyn-deficient cells, and degranulation was not enhanced in ASK mast cells. Regardless, the consequence of Lyn inactivity is highly similar to the effects of Lyn deficiency on mast cell responsiveness. Moreover, these findings clearly confirm the negative role of Lyn in cytokine production. Additional evidence is also afforded by studies on Hck-deficient mast cells (Hong *et al.*, 2007). In this study, the findings point to Hck-mediated negative regulatory control on Lyn activation. Thus, in the absence of Hck, Lyn activity is substantially increased suppressing degranulation and cytokine production when cells encounter a strong stimulus. The mechanism for this effect is less clear since Cbp phosphorylation was enhanced and Fyn activity was normal. Nonetheless, Hck appears to dampen the negative function of Lyn kinase. Collectively these studies argue that, under optimal (strong) conditions of stimulation, Lyn kinase plays a dominant role in controlling the extent of mast cell responses. Thus, the concept of therapeutic targeting of Lyn kinase in allergic disease (Metzger, 1999) should be reconsidered.

3.3. Fyn kinase and PIP$_3$ in mast cell responsiveness

In contrast to Lyn, low strength and high strength stimulation of mast cells demonstrate that Fyn functions to positively regulate mast cell responsiveness through its role in regulating the activation of phosphatidylinositol 3-OH kinase (PI3K) through the adaptor Gab2 (Gu *et al.*, 2001;

Parravicini *et al.*, 2002) and thus its product, PIP_3 (Fig. 3.1). PIP_3 is essential for mast cell responses as reflected by the inhibitory effect of the PI3K inhibitors or genetic manipulation of PI3K activity, on mast cell degranulation and cytokine production (Ali *et al.*, 2004; Barker *et al.*, 1995; Parravicini *et al.*, 2002). As mentioned above, Lyn-deficient mast cells have high levels of PIP_3. In contrast, Fyn-deficient mast cells show a substantial reduction in PIP_3, which correlates with the reduced degranulation response of these cells.

The importance of intracellular PIP_3 levels is further highlighted by studies performed on mast cells derived from the lipid phosphatase SHIP-1-deficient mouse or in which SHIP-2 or another lipid phosphatase PTEN was downregulated by silencing RNA strategies in mouse or human mast cells (HuMC), respectively (Furumoto *et al.*, 2006; Huber *et al.*, 1998a; Leung and Bolland, 2007). In all cases, decreased expression of these phosphatases led to increased PIP_3 levels and enhanced mast cell responses. SHIP-1 and SHIP-2 regulate the production of PIP_3 by dephosphorylating the $5'$ position to generate PI (3,4)-P_2 (Fig. 3.1). PTEN directly opposes PI3K function as it dephosphorylates the $3'$ position of PIP_3, yielding PI(4,5)-P_2 (Fig. 3.1). The increased degranulation and cytokine production seen in SHIP-1- and PTEN-deficient mast cells was associated with increased FcεRI-dependent calcium mobilization (Huber *et al.*, 1998b, 2002). In contrast, SHIP-2-deficient mast cells showed normal calcium levels relative to wild-type cells but had enhanced mast cell degranulation and cytokine production (Leung and Bolland, 2007). This enhanced response was associated with increased microtubule polymerization, which is thought to be important in movement of exocytotic vesicles (Martin-Verdeaux *et al.*, 2003; Nishida *et al.*, 2005), and enhanced Rac 1 activity.

Some additional divergence, which may reflect species differences or distinct roles of PTEN and SHIP, are noteworthy. PTEN deficiency caused a constitutive phosphorylation of Akt and MAP kinases in HuMC (Furumoto *et al.*, 2006) whereas this was not readily apparent in SHIP-1- or SHIP-2-deficient mast cells (Huber *et al.*, 2002). In all cases, Akt and some of the MAP kinases showed increased phosphorylation upon FcεRI stimulation. Cytokine secretion was constitutive in HuMC as both IL-8 and GM-CSF secretion was observed prior to stimulation and was further enhanced after FcεRI engagement (Furumoto *et al.*, 2006). In contrast, SHIP-1- and SHIP-2-deficient cells seemed to require FcεRI stimulation to elicit an enhanced cytokine production (Huber *et al.*, 2002; Leung and Bolland, 2007). This suggests that in mouse mast cells, while SHIP-1 and SHIP-2 also increased PIP_3 levels, these levels may not be sufficient to drive a constitutive response in these cells or PTEN function may diverge from SHIP function in this respect. Regardless, the findings promote the view that PTEN is required for control of PI3K in mast cell homeostasis and activation whereas SHIP-1 and SHIP-2 are primarily active when mast cells are stimulated via FcεRI. Thus, it seems reasonable to conclude that the dominant role of Fyn kinase in promoting PI3K activity places it in control of a key pathway for mast cell responsiveness.

4. REGULATION OF CALCIUM MOBILIZATION IN MAST CELLS

4.1. The kinases

Fyn, Lyn, and Syk are all contributors to calcium responses in mast cells. It is well known that Lyn plays an important role in propagating calcium responses through its ability to phosphorylate the FcεRIγ ITAM leading to Syk activation (reviewed in Rivera and Gilfillan, 2006). Once activated, Syk phosphorylates multiple proteins (such as the adaptor LAT and phospholipase Cγ) that are involved in the regulation of calcium responses (Rivera, 2002). Thus, both Lyn- and Syk-deficient mast cells showed marked decreases in calcium responses (Costello *et al.*, 1996; Kawakami *et al.*, 2000; Odom *et al.*, 2004; Zhang *et al.*, 1996). However, it is important to note that while Lyn-deficient mast cells showed a prolonged delay in calcium responses, there is a modest increase in intracellular calcium with time. In contrast, Syk-deficient mast cells did not show a calcium response. Thus, it seems that, unlike Lyn, Syk is critical for calcium responses. Consistent with this view, Syk activation occurs in the absence of Lyn although it is delayed and reduced relative to wild-type cells.

Recently, we have come to appreciate that Fyn kinase also plays a role in the calcium response (Suzuki and Rivera, unpublished observation). In an early study of Fyn function in mast cells, a slightly more transient calcium response was noted (Parravicini *et al.*, 2002). Using a more sensitive fluorometric analysis, Fyn kinase has now been found to play an important role in the regulation of calcium responses. Loss of Fyn did not affect the mobilization of calcium from intracellular stores but instead reduced the influx of calcium into the cell. This loss of calcium influx is accompanied by a considerable defect in mast cell degranulation. The exact mechanism by which Fyn contributes to calcium influx remains to be determined; however, several possible options exist. One possible mechanism is through Fyn-dependent phosphorylation of plasma membrane calcium channels. Some transient receptor potential channel (TRPC) family members are substrates for Src family kinase, like Fyn (Hisatsune *et al.*, 2004; Vazquez *et al.*, 2004) and the recent finding (Ma *et al.*, 2008) in the mast cell line (RBL-2H3) showing that TRPC channels can contribute to mast cell calcium influx makes this a plausible scenario.

Another possible mechanism could be through the generation of sphingosine-1-phosphate (S1P) by sphingosine kinase 2 (SphK2). SphK1 and SphK2 convert sphingosine (Sph) into S1P (Fig. 3.1), a highly biologically active molecule that functions as an intracellular lipid second messenger (LSM) as well as a ligand for a family of G protein-coupled receptors (GPCRs) expressed on the cell surface of many cell types. The genetic deletion of SphK2 was found to cause a defect in calcium influx (Olivera *et al.*, 2007) independent of the S1P effects on S1P receptors. Additionally, activation of SphK2 was found to be highly dependent on Fyn kinase, as loss of Fyn expression caused a marked defect in SphK2 activation (Olivera *et al.*, 2006). Thus, Fyn kinase might mediate its effect on calcium influx through its role in SphK2 activation. However, what intracellular targets might be modified by SphK2 activity in mast cells is yet to be determined.

4.2. The adaptors

The lipid raft resident transmembrane adaptor molecules LAT1 and LAT2 are known participants in the regulation of mast cell calcium responses (Draberova *et al.*, 2007; Iwaki *et al.*, 2008; Saitoh *et al.*, 2000). How LAT1 regulates calcium in mast cells has been worked out in great detail (Saitoh *et al.*, 2000,, 2003). Briefly, once phosphorylated by Syk, LAT1 recruits phospholipase Cγ (PLCγ) leading to its membrane localization and activation (Fig. 3.1). This interaction is stabilized by cooperative binding of the SH2-containing leukocyte protein of 76 kDa (SLP-76), an adaptor protein that links PLCγ_1 with LAT1-bound adaptors called Gads (Houtman *et al.*, 2005). PLCγ_1 and PLCγ_2 produce diacylglycerol (DAG) and IP$_3$ from PI(4,5)-P$_2$. IP$_3$ is essential for calcium mobilization from intracellular stores as it binds to receptors in the endoplasmic reticulum (ER) eliciting the release of calcium from these stores. The key role of PLCs is manifest through the requirement of calcium and DAG for the activation of classical PKCs, like PKCβ, whose activity has been demonstrated as essential for mast cell degranulation and cytokine production (Fig. 3.1) (Nechushtan *et al.*, 2000; Ozawa *et al.*, 1993). Thus, adaptors such as LAT1 play an important role in regulating the calcium response and the activation of PKC through PLCγ binding and activation. As might be imagined, the binding and activation of PLCγ must be tightly controlled and one mechanism appears to be through the required cooperativity in protein binding and function by LAT1 (Saitoh *et al.*, 2003). This view of cooperativity in protein binding for LAT1-mediated calcium responses is supported by the markedly reduced calcium responses observed in LAT1- and SLP-76-deficient mast cells as well as tyrosine mutants of these adaptors (Pivniouk *et al.*, 1999; Saitoh *et al.*, 2003; Silverman *et al.*, 2006).

In contrast, the mechanism by which LAT2 regulates calcium responses is less clear. Unlike LAT1, LAT2 does not have the YLVV motif to directly bind PLCγ. Nonetheless, it might associate with PLCγ through binding of Grb2, an adaptor molecule that can bind to both LAT2 and PLCγ (Fig. 3.1) (Iwaki *et al.*, 2008). There is still considerable debate on the role of LAT2 in mast cells. LAT2 was found to be required for mast cell degranulation in studies silencing LAT2 expression with RNAi (Tkaczyk *et al.*, 2004). In contrast, genetic deletion of LAT2 caused enhanced mast cell degranulation (Volna *et al.*, 2004; Zhu *et al.*, 2004). Similarly, while LAT2 is not expressed in resting T cells, in activated T cells genetic loss of its induced expression resulted in enhanced calcium responses and increased cytokine production (Zhu *et al.*, 2006). Thus, LAT2 appears to have a dual role as a positive and negative regulator of calcium that may selectively manifest given the difference in the employed strategies for these studies. One might propose a possible competition between LAT1 and LAT2 for lipid raft occupancy or for limiting amounts of key signaling proteins as an explanation for dual roles (Rivera, 2005).

Recent studies help to shed additional light on this matter. While previous studies had demonstrated that LAT2 is a target of Lyn activity (Tkaczyk *et al.*, 2004), recent analysis of the phosphorylation of the individual LAT2 tyrosines revealed their differential phosphorylation depending on the stimulus (Iwaki *et al.*, 2008). *In vitro* experiments have shown that Lyn, Syk, and Kit kinases can phosphorylate overlapping and distinct tyrosine residues in LAT2 (Iwaki *et al.*, 2008). For example, upon phosphorylation of LAT2 by Syk, PLCγ binding could be detected whereas this interaction was not readily detected upon Lyn or Kit phosphorylation of LAT2. The findings showed that the putative Grb2-binding sites at Y193 and Y233 (which are most effectively phosphorylated by Syk) were important for PLCγ association. This suggests that LAT2 may have different roles depending on the stimulus and thus the complete absence of LAT2 (upon genetic deletion) could dominantly manifest an otherwise passive negative role whereas its downregulation (upon RNAi inhibition) may reveal only its positive role as some LAT2 is still available in these cells. This view has recently received additional support from the study of Ca^{2+} regulation by phosphorylated and unphosphorylated LAT2 in mast cells (Draberova *et al.*, 2007). The findings demonstrate that the overexpression as well as the downregulation (by RNAi) of LAT2 expression in the RBL-2H3 mast cell line causes decreased mast cell degranulation. Interestingly, the overexpression of LAT2 was inhibitory for phosphorylation of FcεRI, Syk, and LAT1 whereas its underexpression did not alter the phosphorylation of these proteins suggesting that the mechanisms by which LAT2 decreased mast cell degranulation when over- or underexpressed was different. Most strikingly, Lyn kinase activity was greatly enhanced when LAT2 was overexpressed. This observation appears at odds with the state of decreased FcεRI and Syk phosphorylation. However, since LAT2 phosphorylation was also greatly enhanced by its over-expression, the findings suggest a possible sequestration of Lyn by LAT2 thus effectively blocking its role in FcεRI phosphorylation. Importantly, the study defines a primary role for LAT2 in regulating calcium influx in a manner that is not dependent on its increased tyrosine phosphorylation. While defects in PLCγ₁ phosphorylation and in IP₃ production were observed when LAT2 is either over- or underexpressed, the amount of LAT2 expressed showed a direct correlation to calcium influx rather than to efflux of calcium from intracellular stores (Draberova *et al.*, 2007). Thus, the findings put forward an interesting hypothesis; namely, that LAT1 may be essential for intracellular calcium mobilization through its role in the activation of PLCγ and IP₃ production whereas LAT2 may principally function to regulate calcium influx. This could be feasible regardless of the observation that PLCγ can associate with LAT2 indirectly, since PLCγ₂ was shown to regulate calcium influx independent of its catalytic activity (Patterson *et al.*, 2002; Wang *et al.*, 2000). This of course remains to be directly demonstrated; however, increasing evidence suggests that calcium influx requires its own network of signaling molecules composed of proteins found in the plasma and endoplasmic reticular membranes that communicate to elicit the inward flow of calcium (Smyth *et al.*, 2006).

4.3. The calcium apparatus

Increases in the intracellular concentration of free Ca^{2+} is essential for mast cell degranulation and cytokine responses (reviewed in Blank and Rivera, 2004). Store-operated Ca^{2+} entry requires the depletion of intracellular calcium stores as the trigger to induce Ca^{2+} influx across the plasma membrane through store-operated channels (SOCs) (Putney, 1986) (Fig. 3.2). The demonstration by Hoth and Penner of a calcium release-activated calcium (CRAC) channels in mast cells (Hoth and Penner, 1992) more than 15 years ago has led to an extensive effort to identify these channels and their mechanism of activation (Vig and Kinet, 2007). Major advances have recently been made toward identifying components of this apparatus. The discovery of STIM-1 as an ER-localized calcium sensor (Liou *et al.*, 2005; Roos *et al.*, 2005) and the subsequent discovery of Orai1/CRACM1 (Feske *et al.*, 2006; Vig *et al.*, 2006b) as a key component for calcium entry has rapidly advanced our mechanistic understanding of how SOCs work in response to a receptor stimulus (Fig. 3.2). In this section, we briefly review some of these advances in the context of mast cell signaling and function. More extensive and recent reviews on the topic of calcium regulation are available (Smyth *et al.*, 2006; Vig and Kinet, 2007).

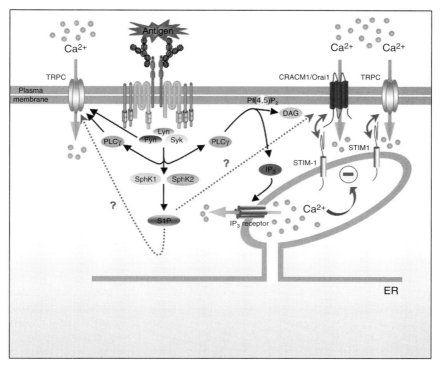

FIGURE 3.2   Calcium regulation in mast cells. Calcium is an important second messenger whose mobilization is precisely controlled in mast cells. Antigen aggregation of FcεRI leads to the activation of Lyn, Fyn, and Syk kinases. These kinases contribute to the calcium response by initiating and/or supporting PLCγ activity. PLCγ catalyzes the hydrolysis of membrane-localized phosphatidylinositol 4,5-bisphosphate (PIP$_2$) to inositol 1,4,5-trisphosphate (IP$_3$) and diacylglycerol (DAG). IP$_3$ binds IP$_3$ receptors in the ER membrane, resulting in the release of Ca^{2+} from intracellular Ca^{2+} stores. DAG regulates the activity of various proteins, such as members of the protein kinase C family. The decrease in the Ca^{2+} concentration in the ER stores is sensed by STIM-1 (a calcium sensor) causing a change in its conformation that allows its translocation proximal to the plasma membrane where it can interact with CRACM1/Orai1. STIM-1 and CRACM1/Orai1 synergize to elicit the influx of calcium. CRACM1/Orai1 appears to encode the *icrac* current channel in mast cells originally described by Hoth and Penner. Recent evidence also demonstrates that STIM-1 and Orai1 cooperate with transient receptor potential channels (TRPC) in mast cells to stimulate nonselective entry of Ca^{2+}. These findings demonstrate an increasing complexity in the regulation of Ca^{2+} entry in mast cells. In other cell types, some TRPCs have also been demonstrated to be targets of Src family kinases, like Fyn, and to be directly regulated by PLCγ. The activation of FcεRI also causes the activation of sphingosine kinases 1 and 2 (SphK1 and SphK2) and the production of sphingosine-1-phosphate (S1P). This sphingolipid regulates Ca^{2+} influx within the cell, but the intracellular target of S1P is unknown. Moreover, the relationship between S1P, STIM-1, and CRACM1/Orai1, if any, remains to be determined. (See Plate 2 in Color Plate Section.)

As outlined above, IP_3 is a key second messenger in the release of calcium from ER stores of mast cells through its engagement of IP_3 receptors which are clustered upon IP_3 binding resulting in a receptor/channel opening and release of stored free Ca^{2+} into the cytosol (Putney *et al.*, 1989) (Fig. 3.2). We now know that this emptying of stores and increase in cytosolic free Ca^{2+} concentration triggers the conformational change in the calcium sensor ER single transmembrane spanning protein (STIM-1, which has a molecular mass of 77 kDa) that causes it to localize proximal to the plasma membrane (although whether it integrates in the plasma membrane is still uncertain) and functions to enhance plasma membrane calcium channel activity (Liou *et al.*, 2005; Peinelt *et al.*, 2006). More recent studies have demonstrated that deficiency of STIM-1 expression in mast cells caused marked loss of FcεRI-mediated degranulation and cytokine production and mice with low levels of STIM-1 showed a reduced anaphylactic response (Baba *et al.*, 2008). The activation of transcription factors such as NFκB and NFAT was markedly reduced whereas early signaling (Syk, LAT, etc.) upon FcεRI stimulation (including IP_3 production) appeared normal. As might be expected, a marked defect in the calcium response was observed in these cells. Strikingly, some residual influx of calcium was observed suggesting a possible STIM-1-independent mechanism contributing to calcium influx in mast cells. Regardless, the findings provide convincing evidence of a key role for STIM-1 in the regulation of mast cell calcium fluxes.

The combined approach of an RNAi screen in Drosophila S2 cells for inhibition of thapsigargin-induced calcium responses or for disruption of an indicator of NFAT nuclear translocation led two groups (Feske *et al.*, 2006; Vig *et al.*, 2006b) to the almost simultaneous discovery of another key component of the calcium apparatus termed CRACM1 or Orai1 (Fig. 3.2). This is a small protein with a molecular mass of 32.7 kDa predicted to span the membrane four times and the presence of both its N- and C-terminal domains in the cytosol has been demonstrated. Mutation of this protein was found in some severe combined immunodeficiency (SCID) patients (Feske *et al.*, 2006) and it was shown to be crucial for calcium entry (Peinelt *et al.*, 2006; Prakriya *et al.*, 2006; Vig *et al.*, 2006b). However, whether CRACM1/Orai1 is itself a calcium channel is still not clear. Nonetheless, a strong functional synergy in calcium entry was found by co-expression of STIM-1 with CRACM1/Orai1 (Peinelt *et al.*, 2006) and it was also found to multimerize (Vig *et al.*, 2006a), which is characteristic of how many ion channels form a functional pore (Fig. 3.2). Recent studies have found that mast cells deficient in CRACM1/Orai1 showed a marked defect in degranulation and cytokine production (Vig *et al.*, 2008). Reconstitution of these deficient cells with CRACM1/Orai1 caused a partial restoration of function. While the deficiency in expression of CRACM1/Orai1 showed a marked defect on mast cell function *in vitro* and *in vivo* (in passive cutaneous anaphylaxis), it did not completely ablate calcium responses. This is consistent with the idea that *icrac*-independent calcium mobilization may contribute to the increased levels of intracellular calcium upon FcεRI stimulation. Regardless, the findings demonstrate that CRACM1/Orai1 is a key component in FcεRI calcium influx and thus makes this protein a potential therapeutic target for intervention in allergic disease.

Additional complexity in the calcium apparatus has been recently recognized by the demonstration that both STIM-1 and CRACM1/Orai1 appear to work synergistically (Peinelt *et al.*, 2006) and associate with other calcium channels, particularly members of the transient receptor potential channels (TRPC) (Fig. 3.2). Recently, the use of silencing RNAi strategy and ectopic expression experiments has revealed that TRPC5 and STIM-1 cooperate to elicit both Ca^{2+} and Sr^{2+} influx in the mast cell line RBL-2H3 (Ma *et al.*, 2008), and this contributes to the enhanced calcium response needed for degranulation. This cooperation was easily distinguished from selective Ca^{2+} entry induced by co-expression of STIM-1 and CRACM1/Orai1. Thus, these findings suggest that calcium channels other than *icrac* are likely contributors to mast cell calcium influx and functional responses. As a tissue resident cell that is found in considerably different microenvironments, one might speculate that the type of calcium channel expressed may differ depending on the microenvironment. This adaptability would afford flexibility in eliciting functional responses in innate and adaptive immunity (Galli, 1997b).

5. SPHINGOSINE-1-PHOSPHATE: A CO-STIMULATORY SIGNAL IN FcεRI-MEDIATED MAST CELL ACTIVATION

As described in the Section 4, S1P has an intrinsic effect on mast cell activation through regulation of calcium responses, but the metabolic breakdown of sphingomyelin generates multiple bioactive lipids, including ceramide (Cer), Sph, and S1P. These lipids have diverse functions and can mediate cell growth, survival, differentiation, calcium homeostasis, and chemotactic motility and show counteracting properties in regulating cell function (Olivera and Rivera, 2005; Spiegel and Milstien, 2003). In mast cells, S1P has been demonstrated to positively regulate responses whereas Sph and Cer inhibit mast cell responsiveness (Olivera and Rivera, 2005; Prieschl *et al.*, 1999). Engagement of FcεRI on mast cells activates the two known mammalian sphingosine kinases (SphK1 and SphK2) resulting in the regulation of Sph levels and the production and secretion of S1P by these cells (Olivera *et al.*, 2006). S1P also mediates its functional effect on mast cells as a ligand for a family of S1P receptors ($S1P_{1-5}$), two of which ($S1P_1$ and $S1P_2$) are expressed on mast cells

(Fig. 3.3). Thus, the role of FcεRI-mediated production of S1P in mast cells is of considerable interest and recent studies delineate an important role for S1P and its receptors in mast cell function and allergic responses.

5.1. Linking FcεRI to sphingosine kinase activation

How SphKs are activated is not completely understood. However, the evidence gathered to date suggests that a complex interplay of protein kinase- and lipid-derived signals are required in an apparent two-step process in which SphKs translocate rapidly to plasma membranes and become active following FcεRI engagement (Olivera and Rivera, 2005; Olivera *et al.*, 2006; Urtz *et al.*, 2004). At the moment, it has been difficult to resolve whether the translocation of SphKs to the plasma membrane is essential for activation or whether it is a consequence of activation. Regardless, under circumstances where the stimulation of SphK by FcεRI is impaired, there is also no translocation of SphK to membranes (Olivera *et al.*, 2006; Urtz *et al.*, 2004). We now know that the FcεRI-mediated translocation and activation of SphKs requires both Fyn and Lyn (Fig. 3.3). While there is an absolute requirement for Fyn activity in activating SphKs, Lyn activity is not essential (Olivera *et al.*, 2006; Urtz *et al.*, 2004). The requirement for Fyn-dependent activation of SphK1 includes a role for Gab2/PI3K/PLD activities (Olivera *et al.*, 2006, unpublished observations). The adaptor Gab2 is required for the activation of PI3K in mast cells and the production of phosphatidylinositols (including PIP_3) is needed for the activation and function of PLD, which in turn produces phosphatidic acid (PA). Both of these lipid messengers can either bind, activate, or induce the translocation of SphKs (Delon *et al.*, 2004; Melendez and Khaw, 2002; Olivera *et al.*, 1996). In contrast to SphK1, the activation of SphK2 requires Fyn, but seems to be less dependent on the activity of Gab2/PI3K (Olivera *et al.*, 2006). Fyn and Lyn also interact with SphKs and this interaction appears to be required for the translocation and activation of SphKs (Olivera *et al.*, 2006; Urtz *et al.*, 2004).

5.2. Linking FcεRI to sphingosine-1-phosphate receptors

Fyn-deficient mast cells are defective in FcεRI-induced degranulation, but degranulation can be partially corrected by the addition of exogenous S1P upon Ag stimulation (Olivera *et al.*, 2006). This suggests that S1P production in mast cells partially contributes to their FcεRI-mediated degranulation (Fig. 3.3). Inhibition of SphK activation, and thus S1P generation, by either competitive analogs of Sph in the RBL tumor mast cell line (Choi *et al.*, 1996) or by antisense SphK mRNA in human mast cells (Melendez and Khaw, 2002) prevented IgE-triggered calcium responses and inhibited degranulation. Thus, the S1P produced by SphK activation in mast cells has both an intracellular and extracellular regulatory role in mast cell

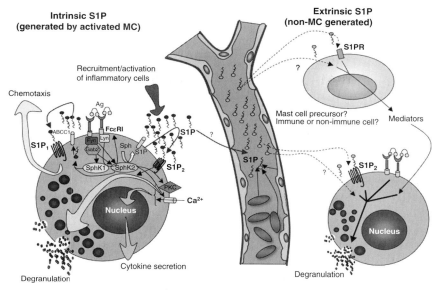

FIGURE 3.3   Role of sphingosine kinases, sphingosine-1-phosphate (S1P), and its receptors in mast cell functions. Mast cell function in its physiological environment is
affected by "extrinsic S1P" generated by cells other than mast cells (right side, dashed
lines with arrows) and "intrinsic S1P" (S1P generated by stimulated mast cells; left side
blue arrows). Crosslinking of the FcɛRI by IgE/antigen in mast cells results in the rapid
activation and translocation of SphK to the plasma membrane and the generation of S1P.
This is mediated by the Src kinases Lyn and Fyn. Fyn is required for SphK1 and SphK2
activation, whereas Lyn is required for the early phase of activity and membrane
translocation (left panel). Fyn-dependent Gab2/PI3K activation, followed by PLD activation is required for SphK1 stimulation, while a not yet determined Fyn-dependent
but Gab2-independent pathway is needed for full SphK2 activation. Our recent studies
implicate SphK2, and not SphK1, in the influx of calcium following IgE receptor independently of IP_3 generation, and thus affecting mast cell functions. S1P is secreted by
activated mast cells to the extracellular media independently of their degranulation
via an ATP-binding cassette (ABC) transporter. Furthermore, S1P is able to rapidly bind
and activate its receptors $S1P_1$ and $S1P_2$ on the plasma membrane. $S1P_1$ induces cyto-
skeletal rearrangements leading to the movement of mast cells toward an antigen
gradient, while transactivation of $S1P_2$ enhances the degranulation response. Mast cell-
secreted S1P can also promote inflammation by activating and recruiting other immune
cells involved in allergic and inflammatory responses. Mast cell granules are illustrated
as blue circles, and the process of degranulation as granules in contact with the plasma
membrane emptying their content (smaller pink and blue dots). Mast cells may be
affected by changes in circulating S1P by different mechanisms. It is possible that changes
in S1P in the circulation affect the priming of $S1P_2$ in mast cells, which are found in close
proximity to blood vessels, enhancing degranulation upon cell activation. Increases or
decreases in extrinsic S1P levels might also induce the differentiation of mast cell
precursors toward a more or less responsive phenotype, respectively. Constant exposure to higher or lower levels of S1P might also indirectly influence mast cell differentiation or function via mediators derived from other immune or nonimmune cells that
respond to the fluctuation in S1P levels. (See Plate 3 in Color Plate Section.)

degranulation. The role of S1P in regulating calcium was discussed above; here we now summarize what is known about its S1P receptor-mediated role in mast cell degranulation.

As previously mentioned, mast cells express two of the five receptors for S1P, $S1P_1$ and $S1P_2$ (Fig. 3.3) (Jolly *et al.*, 2004). The first evidence of a link between the FcεRI and S1P receptors on mast cells was the demonstration that FcεRI-induced S1P formation resulted in the transactivation of these two receptors (Jolly *et al.*, 2004). The S1P receptors couple to different subunits of heterotrimeric G proteins (α_i, α_q, and $\alpha_{12/13}$), and therefore they can trigger diverse signals, including activation of Src kinases, small GTPases, MAPK cascades, phospholipases, PKC, and calcium mobilization (Pyne and Pyne, 2000). $S1P_1$ has been well defined as a chemotatic receptor and in the immune system it is known to be required for thymocyte emigration and lymphocyte recirculation (Matloubian *et al.*, 2004) and outside of the immune system it is required for vascular morphogenesis (Allende and Proia, 2002) among a host of other functions (Spiegel *et al.*, 1996). For mast cells, it was demonstrated that $S1P_1$ is required for their migration toward low concentrations of antigen (Fig. 3.3). In contrast, $S1P_2$ was required for normal degranulation (Fig. 3.3), as downregulation of its expression or deletion of the $S1P_2$ gene in mast cells led to a marked loss (a 50% inhibition) of degranulation (Jolly *et al.*, 2004). Interestingly, $S1P_2$ also appears to exert some control in the function of $S1P_1$, since overexpression of $S1P_2$ inhibited mast cell chemotactic motility. This is consistent with the finding that $S1P_2$ mRNA expression is enhanced as a late consequence of FcεRI engagement, while $S1P_1$ mRNA expression is constitutive. Thus, a gradient of antigen might attract mast cells to their site of action via $S1P_1$. Subsequently, as the mast cells approach higher concentrations of antigen, enhanced $S1P_2$ expression would inhibit migration while promoting degranulation. This is consistent with the requirement of low concentrations of antigen in eliciting the production of chemokines whereas high antigen concentrations are required for mast cell degranulation (Gonzalez-Espinosa *et al.*, 2003). How $S1P_2$ collaborates with FcεRI to elicit a full degranulation of mast cells is not well understood but the findings suggest that the transactivation of this receptor is important for mast cell degranulation. Thus, S1P engagement of $S1P_2$ can be defined as a co-stimulatory signal for FcεRI-mediated mast cell degranulation. The findings also suggest that the extracellular levels of S1P may influence the degranulation of mast cells, a phenomenon for which there is now *in vivo* evidence.

5.3. S1P is an *in vivo* effector of mast cell function

S1P is secreted by mast cells upon FcεRI engagement (reviewed in Olivera and Rivera, 2005). Unlike T cells, B cells, and so on, mast cells secrete a substantial amount of S1P suggesting it is important for mast cell effector

functions. S1P is also highly elevated in the airways of asthmatic individuals (Jolly *et al.*, 2002), and in the joints of arthritic individuals (Kitano
et al., 2006). Both asthma and rheumatoid arthritis are inflammatory conditions in which mast cells have been demonstrated to be important
effector cells, the latter primarily in a mouse model (Lee *et al.*, 2002). This
raises the possibility that S1P is an autocrine and/or a paracrine mediator
involved in the pathophysiology of asthma or other allergic and/or
inflammatory diseases.

In recent studies, we have found a close correlation between circulating levels of S1P and the histamine levels in the plasma following anaphylactic challenge (Olivera *et al.*, 2007). Wild-type mice having high
circulating S1P showed the highest levels of histamine in the plasma,
whereas those with low levels of S1P had reduced plasma histamine.
This correlation is further observed in the *in vivo* anaphylactic responses
of wild-type mice or mice with a genetic deletion of SphK1 or SphK2.
SphK1-null mice had reduced levels of circulating S1P (Allende *et al.*,
2004; Zemann *et al.*, 2006) and these mice were highly resistant to anaphylaxis, suggesting that the level of circulating S1P was a determinant of
mast cell responsiveness. In contrast, SphK2-null mice had enhanced
levels of circulating S1P relative to wild-type mice (Zemann *et al.*, 2006)
and were found to respond normally to an anaphylactic challenge. These
findings pointed to a dominant extrinsic role for S1P in regulating mast
cell responsiveness (Fig. 3.3). This dominant extrinsic role of S1P could
overcome the intrinsic defect of SphK2-null mast cells, which showed
defective degranulation *in vitro*. Moreover, the low circulating levels of
S1P in SphK1-null mice was dominant over the observed normal *in vitro*
degranulation of SphK1-null mast cells manifesting as defective *in vivo*
degranulation. This dominance of circulating S1P levels was further
explored by generating mice that were null for SphK2 with one functional
allele for SphK1. This returned the levels of circulating S1P to normal, as
compared to the SphK2-null mouse, and these mice were now resistant to
an anaphylactic challenge. This relationship between increased circulating levels of S1P and increased *in vivo* mast cell responsiveness is also
manifest in mice differing in genetic backgrounds (see below), or mice
where S1P levels were artificially increased by use of the sphingosine
lyase inhibitor, 2-acetyl-4-tetrahydroxybutylimidazole (THI). In both
cases, increased circulating S1P was associated with increased *in vivo*
mast cell responsiveness (Fig. 3.3) (Olivera and Rivera, unpublished
observation). These findings demonstrate a dominant extrinsic role for
S1P, but it should be noted that S1P generated within the mast cell is also
important in regulating mast cell responsiveness as mast cells that do not
produce S1P, such as Fyn-null mast cells (Olivera *et al.*, 2006), are considerably more defective in degranulation than $S1P_2$-null mast cells (Jolly
et al., 2004). Both intrinsic and extrinsic regulation of mast cell responses
by S1P appears to participate in FcεRI-induced mast cell responses. However, the intracellular target(s) of S1P in these cells remains to be defined.

6. THE ROLE OF GENETICS IN MAST CELL RESPONSIVENESS

The study of the physiological or pathophysiological relevance of mast cell signaling and function has required the use of mouse models. Mice have become the primary mammalian model organism because of their close genetic relationship to humans and the ease by which alterations in their genome allow the study of fundamental biological processes and disease. It has been well established that theses two species share many of the genetic pathways regulating normal and pathological conditions regardless of the obvious physiological and anatomical differences. Since the first use of the mouse as a model for physiological studies, it became clear that not all isogenic backgrounds are appropriate for a given study. Indeed, certain isogenic strains can be poor mimics of human disease or even eclipse the effects of a targeted mutation, which is manifest in another strain (Rivera and Tessarollo, 2008). For example, there is a contrasting degree of airway inflammation seen in the ovalbumin (OVA) model of asthma depending on the mouse strain (Herz *et al.*, 2004). When sensitized to OVA, an aerosol challenge of BALB/c mice with OVA causes a marked increase in lymphocytes, eosinophils, and neutrophils in the bronchoalveolar fluids. This is paralleled by highly elevated amounts of IL-4, IL-5, and TNF and easily measurable airway hyperreactivity. This same model in a different strain, such as C57BL/6, mice showed a much weaker lung inflammation, a lesser cytokine response, and a more modest airway hyperreactivity (Herz *et al.*, 2004). These types of studies argue that, like in humans, the genetic environment is a determinant in allergic responses. However, whether the genetic makeup could directly affect the responsiveness of the mast cell was not known. Here we discuss new evidence supporting the view that differences in genetic makeup contribute to the degree of mast cell responsiveness.

6.1. An apparent contradiction in Lyn function in mast cells is manifest through genetic differences in mouse strains

Various experiments conducted to study the role of Lyn kinase in mast cell activation were apparently contradictory. Thus, the early demonstration that Lyn-null mice were resistant to IgE/Ag-induced passive cutaneous anaphylaxis (Hibbs *et al.*, 1995) appeared to contradict experiments showing that mast cells from another independently generated Lyn-null mouse showed normal mast cell degranulation (Nishizumi and Yamamoto, 1997). Moreover, several later reports showed either normal or enhanced degranulation in Lyn-null mast cells (Hernandez-Hansen *et al.*, 2004; Kawakami *et al.*, 2000; Odom *et al.*, 2004), including an enhanced passive systemic anaphylactic response and increased airway hypersensitivity in Lyn-null mice (Beavitt *et al.*, 2005; Odom *et al.*, 2004), whereas the loss of Lyn function in the mast cell line RBL-2H3 caused loss of mast cell degranulation (Vonakis *et al.*, 2005). While a number of variables differed in these studies, the most identifiable difference in experiments using mouse cells was that they were derived from mice either of the C57BL/6 or mice with a mixed background of 129/SvJ × C57BL/6.

To determine the relative contribution of the difference in genetic background to the published observation, an analysis of the two pure genetic backgrounds (C57BL/6 vs 129/SvJ) was undertaken. Analysis of mast cells derived from C57BL/6 mice carrying a null mutation for Lyn showed a dramatic inhibition of FcεRI-mediated degranulation relative to their wild-type counterparts (Yamashita *et al.*, 2007). In contrast, mast cells from 129/SvJ mice carrying the same mutation showed an enhanced degranulation response when compared to their wild-type counterparts and to all C57BL/6-derived mast cells. These studies identified that in mast cells derived from C57BL/6 mice, Fyn kinase expression was two-fold less than that seen in 129/SvJ-derived mast cells. Since Fyn functions to promote degranulation (Parravicini *et al.*, 2002), ectopic expression in C57BL/6 cells was explored in the context of Lyn deficiency. The increased expression of Fyn caused conversion of the poor responsive C57BL/6 Lyn-null phenotype to a highly responsive 129/SvJ Lyn-null phenotype. Thus, the findings demonstrated that relatively modest differences in gene expression (a 50% decrease) could cause a marked phenotypic change. Interestingly, the role of Fyn and Lyn in human mast cells was also investigated in this study. The silencing of Fyn or Lyn expression in human mast cells caused the inhibition or augmentation of degranulation, respectively (Yamashita *et al.*, 2007). Thus, it appears that human mast cells show a phenotype that is most similar to that of the 129/SvJ mouse rather than the C57BL/6 mouse.

6.2. Genetics influences the *in vivo* environment altering mast cell responsiveness

129/SvJ mouse strains do not breed well, and are reported to have abnormal anatomy and behavior, thus, the backcross newly generated mutant 129/SvJ mice into the C57BL/6 background is widely practiced. C57BL/6 mice are long-lived and are permissive for expression of most mutations (http://jaxmice.jax.org/strain/000664.html) and breed relatively well. However, 129/SvJ and C57BL/6 mouse strains demonstrate a skewing toward T helper 2 (Th2) and T helper 1 (Th1) responses, respectively (O'Neill *et al.*, 2000) and C57BL/6 mice are relatively resistant to tumor development. This limits the use of these strains for certain immunological questions. *In vivo* interrogation of allergic responses must take into consideration the Th1 or Th2 environment present in these mice. Thus, for example, 129/SvJ mice have considerably higher levels of circulating IgE than that found in C57BL/6 mice. Relative to the C57BL/6 mouse, increased levels of IgE in the 129/SvJ mouse enhances the onset and extent of allergic responses because high IgE levels cause increased expression of FcεRI and also cause full occupancy of this receptor, increasing the sensitivity to an allergenic stimulus (Yamashita *et al.*, 2007). More recently, we have found that the 129/SvJ mice also have higher levels of circulating S1P relative to the C57BL/6 mice (Olivera and Rivera, unpublished observation). As outlined above, the increase in S1P in 129/SvJ mice may manifest in increased mast cell responsiveness. Indeed, *in vivo* passive systemic anaphylaxis challenge of 129/SvJ mice showed increased circulating histamine levels when compared to C57BL/6 mice. Thus, while many questions that surround the issue of how genetics influences mast cell responsiveness are still unanswered, the experiments to date reveal the importance of genetics to the mast cell and its environment. In addition, these studies also caution against the extrapolation of experimental results in mice (based on experiments conducted on mast cells or mice from a single genetic strain) as relevant to human disease.

7. SUMMARY AND PERSPECTIVES

7.1. On early signals in the regulation of mast cell activation

The recent advances in our understanding of FcεRI-mediated activation of mast cells have clearly shown an increasing complexity that was previously unappreciated. These findings define a complex relationship of proteins and lipids that serve to target, activate, and regulate the molecular steps required for full activation of a mast cell (Fig. 3.1). At the earliest step in mast cell activation, namely the clustering of FcεRI, one finds that proximity of FcεRI in these clusters is essential in determining if FcεRI becomes phosphorylated and whether the mast cell will become fully activated or not (Sil *et al.*, 2007). This suggests an intrinsic mechanism at the very first step of activation that screens a productive engagement of FcεRI from an unproductive one. Given that the FcεRI is likely to be occupied by IgE of varying antigenic specificities *in vivo*, the requirement for close proximity of FcεRI for its activation would likely avoid the spurious activation of these cells by weak antigens since the proximity of any two receptors is unlikely to be sustained by weak antigens due to their more rapid dissociation (Torigoe *et al.*, 1998).

Other points of regulation in FcεRI-mediated mast cell activation are also revealed in the recent studies. The positive and negative balance is mediated through Src family kinases. Fyn and Lyn, as well as Hck and possibly others (Hernandez-Hansen *et al.*, 2004; Hong *et al.*, 2007; Kawakami *et al.*, 2000; Kitaura *et al.*, 2007; Odom *et al.*, 2004; Xiao *et al.*, 2005), appear to be key components of a network that fine-tunes the extent of a mast cell's response when FcεRI is engaged (Fig. 3.1). In some cases, this appears to be mediated indirectly through regulatory proteins that inactivate Src kinases, like the Cbp/Csk complex (Kitaura *et al.*, 2007; Odom *et al.*, 2004), but we cannot exclude a more direct control through crosstalk of Src family kinases. There is also an increasing body of literature that argues that the cholesterol-enriched plasma membrane microdomains, which are heterogeneous signaling module (Heneberg *et al.*, 2006; Wilson *et al.*, 2000, 2001), also play an important regulatory role (Fig. 3.1). Thus, loss of Lyn from these domains or the failure to cause effective coalescence of these domains and their signaling constituents has marked consequences on the ability of a mast cell to become fully activated and/or may result in selective responses (Heneberg *et al.*, 2006; Kovarova *et al.*, 2006; Young *et al.*, 2003). It is important to note that selective signaling and responses have been shown to occur, such as upon weak stimulation of FcεRI (Gonzalez-Espinosa *et al.*, 2003). This is important as weak stimuli are less effective in coalescing cholesterol-enriched membrane microdomains (Davey *et al.*, 2007, 2008 Heneberg *et al.*, 2006). This is also likely to affect signaling and the function of adaptor molecules, like LAT1 and LAT2, which participate in amplification of various signaling pathways, including calcium mobilization (Draberova *et al.*, 2007; Iwaki *et al.*, 2008; Saitoh *et al.*, 2000) as they are localized within these membrane microdomains. However, how the regulatory role of LAT1 and LAT2 is partitioned in FcεRI-induced signals and mast cell responses remains to be defined. It should be noted that regulation of responses also exist downstream of LAT1 (Torigoe *et al.*, 2007); however, the molecular events involved in the downstream regulatory checkpoint(s) are unknown. Possible candidates for such a role may include the generation of lipid mediators as they can contribute to the overall activation of mast cells by their effects on recruitment and activation of many signaling molecules but can also be dominant in certain pathways leading to selective responses.

7.2. Marrying of lipids in the regulation of calcium and mast cell effector responses

The generation of lipid second messengers has long been recognized to be important in the effector function of mast cells as well as in other cell types (Becker and Hannun, 2005; Rivera and Olivera, 2007; Spiegel *et al.*, 1996). The recent advances have clearly demonstrated that lipid messengers can be dominant in certain signaling pathways and cause cellular responses independent of many of the additional events initiated by FcεRI activation.

A superb example is the finding that downregulation of PTEN expression in human mast cell (by shRNA silencing), which caused increased PIP_3 production, resulted in the activation of Akt and MAP kinases and in mast cell cytokine secretion independently of FcεRI engagement (Furumoto *et al.*, 2006). These findings show that selective mast cell responses depend almost entirely on lipid signals and do not require the additional signals that are elicited by FcεRI engagement. It should be noted that increased or decreased PIP_3 levels (which can be regulated by PTEN) in mast cells is associated with increased or decreased mast cell responsiveness, respectively (Ali *et al.*, 2004; Barker *et al.*, 1999; Furumoto *et al.*, 2006; Gomez *et al.*, 2005; Odom *et al.*, 2004), in many cases similarly affecting calcium responses. How PIP_3 levels might affect calcium responses is not well defined but it is known that many of the proteins, such as PLCγ or SphKs, that regulate calcium also require PIP_3 for their translocation and function. One can imagine that this must be finely tuned particularly since the same substrate, namely $PI(4,5)-P_2$, is used by both PLCγ and PI3K. However, it does not always follow that alterations of PIP_3 levels affect calcium responses (Leung and Bolland, 2007). Regardless, the accumulated data makes a convincing argument for PIP_3 as an essential lipid messenger in FcεRI-mediated mast cell activation (Fig. 3.1) (Ali *et al.*, 2004; Furumoto *et al.*, 2006; Leung and Bolland, 2007).

Mast cell effector functions are not only regulated by lipids, through the modulation of intracellular signaling pathways, but also through autocrine/paracrine mechanisms involving engagement of cell surface receptors on these cells. S1P is such a lipid, whose production and function in mast cells not only causes intracellular effects (such as calcium regulation) but also modulates mast cell chemotaxis and degranulation through the S1P receptors expressed on these cells (Fig. 3.2) (Jolly *et al.*, 2004; Olivera *et al.*, 2006, 2007; Urtz *et al.*, 2004). At the moment, how the intracellular and extracellular effects of S1P cooperate in FcεRI-mediated mast cell activation is a key missing piece in this puzzle. This may be revealed by the identification of the intracellular targets for S1P and its possible partnership in calcium regulation. Now that the molecular elements of the calcium apparatus are rapidly being defined (Smyth *et al.*, 2006; Vig and Kinet, 2007), it will be of particular interest to determine if these are targets of S1P. Given that a previous study suggests that sphingosine is an inhibitor of *icrac*-current (Mathes *et al.*, 1998) and that loss of S1P production selectively affects calcium influx (Olivera *et al.*, 2007), it is possible that SphKs or S1P may play a direct role in regulating calcium channels (Figs. 3.2 and 3.3). However, as the recent studies are demonstrating, an intrinsic defect in calcium responses leading to decreased *in vitro* responsiveness of a mast cell may be overcome by the *in vivo* environment in which they are found (Olivera *et al.*, 2007).

7.3. Translating molecular mechanisms to an *in vivo* environment

Recent studies affirm that *in vitro* analysis of mast cell activation and function (a reductionist approach) does not always translate *in vivo* (Olivera *et al.*, 2007; Yamashita *et al.*, 2007). Multiple factors may be responsible for this failure of translation. These include genetics, the *in vivo* environment, and the heterogeneity of mast cells *in vivo*. It is not surprising that these would be important determinants of *in vivo* mast cell responsiveness and function as this type of plasticity is required of a cell that transcends both innate and adaptive immunity (Galli *et al.*, 2005; Williams and Galli, 2000). What is unexpected, however, is that S1P as an extrinsic regulator of mast cell responsiveness seems to be dominant in the *in vivo* environment and that genetics may contribute to the overall circulating S1P levels, which are closely associated with mast cell responsiveness (Fig. 3.3). Obviously, we need to understand how circulating S1P levels are regulated *in vivo* and what is the source of this S1P, as the data to date suggests a non-mast cell source for circulating S1P (Olivera *et al.*, 2007). Of particular interest is whether one might be able to convert a low mast cell responder mouse (such as a C57BL/6) to a high mast cell responder mouse (129/SvJ or Balb/c) by increasing the circulating S1P levels in the low responder mouse. The importance of such a shift may well relate to human disease given the close association between the circulating levels of S1P and the extent of the response observed during a passive systemic challenge (Olivera *et al.*, 2007). The demonstration that allergen-challenged asthmatics showed high levels of S1P in their lungs relative to controls (Ammit *et al.*, 2001) also argues for a close link to mast cell responsiveness in disease. Further studies are clearly required to define a causal relationship and how genetics may influence circulating S1P levels. Nonetheless, it is clear that the recent advances in our knowledge of the mechanisms underlying FcεRI-mediated mast cell activation have uncovered new complexities. More importantly, they also reveal new areas of investigation with therapeutic potential in disease.

REFERENCES

Ali, K., Bilancio, A., Thomas, M., Pearce, W., Gilfillan, A. M., Tkaczyk, C., Kuehn, N., Gray, A., Giddings, J., Peskett, E., Fox, R., Bruce, I., *et al.* (2004). Essential role for the p110δ phosphoinositide 3-kinase in the allergic response. *Nature* **431,** 10071011.

Allende, M. L., and Proia, R. L. (2002). Sphingosine-1-phosphate receptors and the development of the vascular system. *Biochim. Biophys. Acta.* **1582,** 222–227.

Allende, M. L., Sasaki, T., Kawai, H., Olivera, A., Mi, Y., van Echten-Deckert, G., Hajdu, R., Rosenbach, M., Keohane, C. A., Mandala, S., Spiegel, S., and Proia, R. L. (2004). Mice deficient in sphingosine kinase 1 are rendered lymphopenic by FTY720. *J. Biol. Chem.* **279,** 52487–52492.

Ammit, A. J., Hastie, A. T., Edsall, L. C., Hoffman, R. K., Amrani, Y., Krymskaya, V. P., Kane, S. A., Peters, S. P., Penn, R. B., Spiegel, S., and Panettieri, R. A., Jr. (2001). Sphingosine 1-phosphate modulates human airway smooth muscle cell functions that promote inflammation and airway remodeling in asthma. *Faseb. J.* **15,** 1212–1214.

Baba, Y., Nishida, K., Fujii, Y., Hirano, T., Hikida, M., and Kurosaki, T. (2008). Essential function for the calcium sensor STIM1 in mast cell activation and anaphylactic responses. *Nat. Immunol.* **9,** 81–88.

Barker, S. A., Caldwell, K. K., Hall, A., Martinez, A. M., Pfeiffer, J. R., Oliver, J. M., and Wilson, B. S. (1995). Wortmannin blocks lipid and protein kinase activities associated with PI 3-kinase and inhibits a subset of responses induced by FcεRI cross-linking. *Mol. Biol. Cell* **6,** 1145–1158.

Barker, S. A., Lujan, D., and Wilson, B. S. (1999). Multiple roles for PI 3-kinase in the regulation of PLCγ activity and Ca^{2+} mobilization in antigen-stimulated mast cells. *J. Leukocyte Biol.* **65,** 321–329.

Beavitt, S. J., Harder, K. W., Kemp, J. M., Jones, J., Quilici, C., Casagranda, F., Lam, E., Turner, D., Brennan, S., Sly, P. D., Tarlinton, D. M., Anderson, G. P., *et al.* (2005). Lyn-deficient mice develop severe, persistent asthma: Lyn is a critical negative regulator of Th2 immunity. *J. Immunol.* **175,** 1867–1875.

Becker, K. P., and Hannun, Y. A. (2005). Protein kinase C and phospholipase D: Intimate interactions in intracellular signaling. *Cell Mol. Life Sci.* **62,** 1448–1461.

Benhamou, M., Ryba, N. J. P., Kihara, H., Nishikata, H., and Siraganian, R. P. (1993). Protein-tyrosine kinase p72[syk] in high affinity receptor signaling. Identification as a component of pp72 and associaition with the receptor γ chain after receptor aggregation. *J. Biol. Chem.* **268,** 23318–23324.

Berton, G., Mocsai, A., and Lowell, C. A. (2005). Src and Syk kinases: Key regulators of phagocytic cell activation. *Trends Immunol.* **26,** 208–214.

Blank, U., and Rivera, J. (2004). The Ins and Outs of IgE-dependent mast cell exocytosis. *Trends Immunol.* **25,** 266–273.

Choi, O. H., Kim, J. H., and Kinet, J. P. (1996). Calcium mobilization via sphingosine kinase in signalling by the FcεRI antigen receptor. *Nature* **380,** 634–636.

Costello, P. S., Turner, M., Walters, A. E., Cunningham, C. N., Bauer, P. H., Downward, J., and Tybulewicz, V. L. J. (1996). Critical role for the tyrosine kinase Syk in signalling through the high affinity IgE receptor of mast cells. *Oncogene* **13,** 2595–2605.

Davey, A. M., Krise, K. M., Sheets, E. D., and Heikal, A. A. (2008). Molecular perspective of antigen-mediated mast cell signaling. *J. Biol. Chem.* **283,** 7117–7127.

Davey, A. M., Walvick, R. P., Liu, Y., Heikal, A. A., and Sheets, E. D. (2007). Membrane order and molecular dynamics associated with IgE receptor cross-linking in mast cells. *Biophys. J.* **92,** 343–355.

Delon, C., Manifava, M., Wood, E., Thompson, D., Krugmann, S., Pyne, S., and Ktistakis, N. T. (2004). Sphingosine kinase 1 is an intracellular effector of phosphatidic acid. *J. Biol. Chem.* **279,** 44763–44774.

Draberova, L., Shaik, G. M., Volna, P., Heneberg, P., Tumova, M., Lebduska, P., Korb, J., and Draber, P. (2007). Regulation of Ca^{2+} signaling in mast cells by tyrosine-phosphorylated and unphosphorylated non-T cell activation linker. *J. Immunol.* **179,** 5169–5180.

Feske, S., Gwack, Y., Prakriya, M., Srikanth, S., Puppel, S. H., Tanasa, B., Hogan, P. G., Lewis, R. S., Daly, M., and Rao, A. (2006). A mutation in Orai1 causes immune deficiency by abrogating CRAC channel function. *Nature* **441,** 179–185.

Field, K. A., Holowka, D., and Baird, B. (1997). Compartmentalized activation of the high affinity immunoglobulin E receptor within membrane domains. *J. Biol. Chem.* **272,** 4276–4280.

Furumoto, Y., Brooks, S., Olivera, A., Takagi, Y., Miyagishi, M., Taira, K., Casellas, R., Beaven, M. A., Gilfillan, A. M., and Rivera, J. (2006). Cutting Edge: Lentiviral shRNA silencing of PTEN in human mast cells reveals constitutive signals that promote cytokine secretion and cell survival. *J. Immunol.* **176,** 5167–5171.

Furumoto, Y., Nunomura, S., Terada, T., Rivera, J., and Ra, C. (2004). The FcεRIβ immunoreceptor tyrosine-based activation motif exerts inhibitory control on MAPK and IκB kinase phosphorylation and mast cell cytokine production. *J. Biol. Chem.* **279,** 49177–49187.

Galli, S. J. (1997a). Complexity and redundancy in the pathogenesis of asthma: Reassessing the roles of mast cells and T cells. *J. Exp. Med.* **186,** 343–347.

Galli, S. J. (1997b). The mast cell: A versatile effector cell for a challenging world. *Int. Arch. Allergy Immunol.* **113,** 14–22.

Galli, S. J., Kalesnikoff, J., Grimbaldeston, M. A., Piliponsky, A. M., Williams, C. M., and Tsai, M. (2005). Mast cells as "tunable" effector and immunoregulatory cells: Recent advances. *Annu. Rev. Immunol.* **23,** 749–786.

Gilfillan, A. M., and Tkaczyk, C. (2006). Integrated signalling pathways for mast-cell activation. *Nat. Rev. Immunol.* **6,** 218–230.

Gomez, G., Gonzalez-Espinosa, C., Odom, S., Baez, G., Cid, M. E., Ryan, J. J., and Rivera, J. (2005). Impaired FcεRI-dependent gene expression and defective eicosanoid and cytokine production as a consequence of Fyn-deficiency in mast cells. *J. Immunol.* **175,** 7602–7610.

Gonzalez-Espinosa, C., Odom, S., Olivera, A., Hobson, J. P., Martinez, M. E., Oliveira-Dos-Santos, A., Barra, L., Spiegel, S., Penninger, J. M., and Rivera, J. (2003). Preferential signaling and induction of allergy-promoting lymphokines upon weak stimulation of the high affinity IgE receptor on mast cells. *J. Exp. Med.* **197,** 1453–1465.

Gu, H., Saito, K., Klaman, L. D., Shen, J., Fleming, T., Wang, Y., Pratt, J. C., Lin, G., Lim, B., Kinet, J. P., and Neel, B. G. (2001). Essential role for Gab2 in the allergic response. *Nature* **412,** 186–190.

Heneberg, P., Lebduska, P., Draberova, L., Korb, J., and Draber, P. (2006). Topography of plasma membrane microdomains and its consequences for mast cell signaling. *Eur. J. Immunol.* **36,** 2795–2806.

Hernandez-Hansen, V., Smith, A. J., Surviladze, Z., Chigaev, A., Mazel, T., Kalesnikoff, J., Lowell, C. A., Krystal, G., Sklar, L. A., Wilson, B. S., and Oliver, J. M. (2004). Dysregulated FcεRI signaling and altered Fyn and SHIP activities in Lyn-deficient mast cells. *J. Immunol.* **173,** 100–112.

Herz, U., Renz, H., and Wiedermann, U. (2004). Animal models of type I allergy using recombinant allergens. *Methods* **32,** 271–280.

Hibbs, M. L., Tarlinton, D. M., Armes, J., Grail, D., Hodgson, G., Maglitto, R., Stacker, S. A., and Dunn, A. R. (1995). Multiple defects in the immune system of Lyn-deficient mice, culminating in autoimmune disease. *Cell* **83,** 301–311.

Hisatsune, C., Kuroda, Y., Nakamura, K., Inoue, T., Nakamura, T., Michikawa, T., Mizutani, A., and Mikoshiba, K. (2004). Regulation of TRPC6 channel activity by tyrosine phosphorylation. *J. Biol. Chem.* **279,** 18887–18894.

Hong, H., Kitaura, J., Xiao, W., Horejsi, V., Ra, C., Lowell, C. A., Kawakami, Y., and Kawakami, T. (2007). The Src family kinase Hck regulates mast cell activation by suppressing an inhibitory Src family kinase Lyn. *Blood* **110,** 2511–2519.

Hoth, M., and Penner, R. (1992). Depletion of intracellular calcium stores activates a calcium current in mast cells. *Nature* **355,** 353–356.

Houtman, J. C., Barda-Saad, M., and Samelson, L. E. (2005). Examining multiprotein signaling complexes from all angles. *Febs. J.* **272,** 5426–5435.

Huber, M., Helgason, C. D., Damen, J. E., Liu, L., Humphries, R. K., and Krystal, G. (1998a). The src homology 2-containing inositol phosphatase (SHIP) is the gatekeeper of mast cell degranulation. *Proc. Natl. Acad. Sci. USA* **95,** 11330–11335.

Huber, M., Helgason, C. D., Scheid, M. P., Duronio, V., Humphries, R. K., and Krystal, G. (1998b). Targeted disruption of SHIP leads to Steel factor-induced degranulation of mast cells. *EMBO J.* **17,** 7311–7319.

Huber, M., Kalesnikoff, J., Reth, M., and Krystal, G. (2002). The role of SHIP in mast cell degranulation and IgE-induced mast cell survival. *Immunol. Lett.* **82,** 17–21.

Iwaki, S., Spicka, J., Tkaczyk, C., Jensen, B. M., Furumoto, Y., Charles, N., Kovarova, M., Rivera, J., Horejsi, V., Metcalfe, D. D., and Gilfillan, A. M. (2008). Kit- and FcεRI-induced differential phosphorylation of the transmembrane adaptor molecule NTAL/LAB/LAT2 allows flexibility in its scaffolding function in mast cells. *Cell Signal* **20,** 195–205.

Jahn, T., Leifheit, E., Gooch, S., Sindhu, S., and Weinberg, K. (2007). Lipid rafts are required for Kit survival and proliferation signals. *Blood* **110,** 1739–1747.

Jolly, P., Rosenfeldt, H., Milstien, S., and Spiegel, S. (2002). The roles of sphingosine-1-phosphate in asthma. *Mol. Immunol.* **38,** 1239–1251.

Jolly, P. S., Bektas, M., Olivera, A., Gonzalez-Espinosa, C., Proia, R. L., Rivera, J., Milstien, S., and Spiegel, S. (2004). Transactivation of sphingosine-1-phosphate receptors by FcεRI triggering is required for normal mast cell degranulation and chemotaxis. *J. Exp. Med.* **199,** 959–970.

Kawakami, Y., Kitaura, J., Satterthwaite, A. B., Kato, R. M., Asai, K., Hartman, S. E., Maeda-Yamamoto, M., Lowell, C. A., Rawlings, D. J., Witte, O. N., and Kawakami, T. (2000). Redundant and opposing functions of two tyrosine kinases, Btk and Lyn, in mast cell activation. *J. Immunol.* **165,** 1210–1219.

Kihara, H., and Siraganian, R. P. (1994). Src homology 2 domains of Syk and Lyn bind to tyrosine-phosphorylated subunits of the high affinity IgE receptor. *J. Biol. Chem.* **269,** 22427–22432.

Kitano, M., Hla, T., Sekiguchi, M., Kawahito, Y., Yoshimura, R., Miyazawa, K., Iwasaki, T., Sano, H., Saba, J. D., and Tam, Y. Y. (2006). Sphingosine 1-phosphate/sphingosine 1-phosphate receptor 1 signaling in rheumatoid synovium: Regulation of synovial proliferation and inflammatory gene expression. *Arthritis Rheum.* **54,** 742–753.

Kitaura, J., Kawakami, Y., Maeda-Yamamoto, M., Horejsi, V., and Kawakami, T. (2007). Dysregulation of Src family kinases in mast cells from epilepsy-resistant ASK versus epilepsy-prone EL mice. *J. Immunol.* **178,** 455–462.

Kovarova, M., Tolar, P., Arudchandran, R., Draberova, L., Rivera, J., and Draber, P. (2001). Structure-function analysis of Lyn kinase association with lipid rafts and initiation of early signaling events after Fcε receptor I aggregation. *Mol. Cell Biol.* **21,** 8318–8328.

Kovarova, M., Wassif, C. A., Odom, S., Liao, K., Porter, F. D., and Rivera, J. (2006). Cholesterol-deficiency in a murine model of Smith-Lemli-Opitz Syndrome reveals increased mast cell responsiveness. *J. Exp. Med.* **203,** 1161–1171.

Kraft, S., and Kinet, J. P. (2007). New developments in FcεRI regulation, function and inhibition. *Nat. Rev. Immunol.* **7,** 365–378.

Lee, D. M., Friend, D. S., Gurish, M. F., Benoist, C., Mathis, D., and Brenner, M. B. (2002). Mast cells: A cellular link between autoantibodies and inflammatory arthritis. *Science* **297,** 1689–1692.

Leung, W. H., and Bolland, S. (2007). The inositol 5′-phosphatase SHIP-2 negatively regulates IgE-induced mast cell degranulation and cytokine production. *J. Immunol.* **179,** 95–102.

Liou, J., Kim, M. L., Heo, W. D., Jones, J. T., Myers, J. W., Ferrell, J. E., Jr., and Meyer, T. (2005). STIM is a Ca^{2+} sensor essential for Ca^{2+}-store-depletion-triggered Ca^{2+} influx. *Curr. Biol.* **15,** 1235–1241.

Ma, H. T., Peng, Z., Hiragun, T., Iwaki, S., Gilfillan, A. M., and Beaven, M. A. (2008). Canonical Transient Receptor Potential 5 Channel in conjunction with Orai1 and STIM1 allows Sr2+ entry, optimal influx of Ca^{2+}, and degranulation in a rat mast cell line. *J. Immunol.* **180,** 2233–2239.

Martin-Verdeaux, S., Pombo, I., Iannascoli, B., Roa, M., Varin-Blank, N., Rivera, J., and Blank, U. (2003). Analysis of Munc18-2 compartmentation in mast cells reveals a role for microtubules in granule exocytosis. *J. Cell Sci.* **116,** 325–334.

Mathes, C., Fleig, A., and Penner, R. (1998). Calcium release-activated calcium current (ICRAC) is a direct target for sphingosine. *J. Biol. Chem.* **273**, 25020–25030.

Matloubian, M., Lo, C. G., Cinamon, G., Lesneski, M. J., Xu, Y., Brinkmann, V., Allende, M. L., Proia, R. L., and Cyster, J. G. (2004). Lymphocyte egress from thymus and peripheral lymphoid organs is dependent on S1P receptor 1. *Nature* **427**, 355–360.

Melendez, A. J., and Khaw, A. K. (2002). Dichotomy of Ca^{2+} signals triggered by different phospholipid pathways in antigen stimulation of human mast cells. *J. Biol. Chem.* **277**, 17255–17262.

Metzger, H. (1978). The IgE-mast cell system as a paradigm for the study of antibody mechanisms. *Immunol. Rev.* **41**, 186–199.

Metzger, H. (1999). It's spring, and thoughts turn to … allergies. *Cell* **97**, 287–290.

Mitra, P., Oskeritzian, C. A., Payne, S. G., Beaven, M. A., Milstien, S., and Spiegel, S. (2006). Role of ABCC1 in export of sphingosine-1-phosphate from mast cells. *Proc. Natl. Acad. Sci. USA* **103**, 16394–16399.

Nadler, M. J., Matthews, S. A., Turner, H., and Kinet, J. P. (2000). Signal transduction by the high-affinity immunoglobulin E receptor FcεRI: coupling form to function. *Adv. Immunol.* **76**, 325–355.

Nechushtan, H., Leitges, M., Cohen, C., Kay, G., and Razin, E. (2000). Inhibition of degranulation and interleukin-6 production in mast cells derived from mice deficient in protein kinase Cβ. *Blood* **95**, 1752–1757.

Nishida, K., Yamasaki, S., Ito, Y., Kabu, K., Hattori, K., Tezuka, T., Nishizumi, H., Kitamura, D., Goitsuka, R., Geha, R. S., Yamamoto, T., Yagi, T., and Hirano, T. (2005). FcεRI-mediated mast cell degranulation requires calcium-independent microtubule-dependent translocation of granules to the plasma membrane. *J. Cell Biol.* **170**, 115–126.

Nishizumi, H., and Yamamoto, T. (1997). Impaired tyrosine phosphorylation and Ca^{2+} mobilization, but not degranulation, in Lyn-deficient bone marrow-derived mast cells. *J. Immunol.* **158**, 2350–2355.

O'Neill, S. M., Brady, M. T., Callanan, J. J., Mulcahy, G., Joyce, P., Mills, K. H., and Dalton, J. P. (2000). Fasciola hepatica infection downregulates Th1 responses in mice. *Parasite Immunol.* **22**, 147–155.

Odom, S., Gomez, G., Kovarova, M., Furumoto, Y., Ryan, J. J., Wright, H. V., Gonzalez-Espinosa, C., Hibbs, M. L., Harder, K. W., and Rivera, J. (2004). Negative regulation of immunoglobulin E-dependent allergic responses by Lyn kinase. *J. Exp. Med.* **199**, 1491–1502.

Olivera, A., Mizugishi, K., Tikhonova, A., Ciaccia, L., Odom, S., Proia, R. L., and Rivera, J. (2007). The sphingosine kinase-sphingosine-1-phosphate axis is a determinant of mast cell function and anaphylaxis. *Immunity* **26**, 287–297.

Olivera, A., and Rivera, J. (2005). Sphingolipids and the balancing of immune cell function: lessons from the mast cell. *J. Immunol.* **174**, 1153–1158.

Olivera, A., Rosenthal, J., and Spiegel, S. (1996). Effect of acidic phospholipids on sphingosine kinase. *J. Cell Biochem.* **60**, 529–537.

Olivera, A., Urtz, N., Mizugishi, K., Yamashita, Y., Gilfillan, A. M., Furumoto, Y., Gu, H., Proia, R. L., Baumruker, T., and Rivera, J. (2006). IgE-dependent activation of sphingosine kinases 1 and 2 and secretion of sphingosine 1-phosphate requires Fyn kinase and contributes to mast cell responses. *J. Biol. Chem.* **281**, 2515–2525.

On, M., Billingsley, J. M., Jouvin, M. H., and Kinet, J. P. (2004). Molecular dissection of the FcεRβ signaling amplifier. *J. Biol. Chem.* **279**, 45782–45790.

Ozawa, K., Szallasi, Z., Kazanietz, M., Blumberg, P., Mischak, H., Mushinski, J., and Beaven, M. (1993). Ca^{2+}-dependent and Ca^{2+}-independent isozymes of protein kinase C mediate exocytosis in antigen-stimulated Rat Basophilic RBL-2H3 cells. *J. Biol. Chem.* **268**, 1749–1756.

Parravicini, V., Gadina, M., Kovarova, M., Odom, S., Gonzalez-Espinosa, C., Furumoto, Y., Saitoh, S., Samelson, L. E., O'Shea, J. J., and Rivera, J. (2002). Fyn kinase initiates complementary signals required for IgE-dependent mast cell degranulation. *Nat. Immunology* **3**, 741–748.

Patterson, R. L., van Rossum, D. B., Ford, D. L., Hurt, K. J., Bae, S. S., Suh, P. G., Kurosaki, T., Snyder, S. H., and Gill, D. L. (2002). Phospholipase C-gamma is required for agonist-induced Ca^{2+} entry. *Cell* **111**, 529–541.

Peinelt, C., Vig, M., Koomoa, D. L., Beck, A., Nadler, M. J., Koblan-Huberson, M., Lis, A., Fleig, A., Penner, R., and Kinet, J. P. (2006). Amplification of CRAC current by STIM1 and CRACM1 (Orai1). *Nat. Cell Biol.* **8**, 771–773.

Pivniouk, V. I., Martin, T. R., Lu-Kuo, J. M., Katz, H. R., Oettgen, H. C., and Geha, R. S. (1999). SLP-76 deficiency impairs signaling via the high-affinity IgE receptor in mast cells. *J. Clin. Invest.* **103**, 1737–1743.

Pombo, I., Rivera, J., and Blank, U. (2003). Munc18-2/syntaxin3 complexes are spatially separated from syntaxin3-containing SNARE complexes. *FEBS Lett.* **550**, 144–148.

Prakriya, M., Feske, S., Gwack, Y., Srikanth, S., Rao, A., and Hogan, P. G. (2006). Orai1 is an essential pore subunit of the CRAC channel. *Nature* **443**, 230–233.

Pribluda, V. S., Pribluda, C., and Metzger, H. (1994). Transphosphorylation as the mechanism by which the high-affinity receptor for IgE is phosphorylated upon aggregation. *Proc. Natl. Acad. Sci. USA* **91**, 11246–11250.

Prieschl, E. E., Csonga, R., Novotny, V., Kikuchi, G. E., and Baumruker, T. (1999). The balance between sphingosine and sphingosine-1-phosphate is decisive for mast cell activation after Fcε receptor I triggering. *J. Exp. Med.* **190**, 1–8.

Puri, N., and Roche, P. A. (2006). Ternary SNARE complexes are enriched in lipid rafts during mast cell exocytosis. *Traffic* **7**, 1482–1494.

Putney, J. W., Jr. (1986). A model for receptor-regulated calcium entry. *Cell Calcium.* **7**, 1–12.

Putney, J. W., Jr., Takemura, H., Hughes, A. R., Horstman, D. A., and Thastrup, O. (1989). How do inositol phosphates regulate calcium signaling? *FASEB Journal* **3**, 1899–1905.

Pyne, S., and Pyne, N. (2000). Sphingosine 1-phosphate signalling via the endothelial differentiation gene family of G-protein-coupled receptors. *Pharmacol. Ther.* **88**, 115–131.

Rivera, J. (2002). Molecular adapters in FcεRI signaling and the allergic response. *Curr. Opin. Immunol.* **14**, 688–693.

Rivera, J. (2005). NTAL/LAB and LAT: A balancing act in mast cell activation and function. *Trends. Immunol.* **26**, 119–122.

Rivera, J., and Beaven, M. A. (1997). *In* "Protein Kinase C", (P. J. Parker, and and L. V. Dekker, eds.), pp. 133–166. Landes Bioscience, Austin, TX.

Rivera, J., and Gilfillan, A. M. (2006). Molecular regulation of mast cell activation. *J. Allergy. Clin. Immunol.* **117**, 1214–1225; quiz 1226.

Rivera, J., Gonzalez-Espinosa, C., Kovarova, M., and Parravicini, V. (2002). The Architecture of IgE-dependent mast cell signaling. A complex story. *Allergy Clin. Immunol. Int.* **14**, 25–36.

Rivera, J., and Olivera, A. (2007). Src family kinases and lipid mediators in control of allergic inflammation. *Immunol. Rev.* **217**, 255–268.

Rivera, J., and Tessarollo, L. (2008). Genetic background and the dilemma of translating mouse studies to humans. *Immunity* **28**, 1–4.

Roos, J., DiGregorio, P. J., Yeromin, A. V., Ohlsen, K., Lioudyno, M., Zhang, S., Safrina, O., Kozak, J. A., Wagner, S. L., Cahalan, M. D., Velicelebi, G., and Stauderman, K. A. (2005). STIM1, an essential and conserved component of store-operated Ca^{2+} channel function. *J. Cell Biol.* **169**, 435–445.

Saitoh, S., Arudchandran, R., Manetz, T. S., Zhang, W., Sommers, C. L., Love, P. E., Rivera, J., and Samelson, L. E. (2000). LAT is essential for FcεRI-mediated mast cell activation. *Immunity* **12**, 525–535.

Saitoh, S., Odom, S., Gomez, G., Sommers, C. L., Young, H. A., Rivera, J., and Samelson, L. E. (2003). The four distal tyrosines are required for LAT-dependent signaling in FcεRI-mediated mast cell activation. *J. Exp. Med.* **198,** 831–843.

Sheets, E. D., Holowka, D., and Baird, B. (1999). Critical role for cholesterol in Lyn-mediated tyrosine phosphorylation of FcεRI and their association with detergent-resistant membranes. *J. Cell Biol.* **145,** 877–887.

Sil, D., Lee, J. B., Luo, D., Holowka, D., and Baird, B. (2007). Trivalent ligands with rigid DNA spacers reveal structural requirements for IgE receptor signaling in RBL mast cells. *ACS Chem. Biol.* **2,** 674–684.

Silverman, M. A., Shoag, J., Wu, J., and Koretzky, G. A. (2006). Disruption of SLP-76 interaction with Gads inhibits dynamic clustering of SLP-76 and FcεRI signaling in mast cells. *Mol. Cell Biol.* **26,** 1826–1838.

Simons, K., and Toomre, D. (2000). Lipid rafts and signal transduction. *Nat. Rev. Mol. Cell Biol.* **1,** 31–39.

Siraganian, R. P., Zhang, J., Suzuki, K., and Sada, K. (2002). Protein tyrosine kinase Syk in mast cell signaling. *Mol. Immunol.* **38,** 1229–1233.

Smyth, J. T., Dehaven, W. I., Jones, B. F., Mercer, J. C., Trebak, M., Vazquez, G., and Putney, J. W., Jr. (2006). Emerging perspectives in store-operated Ca(2+) entry: Roles of Orai, Stim and TRP. *Biochim. Biophys. Acta.* **1763,** 1147–1160.

Spiegel, S., Foster, D., and Kolesnick, R. (1996). Signal transduction through lipid second messengers. *Curr. Opin. Cell Biol.* **8,** 159–167.

Spiegel, S., and Milstien, S. (2003). Sphingosine-1-phosphate: An enigmatic signalling lipid. *Nat. Rev. Mol. Cell Biol.* **4,** 397–407.

Swieter, M., Berenstein, E. H., and Siraganian, R. P. (1995). Protein tyrosine phosphatase activity associates with the high affinity IgE receptor and dephosphorylates the receptor subunits, but not Lyn or Syk. *J. Immunol.* **155,** 5330–5336.

Tkaczyk, C., Horejsi, V., Iwaki, S., Draber, P., Samelson, L. E., Satterthwaite, A. B., Nahm, D.-H., Metcalfe, D. D., and Gilfillan, A. M. (2004). NTAL phosphorylation is a pivotal link between the signaling cascades leading to human mast cell degranulation following KIT activation and FcεRI aggregation. *Blood* **104,** 207–214.

Tkaczyk, C., Iwaki, S., Metcalfe, D. D., and Gilfillan, A. M. (2005). Roles of adaptor molecules in mast cell activation. *Chem. Imunnol. Allergy* **87,** 43–58.

Torigoe, C., Faeder, J. R., Oliver, J. M., and Goldstein, B. (2007). Kinetic proofreading of ligand-FcεRI interactions may persist beyond LAT phosphorylation. *J. Immunol.* **178,** 3530–3535.

Torigoe, C., Inman, J. K., and Metzger, H. (1998). An unusual mechanism for ligand antagonism. *Science* **281,** 568–572.

Urtz, N., Olivera, A., Bofill-Cardona, E., Csonga, R., Billich, A., Mechtcheriakova, D., Bornancin, F., Woisetschlager, M., Rivera, J., and Baumruker, T. (2004). Early activation of sphingosine kinase in mast cells and recruitment to FcεRI are mediated by its interaction with Lyn kinase. *Mol. Cell Biol.* **24,** 8765–8777.

Vazquez, G., Wedel, B. J., Kawasaki, B. T., Bird, G. S., and Putney, J. W., Jr. (2004). Obligatory role of Src kinase in the signaling mechanism for TRPC3 cation channels. *J. Biol. Chem.* **279,** 40521–40528.

Vig, M., Beck, A., Billingsley, J. M., Lis, A., Parvez, S., Peinelt, C., Koomoa, D. L., Soboloff, J., Gill, D. L., Fleig, A., Kinet, J. P., and Penner, R. (2006a). CRACM1 multimers form the ion-selective pore of the CRAC channel. *Curr. Biol.* **16,** 2073–2079.

Vig, M., DeHaven, W. I., Bird, G. S., Billingsley, J. M., Wang, H., Rao, P. E., Hutchings, A. B., Jouvin, M. H., Putney, J. W., and Kinet, J. P. (2008). Defective mast cell effector functions in mice lacking the CRACM1 pore subunit of store-operated calcium release-activated calcium channels. *Nat. Immunol.* **9,** 89–96.

Vig, M., and Kinet, J. P. (2007). The long and arduous road to CRAC. *Cell Calcium* **42,** 157–162.

Vig, M., Peinelt, C., Beck, A., Koomoa, D. L., Rabah, D., Koblan-Huberson, M., Kraft, S., Turner, H., Fleig, A., Penner, R., and Kinet, J. P. (2006b). CRACM1 is a plasma membrane protein essential for store-operated Ca2+ entry. *Science* **312**, 1220–1223.

Volna, P., Lebduska, P., Draberova, L., Simova, S., Heneberg, P., Boubelik, M., Bugajev, V., Malissen, B., Wilson, B. S., Horejsi, V., Malissen, M., and Draber, P. (2004). Negative regulation of mast cell signaling and function by the adaptor LAB/NTAL. *J. Exp. Med.* **200**, 1001–1013.

Vonakis, B. M., Gibbons, S. P., Jr., Rotte, M. J., Brothers, E. A., Kim, S. C., Chichester, K., and MacDonald, S. M. (2005). Regulation of rat basophilic leukemia-2H3 mast cell secretion by a constitutive Lyn kinase interaction with the high affinity IgE receptor (FcɛRI). *J. Immunol.* **175**, 4543–4554.

Wang, D., Feng, J., Wen, R., Marine, J. C., Sangster, M. Y., Parganas, E., Hoffmeyer, A., Jackson, C. W., Cleveland, J. L., Murray, P. J., and Ihle, J. N. (2000). Phospholipase Cγ2 is essential in the functions of B cell and several Fc receptors. *Immunity* **13**, 25–35.

Williams, C. M., and Galli, S. J. (2000). The diverse potential effector and immunoregulatory roles of mast cells in allergic disease. *J. Allergy Clin. Immunol.* **105**, 847–859.

Wilson, B. S., Pfeiffer, J. R., and Oliver, J. M. (2000). Observing FcɛRI signaling from the inside of the mast cell membrane. *J. Cell Biol.* **149**, 1131–1142.

Wilson, B. S., Pfeiffer, J. R., Surviladze, Z., Gaudet, E. A., and Oliver, J. M. (2001). High resolution mapping of mast cell membranes reveals primary and secondary domains of FcɛRI and LAT. *J. Cell Biol.* **154**, 645–658.

Xiao, W., Nishimoto, H., Hong, H., Kitaura, J., Nunomura, S., Maeda-Yamamoto, M., Kawakami, Y., Lowell, C. A., Ra, C., and Kawakami, T. (2005). Positive and negative regulation of mast cell activation by Lyn via the FcɛRI. *J. Immunol.* **175**, 6885–6892.

Yamashita, Y., Charles, N., Furumoto, Y., Odom, S., Yamashita, T., Gilfillan, A. M., Constant, S., Bower, M. A., Ryan, J. J., and Rivera, J. (2007). Cutting Edge: Genetic variation influences FcɛRI-induced mast cell activation and allergic responses. *J. Immunol.* **179**, 740–743.

Young, R. M., Holowka, D., and Baird, B. (2003). A lipid raft environment enhances Lyn kinase activity by protecting the active site tyrosine from dephosphorylation. *J. Biol. Chem.* **278**, 20746–20752.

Zemann, B., Kinzel, B., Muller, M., Reuschel, R., Mechtcheriakova, D., Urtz, N., Bornancin, F., Baumruker, T., and Billich, A. (2006). Sphingosine kinase type 2 is essential for lymphopenia induced by the immunomodulatory drug FTY720. *Blood* **107**, 1454–1458.

Zhang, J., Berenstein, E. H., Evans, R. L., and Siraganian, R. P. (1996). Transfection of Syk protein tyrosine kinase reconstitutes high affinity IgE receptor-mediated degranulation in a Syk-negative variant of rat basophilic leukemia RBL-2H3 cells. *J. Exp. Med.* **184**, 71–79.

Zhang, J., Kimura, T., and Siraganian, R. P. (1998). Mutations in the activation loop tyrosines of protein tyrosine kinase Syk abrogate intracellular signaling but not kinase activity. *J. Immunol.* **161**, 4366–4374.

Zhu, M., Koonpaew, S., Liu, Y., Shen, S., Denning, T., Dzhagalov, I., Rhee, I., and Zhang, W. (2006). Negative regulation of T cell activation and autoimmunity by the transmembrane adaptor protein LAB. *Immunity* **25**, 757–768.

Zhu, M., Liu, Y., Koonpaew, S., Ganillo, O., and Zhang, W. (2004). Positive and negative regulation of FcɛRI-mediated signaling by the adaptor protein LAB/NTAL. *J. Exp. Med.* **200**, 991–1000.

CHAPTER 4

B Cells and Autoantibodies in the Pathogenesis of Multiple Sclerosis and Related Inflammatory Demyelinating Diseases

Katherine A. McLaughlin[*,†] and
Kai W. Wucherpfennig[*,†,‡]

Contents

Abstract

Multiple sclerosis (MS) is a chronic inflammatory demyelinating disease of the central nervous system (CNS). The mainstream view is that MS is caused by an autoimmune attack of the CNS myelin by myelin-specific CD4 T cells, and this perspective is

* Department of Cancer Immunology and AIDS, Dana-Farber Cancer Institute, Boston, Massachusetts 02115
† Program in Immunology, Harvard Medical School, Boston, Massachusetts 02115
‡ Department of Neurology, Harvard Medical School, Boston, Massachusetts 02115

Advances in Immunology, Volume 98
ISSN 0065-2776, DOI: 10.1016/S0065-2776(08)00404-5

supported by extensive work in the experimental autoimmune encephalomyelitis (EAE) model of MS as well as immunological and genetic studies in humans. However, it is important to keep in mind that other cell populations of the immune system are also essential in the complex series of events leading to MS, as exemplified by the profound clinical efficacy of B cell depletion with Rituximab. This review discusses the mechanisms by which B cells contribute to the pathogenesis of MS and dissects their role as antigen-presenting cells (APCs) to T cells with matching antigen specificity, the production of proinflammatory cytokines and chemokines, as well as the secretion of autoantibodies that target structures on the myelin sheath and the axon. Mechanistic dissection of the interplay between T cells and B cells in MS may permit the development of B cell based therapies that do not require depletion of this important cell population.

1. MS AND RELATED INFLAMMATORY DEMYELINATING CNS DISEASES

Multiple sclerosis (MS) is the most common neurological disease in young adults, affecting over 250,000 individuals in the United States and up to 1.2 million worldwide. It is believed to result from an autoimmune attack on protein components of myelin, the insulation which allows for rapid conductance of electrical signals along axons. MS is characterized by discrete regions of central nervous system (CNS) inflammation, lymphocyte infiltration, demyelination, axonal damage, and ultimately the death of myelin-producing oligodendrocytes. Depending on the localization of these plaques, MS patients suffer from a wide variety of symptoms, including weakness, sensory disturbances, ataxia, and visual impairment. Magnetic resonance imaging (MRI) allows visualization of active lesions in the absence of clinical symptoms, and has become a valuable tool for both diagnosis and monitoring of disease activity. A diagnosis of MS requires multiple episodes of demyelination separated in space and time (Poser *et al.*, 1983), which may be supported by clinical evidence of multiple demyelinating events or the presence of multiple lesions of different age on MRI (McDonald *et al.*, 2001). On the basis of the temporal pattern of demyelinating events and accumulation of permanent disability, MS is classified into relapsing-remitting (RR), secondary progressive (SP), and primary progressive (PP) types (Hauser and Oksenberg, 2006; Noseworthy *et al.*, 2000). About 80–85% of MS patients initially experience a relapsing-remitting course (RRMS) and the intervals between and duration of relapses are highly variable, not necessarily correlating with the presence of lesions by MRI because most lesions are clinically silent (Goodin, 2006). Over time, many RRMS patients develop a progressive worsening of symptoms between relapses, known as SPMS. Cumulative axonal loss may be an important contributor to the progressive decline in neurological function (Hauser and Oksenberg, 2006; Sospedra and Martin, 2005; Trapp *et al.*, 1998). Within 15 years of diagnosis, 50–60% of RRMS patients cannot walk unassisted, and 70% are limited in performing activities central to daily life (Hauser and Oksenberg, 2006). The remaining 15–20% of patients experience a progressive pattern of disease (PPMS), characterized by a gradual accumulation of symptoms and concurrent decline in function. The progression of disability occurs despite a lower frequency of active (gadolinium-enhancing) lesions by MRI relative to RRMS (Thompson *et al.*, 1997).

MS is a complex disease, with contributions from multiple genetic and environmental factors. The frequency of MS varies between ethnic populations (Rosati, 2001), and descendents of Northern Europeans have an increased prevalence relative to African, Asian, and Native American populations. The relative risk of developing MS is significantly higher for individuals with an affected first- or second-degree relative (Chataway *et al.*, 1998; Hauser and Oksenberg, 2006). The MHC locus on human chromosome 6 shows the strongest linkage to disease susceptibility, in particular the MHC class II region and a specific MHC class II haplotype (DRB1*1501-DQB1*0602) (Haines *et al.*, 1996; Sawcer *et al.*, 1996). Recent genome-wide studies have also identified single nucleotide polymorphisms in the IL-2 receptor alpha and IL-7 receptor alpha chains (The International Multiple Sclerosis Genetics Consortium, 2007), again indicating that genetic variations of immune response genes contribute to MS susceptibility. Women are affected by MS twice as frequently as men, which may be due to genetic or hormonal factors (Czlonkowska *et al.*, 2005).

A large body of work in the experimental autoimmune encephalomyelitis (EAE) model of MS indicates that CD4 T cells specific for myelin antigens are essential for the development of inflammatory, demyelinating CNS lesions. EAE can be induced by immunization with myelin proteins or peptides in complete Freund's adjuvant (CFA) or by transfer of myelin-specific T cell clones (Stromnes and Goverman, 2006a,b). Linkage of disease susceptibility to the MHC class II locus in MS indicates that peptide presentation to CD4 T cells is also important in the human disease. Even though disease can be transferred with purified T cells with a single antigen specificity, it is important to keep in mind that other cell populations of the immune system also play an essential role, as exemplified by the clinical trials with Rituximab which depletes B cells. The function of B cells in the chronic inflammatory process in MS will be discussed in detail.

1.1. Related demyelinating CNS diseases

Neuromyelitis optica (NMO), acute disseminated encephalomyelitis (ADEM), and clinically isolated demyelinating syndromes (CIS) are part of the spectrum of inflammatory demyelinating CNS diseases. CIS is an isolated demyelinating event, and 50–95% of CIS cases are diagnosed with MS following a second demyelinating episode (Miller, 2004). Initiation of therapy with interferon beta at the time of CIS has been shown to significantly prevent or delay progression to MS, especially in patients with multiple MRI lesions who may be at higher risk (Kappos et al., 2007; Miller, 2004; Thrower, 2007).

NMO (also known as Devic's disease) primarily affects the optic nerve and spinal cord, leading to the defining symptoms of optic neuritis and transverse myelitis (Wingerchuk et al., 2006). Longitudinally extensive spinal cord lesions are commonly found in NMO, although clinically silent brain lesions may also be seen on MRI. NMO is found in populations with a low incidence of MS, and affects women up to nine times more often than men. The majority of cases follow a relapsing course with incomplete recovery between episodes, but NMO has a worse long-term prognosis than does MS. More than 50% of patients are blind in at least one eye or cannot walk unassisted five years following diagnosis (Wingerchuk et al., 2007). The recent discovery of a specific autoantibody in approximately 75% of NMO patients has aided in diagnosis and provided a therapeutic target, as discussed later in this chapter.

ADEM, as the name implies, is an acute and often rapidly progressing demyelinating condition and is more common in children than adults (Tenembaum et al., 2007). Patients with ADEM typically have a variety of neurological symptoms and diagnosis requires the presence of encephalopathy, defined as a change in behavior or consciousness (Krupp et al., 2007; Tenembaum et al., 2007). ADEM often occurs within 4 weeks of infection or vaccination, although a causative relationship with infection can be difficult to establish. While the majority of ADEM patients experience a monophasic disease course with partial or complete recovery, some have a recurrence of the same symptoms and lesion location more than 3 months following the initial event (recurrent ADEM) or experience new lesions and symptoms (multiphasic ADEM) (Krupp et al., 2007). A minority of individuals with ADEM can later develop MS, although the reported frequency of this occurrence is highly variable (0–28%) (Dale et al., 2000; Leake et al., 2004; Mikaeloff et al., 2004, 2007; Tenembaum et al., 2002), requires many years of follow-up, and is dependent on the criteria used to define MS. For instance, Poser's criteria (Poser et al., 1983) would consider multiphasic ADEM as MS, as both are a recurring demyelinating events separated in space and time. The overlapping symptoms and lack of specific biomarkers can cause difficulties in distinguishing between MS and ADEM, especially in pediatric populations (Belman *et al.*, 2007; Krupp *et al.*, 2007; Wingerchuk, 2003).

2. THERAPEUTIC DEPLETION OF B CELLS IN MS WITH RITUXIMAB

Recent clinical trials with Rituximab have shown that B cells play an important role in the pathogenesis of MS. Rituximab (marketed by Genentech as Rituxan®) is a monoclonal antibody directed against CD20, a transmembrane protein expressed on the surface of B cells, but absent from fully differentiated plasma cells (Sabahi and Anolik, 2006). Administration of this antibody rapidly depletes CD20-expressing cells from the circulation via complement-mediated lysis and cell-mediated cytotoxicity (Reff *et al.*, 1994), and B cell numbers remain low for 3–12 months following treatment. Rituximab has been in use as a treatment for non-Hodgkin's B cell lymphomas since FDA approval in 1997 and has shown an excellent safety profile (Rastetter *et al.*, 2004). Most adverse events are related to the antibody infusion process, while recurrent infections occur at a similar frequency as in placebo control groups and opportunistic infections are rare. The preservation of antibody secretion by plasma cells may explain the low incidence of infection.

The efficacy and safety profile of Rituximab prompted its clinical testing in complex antibody-associated autoimmune diseases, initially rheumatoid arthritis (RA) (Edwards *et al.*, 2004). A phase III trial reported the combined use of methotrexate and Rituximab in patients with established RA who had previously failed tumor necrosis factor (TNF)-directed therapies (Cohen *et al.*, 2006). Treatment with Rituximab resulted in a significant reduction of disease severity compared to methotrexate alone, while patients in the control group experienced a worsening of RA symptoms. The response was not associated with a reduction in rheumatoid factor antibodies, and the mean serum levels of IgM, IgG, and IgA generally stayed within normal limits through the trial period. A follow-up study demonstrated the safety of subsequent infusions following B cell repopulation, with no increase in the infection rate (Keystone *et al.*, 2007). On the basis of these results, Rituximab was approved in 2006 for use in RA patients who do not respond to TNF-antagonists (Sabahi and Anolik, 2006). Rituximab has also shown benefit in several other autoimmune diseases, including autoantibody-associated neuropathies, immune thrombocytopenia and systemic lupus erythematosus (Edwards and Cambridge, 2006; Renaud *et al.*, 2003, 2006; Sabahi and Anolik, 2006; Tanaka *et al.*, 2007).

A recent phase II randomized, placebo-controlled multicenter trial of Rituximab in RRMS showed substantial clinical benefit (Hauser *et al.*, 2008).

Following the single infusion of antibody, depletion of B cells was rapid and persisted for greater than 24 weeks. The total number of lesions and number of new lesions were dramatically reduced in the treatment compared to control group, with 89.4% of patients in the treatment group experiencing one or fewer total lesions over the course of the trial. About 84.8% of patients receiving Rituximab had no new lesions, and although highly variable, the volume of lesions was decreased in the treatment group but increased in controls. The decrease in lesion activity was accompanied by a reduction in clinical relapses. The infection rate was similar in both groups, and no significant opportunistic infections were observed. These promising results are certain to be followed up with a large phase III trial to address safety and long-term efficacy issues. Smaller preceding trials support these findings. A report of one RRMS patient receiving two doses of Rituximab showed effective depletion of B cells for six months in both peripheral blood and CSF, and a complete absence of relapses and new lesions for nine months (Stuve et al., 2005). A small trial in PPMS demonstrated rapid and effective peripheral B cell depletion, but an overall increase in CSF B cell frequencies after 2–18 months (Monson et al., 2005). The majority of these cells were plasma cells, which are not targeted by Rituximab. In a larger study using Ritux-imab as an add-on therapy for RRMS patients with suboptimal responses to current treatment, B and T cell numbers were reduced in CSF following treatment, but no significant changes were observed in the number of oligoclonal bands, CSF IgG concentration or disability scores (Cross et al., 2006). No serious complications or infections were observed in any of these trials. Rituximab may also be useful for the treatment of other inflammatory demyelinating CNS diseases. Treatment with Rituximab led to significant neurological improvement in seven of eight NMO patients, including recovery of walking ability and regain of bladder and bowel control (Cree et al., 2005), although the possibility of spontane-ous recovery was not addressed with a control group in this open label trial.

3. WHICH B CELL FUNCTIONS ARE CRITICAL IN THE PATHOGENESIS OF MS?

It is obvious why Rituximab is an effective treatment for lymphoma – it eliminates the transformed cells. It is more difficult to dissect why treat-ment with Rituximab has such a profound effect on MS. The contribution of B cells may go beyond antibody production – B cells are potent APCs and the B cell – T cell interaction shapes the ensuing T cell response through expression of costimulatory molecules as well as production of cytokines and chemokines.

3.1. B cells as antigen presenting cells

Unlike other professional APCs, B cells have the unique capability to efficiently capture even minute amounts of antigen through the B cell receptor (BCR) (Fig. 4.1). B cells are thus the most efficient APCs for T cells with the same antigen specificity (Lanzavecchia, 1985). Antigen presentation by B cells is closely linked to activation. When crosslinked by a bound protein, the BCR signaling subunits (Ig-α and Ig-β) trigger receptor ubiquitination, internalization and targeting to an endocytic compartment optimized for peptide loading onto class II MHC (Drake *et al.*, 2006; West *et al.*, 1994). There, the BCR-associated protein antigen is degraded into peptides for presentation by MHC class II molecules. Intracellular trafficking of MHC class II molecules is also modulated by BCR ligation to promote maximal loading of antigen-derived peptides, even when present in low abundance (Vascotto *et al.*, 2007). Activation also induces recruitment of HLA-DM to this compartment (Lankar *et al.*, 2002) which

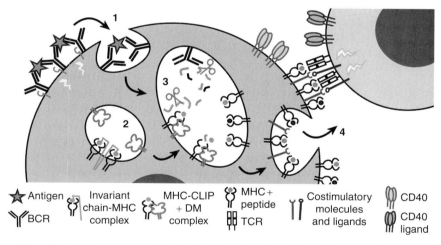

FIGURE 4.1   Antigen presentation by B cells to T cells with matching specificity. Crosslinking of the BCR by antigen induces signaling events leading to B cell activation and receptor endocytosis (1). MHC class II molecules assemble in the ER with the invariant chain (2). The internalized antigen is targeted to an intracellular compartment enriched in MHC (3), where proteases break down the antigen into peptides. In this compartment, the invariant is proteolytically cleaved such that only the CLIP peptide remains bound. DM facilitates the exchange of the MHC-associated CLIP for other peptides, including those derived from the endocytosed antigen. MHC molecules bearing high affinity peptides leave the loading compartment and are exported to the cell surface with co-stimulatory molecules (4). Antigen-derived peptide-MHC complexes can then activate specific CD4 T cells. Initial T cell activation leads to expression of CD40 ligand, which provides signals required for B cell survival and differentiation into memory cells through CD40 on the B cell surface. (See Plate 4 in Color Plate Section.)

catalyzes the exchange of low affinity MHC-associated peptides with higher affinity peptides; these complexes are subsequently transported to the cell surface for recognition by CD4 T cells. Costimulatory molecules including CD80, CD86, and ICOS ligand are also upregulated on the B cell surface following activation (Cambier *et al.*, 1994; Shilling *et al.*, 2006), and promote T cell activation upon peptide–MHC binding. Antigen-specific B and T cells form long-lasting immune synapses characterized by accumulation of peptide–MHC complexes at the site of cell-cell contact (Gordy *et al.*, 2004). The coordination of antigen uptake, class II loading, and costimulatory molecule expression allows antigen-specific B cells to activate cognate T cells far more efficiently than nonspecific B cells or monocytes (Lanzavecchia, 1985), and antigen-specific B cells are required to prime CD4 T cells with low doses of antigen (Rivera *et al.*, 2001). In addition to directly presenting antigens to autoreactive T cells, live B cells can also transfer antigen to macrophages by cell–cell contact, but the precise mechanism of this transfer is unknown (Harvey *et al.*, 2007; Townsend and Goodnow, 1998).

In the context of autoimmunity, which can be dominated by low-affinity T cells and limiting amounts of antigen, the potent APC function of B cells may be harmful. The immunodominant antibody epitope of myelin basic protein (MBP) colocalizes with a major T cell epitope of the antigen (Wucherpfennig *et al.*, 1997), and binding to the BCR may thus protect the immunodominant epitope of MBP from degradation and promote its presentation to T cells. Similarly, the major epitope recognized by high-affinity insulin-specific autoantibodies is located in the same segment of the target antigen as the T cell epitope. The presence of autoantibodies with this specificity is highly predictive of later development of type 1 diabetes (Achenbach *et al.*, 2004; Kent *et al.*, 2005).

In an immune response to a pathogen, B cells are activated in a lymph node by antigen binding to the BCR, migrate to follicles, and continue to undergo antigen-driven maturation by forming germinal centers (MacLennan, 1994). Follicular dendritic cells display native proteins in the form of immune complexes to B cells, which take up the antigen though the BCR, process it and present antigen-derived peptides to CD4 T cells (McHeyzer-Williams *et al.*, 2006). In turn, T cells recognizing peptide-MHC complexes on the B cell become activated, upregulate expression of stimulatory molecules including CD40 ligand (CD154), and secrete cytokines, providing antigen-dependent survival and differentiation signals to the B cell. After acquiring T cell help, B cells rapidly undergo several rounds of proliferation in the germinal center, during which time the immunoglobulin variable sequences are mutated (Klein and Dalla-Favera, 2008; MacLennan, 1994). The resulting daughter cells once again sample antigens on follicular dendritic cells (FDCs), present peptides to T cells, and obtain survival signals.

Because the germinal center reaction is highly competitive, B cells containing rearranged receptors with lower affinity for antigen do not receive the required signals from T cells and undergo death by apoptosis. Cells with higher affinity for antigen proliferate and leave the follicle as memory B cells or plasmablasts, the precursors of long-lived plasma cells. Although capable of establishing residence in the bone marrow, newly generated plasma cells can also traffic to sites of inflammation by chemokine and integrin dependent homing (Kunkel and Butcher, 2003). The process of immunoglobulin diversification in germinal centers can give rise to self-reactive B cells, which should undergo apoptosis in the absence of cognate T cell interactions. However, the mechanisms of central and peripheral T cell tolerance are not perfect, and cells which escape can promote the survival and maturation of autoreactive B cells. The control of self-reactive B and T cell activation are closely linked in the germinal center reaction, making it an attractive target for treatment of autoimmune diseases.

There is substantial evidence for such antigen-driven B cell responses in MS. Structures similar to lymph node B cell follicles, termed tertiary lymphoid organs, develop in the target organs of inflammatory autoimmune diseases and are induced by a positive feedback loop of tissue chemokine expression, lymphocyte recruitment and activation, and cytokine production (Aloisi and Pujol-Borrell, 2006). Follicles containing B cells and FDCs can be found in the meninges of patients with secondary-progressive MS, and their presence is associated with a younger age of onset and more severe pathology (Magliozzi *et al.*, 2007; Serafini *et al.*, 2004). Similar structures have also been found in mice with progressive relapsing EAE (Magliozzi *et al.*, 2004). The formation of these ectopic follicles indicates that B cells migrate to the brain, are activated locally and present antigens, and differentiate into memory B cells or plasmablasts within the CNS, rather than being activated in the periphery and homing to the CNS in a fully mature state. The establishment of germinal centers in the brain may reflect differences in the immune response in RRMS and SPMS (Lyons *et al.*, 1999; Magliozzi *et al.*, 2007).

An important role for antigen presentation by B cells has also been demonstrated in mouse models of autoimmunity. Mice with a mutated immunoglobulin heavy chain (mIgM mice) express membrane-bound BCR but cannot secrete antibodies. When bred onto autoimmune-prone backgrounds such as NOD or MRL/lpr, mIgM mice developed type 1 diabetes and lupus nephritis, respectively (Chan *et al.*, 1999a; Wong *et al.*, 2004), while B cell deficient mice are resistant to the development of these autoimmune diseases (Chan and Shlomchik, 1998; Chan *et al.*, 1999b; Serreze *et al.*, 1996).

A strong B cell dependence has also been found in particular EAE models in which myelin oligodendrocyte glycoprotein (MOG) is used as the target antigen. These studies showed that B cells are required for the development of severe EAE following immunization with whole MOG protein, but are dispensable following MOG peptide immunization (Hjelmstrom *et al.*, 1998; Svensson *et al.*, 2002). The models involving immunization with whole proteins may be more relevant to the human disease because APCs in MS lesions phagocytose large fragments of myelin containing intact proteins. The differences between protein and peptide immunization may at least in part be due to the dose of the immunogen: smaller molar quantities are typically administered when intact proteins are utilized and the efficient antigen presentation function of B cells may then be critical. Furthermore, T cell–B cell collaboration can result in the production of conformation-sensitive antibodies which can induce demyelination. In contrast, peptide immunization only induces antibodies against this linear epitope and such antibodies do not bind to native MOG (von Budingen *et al.*, 2004).

The synergistic action of myelin-specific T cells and B cells has also been highlighted by recently developed spontaneous models of CNS autoimmunity. Mice that only express a transgenic MOG-specific TCR develop isolated optic neuritis in the absence of classical EAE symptoms (Bettelli *et al.*, 2003). Breeding of these TCR transgenic mice with an anti-MOG heavy chain knock-in results in a high frequency of MOG-specific T cells and B cells and an aggressive autoimmune disease in which lesions are largely confined to the optic nerves and spinal cord, as seen in NMO (Bettelli *et al.*, 2006; Krishnamoorthy *et al.*, 2006).

3.2. B cells as a source of cytokines

The costimulatory molecules and cytokines present during antigen recognition by a naive T cell determine whether that cell becomes activated and what effector phenotype it acquires. As shown in Figure 4.2, activated B cells can be induced to produce a variety of immunomodulatory cytokines and growth factors (Pistoia, 1997). B cells can differentiate into effector subsets with distinct cytokine profiles, both *in vitro* and *in vivo*. Production of type I cytokines, including interferon-gamma (IFN-γ) and IL-12, induces T cells to adopt a Th1 phenotype characterized by expression of IFN-γ. This type of response is optimized for the clearance of intracellular pathogens and leads to activation of cytotoxic CD8 T cells, NK cells, and production of complement-fixing antibody isotypes, which can all contribute to tissue damage in autoimmune diseases. TNFα production by B cells can also amplify Th1 differentiation and IFNγ production by T cells (Menard *et al.*, 2007). B cell derived IL-6, in combination with TGFβ, promotes T cell differentiation into highly pathogenic Th17 cells, which secrete high levels of IL-17 and other proinflammatory cytokines (Bettelli *et al.*, 2007). Secretion of type 2 cytokines by B cells is

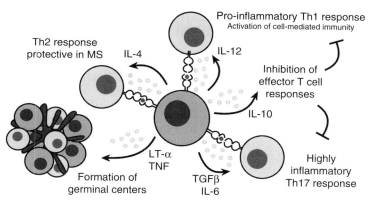

FIGURE 4.2   B cell-derived cytokines direct the ensuing immune response. Antigen-specific immune responses are perpetuated by collaboration of T and B cells in germinal centers, the formation of which is dependent on production of lymphotoxin α and TNF by activated B cells. The cytokines present during antigen presentation to T cells determine their effector phenotype. In the presence of IL-4, T cells follow the Th2 maturation pathway, which is involved in allergic responses and protective in MS. Expression of IL-12 or TGFβ and IL-6 by APCs lead to differentiation of proinflammatory Th1 and Th17 cells, respectively, which are pathogenic in animal models of MS. These proinflammatory cells can be inhibited by IL-10, produced by APCs and a subset of regulatory T cells. (See Plate 5 in Color Plate Section.)

associated with development of Th2 cells, which modulate inflammation by production of cytokines that promote the clearance of extracellular pathogens (Harris *et al.*, 2000). Although characteristic of allergies, Th2 responses are protective in MS and animal models, and are promoted by approved MS therapies, including interferon β and glatiramer acetate (Sospedra and Martin, 2005).

B cell-derived cytokines are also critical for the establishment of germinal centers in lymph nodes and inflamed tissues. In mice, expression of lymphotoxin (LT) and TNF by B cells is required for the maturation of follicular dendritic cells and organization of germinal centers in the spleen (Endres *et al.*, 1999; Fu *et al.*, 1998; Gonzalez *et al.*, 1998). The induction of lymphoid follicles in the gut and at sites of inflammation is also dependent on provision of LT by B cells (Gommerman and Browning, 2003; Lorenz *et al.*, 2003). Ectopic germinal centers are formed in the CNS of MS patients, and both TNF and LT have been detected in lesions (Selmaj *et al.*, 1991). Increased levels of serum TNF and LT have also been detected in MS patients, although other studies have not found a difference between MS and controls (Ledeen and Chakraborty, 1998). Monoclonal antibody inhibitors of TNF are approved therapies for RA, but MS patients receiving treatment had an increased attack rate relative to controls, and a clinical trial was discontinued prematurely due to this negative outcome (The Lenercept Multiple Sclerosis Study Group, 1999).

The context of activation plays an important role in determining the cytokine profile of a B cell: activation through the BCR with T cell help (CD40 ligation) leads to production of proinflammatory cytokines, while ligation of CD40 in the absence of specific BCR activation promotes secretion of the immunosuppressive cytokine IL-10 (Duddy *et al.*, 2004). IL-10 has potent antiinflammatory properties and can inhibit antigen presentation to CD4 T cells, indirectly downregulating their activation, cytokine production, and proliferation (Roncarolo *et al.*, 2006). Production of IL-10 by B cells is essential for recovery from MOG peptide-induced EAE (Fillatreau *et al.*, 2002), and overexpression of IL-10 protects mice from EAE (Cua *et al.*, 1999). B cells from MS patients produce less IL-10 than B cells from healthy individuals, but high levels of IL-10 and TGF-β are produced by newly generated B cells following Rituximab treatment (Duddy *et al.*, 2007).

The antigen-driven B cell maturation process in the MS brain is made possible by local changes in the immune environment. Expression of the lymphocyte adhesion molecule LFA-1 is increased in active MS plaques relative to noninflammatory neurological diseases (Cannella and Raine, 1995), signifying the active migration of cells into the plaque. Under normal conditions, astrocytes maintain an immunosuppressive cytokine environment in the CNS, with detectable levels of IL-10 and TGFβ but no TNF (Hickey, 2001). In MS, the balance of cytokines is tipped towards inflammation rather than tolerance, with increased levels of IL-1, IL-2, TNF, and IFNγ (Cannella and Raine, 1995). Normally secreted at low levels by astrocytes, the potent B cell activating factor BAFF is upregulated in MS, to levels comparable to secondary lymphoid tissues (Krumbholz *et al.*, 2005). Autoreactive B cells compete poorly with nonautoreactive cells for BAFF-induced survival signals, and mice engineered to overexpress BAFF develop B cell-associated autoimmunity (Lesley *et al.*, 2004). Regulation of local BAFF concentration can be an effective method of preventing autoimmunity, and loss of this control contributes to autoreactive B cell proliferation and survival in the brains of MS patients.

3.3. Autoantibody production by B cells

B cells are unique in their ability to produce antibodies, which circulate throughout the body and target specific antigen-bearing targets for clearance. Antibodies to self proteins are directly responsible for the pathology in both systemic and organ-specific autoimmune diseases, and have multiple mechanisms of causing tissue damage. In some cases, autoantibodies are directed to a cell-surface receptor, such as the acetylcholine receptor in myasthenia gravis or the thyroid stimulating hormone receptor in Grave's disease (Kohn and Harii, 2003; Vincent, 2002). Antibodies can also form immune complexes with circulating autoantigens, which accumulate in the kidneys and joint synovium in patients with SLE and RA (Davidson and Aranow, 2006; Dorner *et al.*, 2004; Rahman and Isenberg, 2008; Singh, 2005). Other autoantibodies bind to circulating cells such as platelets or erythrocytes, triggering complement-mediated lysis or phagocytosis by Fc-receptor bearing cells (Elson and Barker, 2000). Much effort has been placed on determining the specificities of antibodies found in the lesions, serum, and CSF of patients with MS and related demyelinating diseases, and a wide variety of antibody targets on myelin, axons, and astrocytes have been studied (Fig. 4.3).

A common finding in MS patients is the intrathecal synthesis of IgG: oligoclonal IgG bands are detectable in more than 95% of MS patients and increased CSF concentrations of IgG relative to albumin are found in up to 70% of cases (Link and Huang, 2006). The oligoclonal nature of these antibodies suggests an antigen-driven process, and determining the specificities of such antibodies could thus provide insights into the causes of MS. Several candidate autoantigens have been proposed as targets, as well as antigens from infectious agents, but there is currently no agreement on the specificities or significance of oligoclonal bands in MS, although they are the only immunological feature used to support the diagnosis (Cepok *et al.*, 2005; Correale and de los Milagros Bassani Molinas, 2002; Franciotta *et al.*, 2005).

Clonally expanded B cells have been detected by single-cell PCR in the cerebrospinal fluid and lesions of patients with MS and optic neuritis, and one study found expansion of MBP-specific B cells in the CSF (Lambracht-Washington *et al.*, 2007). The immunoglobulin gene rearrangements found in CSF B cells are rarely observed in the periphery (Colombo *et al.*, 2000), and memory B cells and plasma cells with similar rearrangements have been detected in the same patient (Haubold *et al.*, 2004). These cells carry mutations typical of germinal center-derived memory B cells (Harp *et al.*, 2007), with overrepresentation of specific immunoglobulin variable regions, especially V_H4 (Baranzini *et al.*, 1999; Owens *et al.*, 1998). The variable region sequences of MS lesion-resident B cells are mutated to a similar extent as those in B cells undergoing an antigen-driven response to CNS measles virus infection (Smith-Jensen *et al.*, 2000). Together, these findings provide support for a local antigen-dependent B cell maturation process in the CNS of MS patients.

Histological analysis of MS tissue has demonstrated deposition of antibodies and complement on the myelin sheath of a substantial fraction of lesions, but also indicated significant heterogeneity in the composition of demyelinating lesions. Lucchinetti and colleagues have classified MS lesion biopsies into four categories based on the observed inflammatory

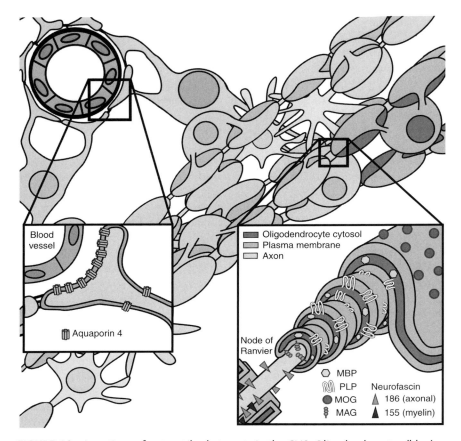

FIGURE 4.3 Locations of autoantibody targets in the CNS. Oligodendrocytes (blue) ensheath axons with the spiral myelin membrane, a specialized extension of the plasma membrane. Each oligodendrocyte produces a segmental myelin structure for several surrounding axons. MBP and PLP are the primary protein components of myelin, but are located within the many layers and therefore unavailable to antibody binding in noninjured myelin. Although less abundant, MOG is specifically expressed on the outer surface of myelin, making it a target for demyelinating antibodies. Antibodies to neurofascin, found at the myelin-axon junction and exposed on axons at the nodes of Ranvier, can contribute to both demyelination and axonal damage. Specific lipids and glycolipids can be released from the membranes of damaged oligodendrocytes, and are also recognized by antibodies in MS. The foot processes of astrocytes (green) form the blood-brain barrier which restricts access of molecules from the circulatory system to the CNS. Aquaporin-4, a water channel, is enriched in these structures, and is the target of autoantibodies specifically found in NMO. (See Plate 6 in Color Plate Section.)

and neurodegenerative characteristics (Lassmann *et al.*, 2001; Lucchinetti *et al.*, 2000). The first two patterns are characterized by demarcated regions of infiltrating T cells and activated macrophages, centered around small veins. Areas of demyelination, with a simultaneous loss of multiple myelin proteins including MBP, MOG, proteolipid protein (PLP), and myelin-associated glycoprotein (MAG), and remyelination were observed in both patterns. The presence (pattern II) or absence (pattern I) of antibody and complement deposition distinguishes these two types of lesions, suggestive of different levels of B cell involvement. Infiltrating T cells and macrophages were also found in pattern III and IV lesions, but other features pointed to an oligodendrocyte defect with a preferential loss of MAG (pattern III) or nearly complete loss of oligodendrocytes (pattern IV) and complete lack of remyelination. Although different between patients, the patterns observed in multiple lesions from the same patient were always identical. Pattern II lesions were the most common, observed in 115 areas from 16/27 patients and including all clinical subtypes of MS. Pattern III lesions were primarily found in acute disease, with a duration of less than 2 months before biopsy, and although rare, pattern IV was only found in primary-progressive MS. The IgG found in pattern II lesions may directly contribute to disease progression, as patients with pattern II lesions, but not pattern I or III, respond to therapeutic plasma exchange (Keegan et al., 2005). Development of novel methods for determining the composition of lesions may enable classification of MS based on the mechanism of pathogenesis rather than clinical presentation and facilitate individualized treatments.

Antibodies specific for myelin (Genain et al., 1999; O'Connor et al., 2005; Warren and Catz, 1993) and axonal proteins (Mathey et al., 2007; Zhang et al., 2005) have been eluted from MS lesions. The terminal complement complex C9neo is also found in active lesions containing IgG (Prineas and Graham, 1981; Storch et al., 1998). Histological analysis of lesions has also shown active phagocytosis of antibody-decorated myelin by macrophages and opsonized myelin fragments have been detected within endocytic vesicles (Prineas and Graham, 1981; Storch et al., 1998). Although IgG-containing lesions can be found in all types of MS, different antigens may be targeted in different subtypes.

MOG has been extensively studied as a target for autoantibodies in MS because it is selectively expressed by oligodendrocytes in the CNS. Unlike intracellular antigens such as MBP and PLP, MOG is localized to the surface of oligodendrocytes and the myelin sheath and therefore available for antibody binding in the absence of prior tissue damage (Brunner et al., 1989). MOG was originally discovered by affinity purification with the well-characterized 8–18C5 antibody (Linington et al., 1984), which recognizes a conformational epitope and can exacerbate demyelination when administered to animals with mild EAE (Breithaupt et al., 2003; Linington et al., 1988). Administration of antibodies to MOG to a healthy mouse cannot cause disease, as the blood–brain barrier must be compromised to allow access to the target antigen. The conformation-specific antibodies produced following immunization with refolded MOG protein can directly contribute to demyelination in both rodents and nonhuman primates and also exacerbate EAE by passive transfer (Genain *et al.*, 1995; Lyons *et al.*, 2002; von Budingen *et al.*, 2002). However, antibodies to linear epitopes or peptides cannot bind MOG on the cell surface and are not pathogenic (Brehm *et al.*, 1999; von Budingen *et al.*, 2004).

The ability of conformation-sensitive antibodies to exacerbate demyelination in animal models has made MOG an attractive candidate antigen for autoantibody-mediated injury in MS. A provocative report that antibodies to MOG are present in patients with CIS and correlate with an increased risk of progression to MS (Berger *et al.*, 2003) has not been reproduced by several groups, though antibodies may be associated with a higher MRI lesion load (Kuhle *et al.*, 2007; Lim *et al.*, 2005; Pelayo *et al.*, 2007). Increased frequencies of antibodies to MOG have been reported in the serum and CSF of MS patients relative to controls (Kennel De March *et al.*, 2003), but other studies have found similar levels of anti-MOG in MS and other neurological diseases or controls (Brokstad *et al.*, 1994; Lampasona *et al.*, 2004; Mantegazza *et al.*, 2004; Markovic *et al.*, 2003; O'Connor *et al.*, 2007; Reindl *et al.*, 1999). Although detectable in some cases, the reported frequency of anti-MOG in serum and CSF from MS patients varies considerably between studies. This variability is primarily attributable to differences in the assay methodologies and antigens utilized. Antibodies that bind to native MOG are most relevant to the disease, but are not detectable with all assays. Many of these studies used Western blots, which only detect antibodies to denatured antigen, or ELISAs, which cannot distinguish between responses to folded and unfolded proteins (Mathey *et al.*, 2004). Furthermore, the results obtained with Western blots and ELISAs do not necessarily correlate, even with the same sera (Pittock *et al.*, 2007). New methods have therefore been developed that enable specific detection of antibodies to properly folded and glycosylated MOG protein. The first method involves labeling of MOG transfectants with serum or CSF antibodies (Brehm *et al.*, 1999; Lalive *et al.*, 2006; O'Connor *et al.*, 2007; Zhou *et al.*, 2006), but the frequency of anti-MOG in MS still varies considerably between these studies, suggesting heterogeneity in the studied patient populations. The second method is a sensitive radioimmunoassay with a tetrameric version of folded and glycosylated MOG protein. The tetrameric nature of the antigen enables bivalent binding by IgG antibodies and thus increases the avidity of the interaction. With this assay, MOG antibodies were detected in nearly 20% of ADEM patients but only infrequently in adult-onset MS patients (O'Connor *et al.*, 2007). These results suggest that antibodies to MOG are more prevalent in ADEM than in adult-onset MS.

Antibodies to other myelin components including MBP, PLP, MAG, and cyclic nucleotide phosphodiesterase (CNP) have each been detected in some studies of MS serum or CSF (Hafler *et al.*, 2005). The frequency of these antibodies in MS and controls, like anti-MOG, varies considerably between studies, and none are useful as a diagnostic or prognostic marker. Furthermore, the contribution of antibodies to intracellular proteins such as MBP to the initiation of myelin damage is likely to be minimal, as they are only exposed once the demyelinating process has begun, but MBP-specific B cells may nevertheless be relevant as APC. The myelin sheath contains a variety of lipids, some of which may also be recognized by autoreactive B cells. Using arrays of 50 unique brain and microbial lipids, Kanter *et al.* have defined an antibody response to sulfatide in MS CSF, and addition of sulfatide to a myelin peptide immunization worsens the course of EAE in mice (Kanter *et al.*, 2006).

Axonal degeneration appears to be an important factor in the cumulative neurological disability that develops over time in MS patients. A recent study demonstrated that an axonal protein is targeted by autoantibodies in a subset of MS patients, suggesting that the B cell response may not only contribute to demyelination but also to axonal damage. Approximately one-third of MS sera contained antibodies to neurofascin, an adhesion molecule expressed by oligodendrocytes and neurons which localizes to the myelin–axon interface at the nodes of Ranvier (Mathey *et al.*, 2007). Antibodies to neurofascin caused rapid worsening of EAE following systemic administration due to complement-mediated damage of the axon segments exposed at the nodes. High-throughput proteomic methods are likely to identify yet other autoantibody targets in MS.

The recent discovery of an autoantibody associated with NMO demonstrates that different inflammatory demyelinating diseases can be distinguished using antibodies as biomarkers. Serum IgG reactivity against the blood–brain barrier of mouse CNS tissue was found in the majority of patients with NMO (73%) or an optic-spinal form of MS (58%), but rarely in MS or controls, with 91–100% specificity for optic-spinal conditions (Lennon *et al.*, 2004). This "NMO-IgG" was also detectable in individuals with longitudinally extensive myelitis or recurrent optic neuritis, who are at a high risk of developing NMO. A single autoantigen, aquaporin-4 (Aqp4), was later found to be targeted by the autoantibody in NMO (Lennon *et al.*, 2005). Aquaporins are a family of water channels responsible for maintaining fluid balance in the renal medulla and gastric mucosa, and are also expressed by astrocytes in the CNS, localizing to the foot processes at the blood–brain barrier (Fig. 4.3). Serum antibodies from NMO patients showed identical reactivity to a monoclonal anti-Aqp4, and no reactivity was observed in tissue sections from Aqp4-null mice (Lennon *et al.*, 2005; Paul *et al.*, 2007; Takahashi *et al.*, 2007). The high specificity and sensitivity of anti-Aqp4 assays were confirmed in larger studies of patients with NMO and other neurological or autoimmune diseases including pediatric and adult-onset MS (Banwell *et al.*, 2008; Paul *et al.*, 2007; Takahashi *et al.*, 2007).

Antibody titer in NMO patients was found to correlate with disease severity (Takahashi *et al.*, 2007), and anti-Aqp4 status has since been incorporated into diagnostic criteria for NMO (Wingerchuk *et al.*, 2006). Antibodies to aquaporin play an active role in the progression of NMO, as Aqp4 is expressed at high levels in the spinal cord and optic nerve (Nielsen *et al.*, 1997) and demyelinated brain lesions are found more often in regions of high aquaporin expression (Pittock *et al.*, 2006). Aqp4 protein is lost in both active and inactive NMO lesions (Roemer *et al.*, 2007). When incubated with serum or purified IgG from NMO patients, cells transfected with Aqp4 rapidly internalize the antigen and can also undergo complement-mediated lysis (Hinson *et al.*, 2007). Patients with NMO respond positively to plasmapheresis (Weinshenker *et al.*, 1999), although the effect may be attributable to a change in the balance of cytokines rather than merely depletion of autoantibodies. Because autoantibody levels typically do not change following treatment, the benefit of Rituximab in NMO and MS is likely due to disruption of antibody-independent B cell functions, such as antigen presentation and cytokine production within the CNS.

4. CONCLUSIONS AND FUTURE DIRECTIONS

The significant clinical and pathogenic heterogeneity of MS and related diseases make it challenging to define the mechanisms of demyelination in an individual, and biomarkers that reflect pathogenetic mechanisms may thus be valuable for individualized treatment. Autoantibodies circulating in the blood and CSF are attractive candidates for the development of biomarkers, and have already proven useful in differentiating NMO and MS. Autoantibodies to multiple islet antigens are highly predictive of progression to type I diabetes (Verge *et al.*, 1996; Ziegler *et al.*, 1999), and measurement of autoantibodies may also become clinically useful in MS and related demyelinating diseases once more antigens are identified and characterized. The MS community has traditionally focused on well-defined candidate autoantigens such as MBP, PLP, and MOG, but recent systematic approaches for the identification of novel antigens have yielded several new protein and lipid targets. Definition of the repertoire of autoantibodies may allow classification of demyelinating diseases based on pathogenetic mechanisms rather than clinical presentation, facilitating individualized treatment or prediction of disease progression. For example, the presence of Aqp4 antibodies in NMO patients predicts a better response to plasma exchange or intravenous immunoglobulin. Antibodies may be a key component of the primary immune response responsible for initiation of tissue damage in MS, or may be merely reflect tissue damage. *In vivo* studies with purified or recombinant antibodies will be necessary to distinguish between these possibilities.

Now that clinical trials with Rituximab have proven that B cells play an important role in the pathogenesis of MS, more selective approaches can be tested that target either defined B cell populations or functions. Because plasma cells are preserved and autoantibody titers do not decrease in all patients following B cell depletion, the beneficial effects of Rituximab in MS suggest that other B cell functions are critical, such as antigen presentation to T cells with a matching antigen specificity and/or cytokine and chemokine production. Rather than depleting B cells systemically, it may be preferable to decrease levels of prosurvival factors for which autoreactive cells must compete or block the cytokines required for perpetuation of an autoimmune response. A soluble version of the BAFF receptor has successfully been used to treat and prevent MOG-induced EAE (Huntington *et al.*, 2006). Recovery was associated with decreased anti-MOG titer and alterations in T cell cytokine production with increased levels of TGFβ and a reduction of IFNγ. Small interfering RNAs (siRNAs) have been targeted to specific cells in the periphery and CNS as complexes with soluble antibodies or antibody-decorated liposomes (Kumar *et al.*, 2007; Song *et al.*, 2005). siRNAs could thus be used to target antigen presenting molecules or cytokines in either the entire B cell pool, or cells with a defined activation/differentiation state.

B cells are now recognized to play a central role in MS and many other chronic inflammatory diseases. Although antibodies to myelin and neurons are likely to contribute to demyelination, the capacity of B cells to activate antigen-specific T cells or promote a proinflammatory cytokine environment is critical for disease progression. The complex interplay between B cells, T cells and other cells of the immune system may thus offer a number of other approaches to therapy that preserve the B cell repertoire.

REFERENCES

Achenbach, P., Koczwara, K., Knopff, A., Naserke, H., Ziegler, A., and Bonifacio, E. (2004). Mature high-affinity immune responses to (pro)insulin anticipate the autoimmune cascade that leads to type 1 diabetes. *J. Clin. Invest.* **114**(4), 589–597.

Aloisi, F., and Pujol-Borrell, R. (2006). Lymphoid neogenesis in chronic inflammatory diseases. *Nat. Rev. Immunol.* **6**(3), 205–217.

Banwell, B., Tenembaum, S., Lennon, V. A., Ursell, E., Kennedy, J., Bar-Or, A., Weinshenker, B. G., Lucchinetti, C. F., and Pittock, S. J. (2008). Neuromyelitis optica-IgG in childhood inflammatory demyelinating CNS disorders. *Neurology* **70**(5), 344–352.

Baranzini, S. E., Jeong, M. C., Butunoi, C., Murray, R. S., Bernard, C. C. A., and Oksenberg, J. R. (1999). B cell repertoire diversity and clonal expansion in multiple sclerosis brain lesions. *J. Immunol.* **163**(9), 5133–5144.

Belman, A. L., Chitnis, T., Renoux, C., Waubant, E., and the International Pediatric MS Study Group. (2007). Challenges in the classification of pediatric multiple sclerosis and future directions. *Neurology* **68**(16, Suppl. 2), S70–S74.

Berger, T., Rubner, P., Schautzer, F., Egg, R., Ulmer, H., Mayringer, I., Dilitz, E., Deisenhammer, F., and Reindl, M. (2003). Antimyelin antibodies as a predictor of clinically definite multiple sclerosis after a first demyelinating event. *N. Engl. J. Med.* **349**(2), 139–145.

Bettelli, E., Baeten, D., Jäger, A., Sobel, R. A., and Kuchroo, V. K. (2006). Myelin oligodendrocyte glycoprotein-specific T and B cells cooperate to induce a Devic-like disease in mice. *J. Clin. Invest.* **116**(9), 2393–2402.

Bettelli, E., Korn, T., and Kuchroo, V. K. (2007). Th17: The third member of the effector T cell trilogy. *Curr. Opin. Immunol.* **19**(6), 652–657.

Bettelli, E., Pagany, M., Weiner, H. L., Linington, C., Sobel, R. A., and Kuchroo, V. K. (2003). Myelin oligodendrocyte glycoprotein-specific T cell receptor transgenic mice develop spontaneous autoimmune optic neuritis. *J. Exp. Med.* **197**(9), 1073–1081.

Brehm, U., Piddlesden, S. J., Gardinier, M. V., and Linington, C. (1999). Epitope specificity of demyelinating monoclonal autoantibodies directed against the human myelin oligodendrocyte glycoprotein (MOG). *J. Neuroimmunol.* **97**(1–2), 9–15.

Breithaupt, C., Schubart, A., Zander, H., Skerra, A., Huber, R., Linington, C., and Jacob, U. (2003). Structural insights into the antigenicity of myelin oligodendrocyte glycoprotein. *Proc. Natl. Acad. Sci.* **100**(16), 9446–9451.

Brokstad, K. A., Page, M., Nyland, H., and Haaheim, L. R. (1994). Autoantibodies to myelin basic protein are not present in the serum and CSF of MS patients. *Acta Neurol. Scand.* **89**(6), 407–411.

Brunner, C., Lassmann, H., Waehneldt, T. V., Matthieu, J.-M., and Linington, C. (1989). Differential ultrastructural localization of myelin basic protein, myelin/oligodendroglial glycoprotein, and 2′,3′-cyclic nucleotide 3′-phosphodiesterase in the CNS of adult rats. *J. Neurochem.* **52**(1), 296–304.

Cambier, J. C., Pleiman, C. M., and Clark, M. R. (1994). Signal transduction by the B cell antigen receptor and its coreceptors. *Annu. Rev. Immunol.* **12**(1), 457–486.

Cannella, B., and Raine, C. S. (1995). The adhesion molecule and cytokine profile of multiple sclerosis lesions. *Ann. Neurol.* **37**(4), 424–435.

Cepok, S., Zhou, D., Srivastava, R., Nessler, S., Stei, S., Bussow, K., Sommer, N., and Hemmer, B. (2005). Identification of Epstein-Barr virus proteins as putative targets of the immune response in multiple sclerosis. *J. Clin. Invest.* **115**(5), 1352–1360.

Chan, O., and Shlomchik, M. J. (1998). A new role for B cells in systemic autoimmunity: B cells promote spontaneous T cell activation in MRL-lpr/lpr mice. *J. Immunol.* **160**(1), 51–59.

Chan, O. T. M., Hannum, L. G., Haberman, A. M., Madaio, M. P., and Shlomchik, M. J. (1999a). A novel mouse with B cells but lacking serum antibody reveals an antibody-independent role for B cells in murine lupus. *J. Exp. Med.* **189**(10), 1639–1648.

Chan, O. T. M., Madaio, M. P., and Shlomchik, M. J. (1999b). B cells are required for lupus nephritis in the polygenic, Fas-intact MRL model of systemic autoimmunity. *J. Immunol.* **163**(7), 3592–3596.

Chataway, J., Feakes, R., Coraddu, F., Gray, J., Deans, J., Fraser, M., Robertson, N., Broadley, S., Jones, H., Clayton, D., Goodfellow, P., Sawcer, S., *et al.* (1998). The genetics of multiple sclerosis: Principles, background and updated results of the United Kingdom systematic genome screen. *Brain* **121**(10), 1869–1887.

Cohen, S. B., Emery, P., Greenwald, M. W., Dougados, M., Furie, R. A., Genovese, M. C., Keystone, E. C., Loveless, J. E., Burmester, G.-R., Cravets, M. W., Hessey, E. W., Shaw, T., *et al.* (2006). Rituximab for rheumatoid arthritis refractory to anti-tumor necrosis factor therapy: Results of a multicenter, randomized, double-blind, placebo-controlled, phase III trial evaluating primary efficacy and safety at twenty-four weeks. *Arthritis Rheum.* **54**(9), 2793–2806.

Colombo, M., Dono, M., Gazzola, P., Roncella, S., Valetto, A., Chiorazzi, N., Mancardi, G. L., and Ferrarini, M. (2000). Accumulation of clonally related B Lymphocytes in the cerebrospinal fluid of multiple sclerosis patients. *J. Immunol.* **164**(5), 2782–2789.

Correale, J., and de los Milagros Bassani Molinas, M. A. (2002). Oligoclonal bands and antibody responses in multiple sclerosis. *J. Neurol.* **249**(4), 375–389.

Cree, B. A. C., Lamb, S., Morgan, K., Chen, A., Waubant, E., and Genain, C. (2005). An open label study of the effects of rituximab in neuromyelitis optica. *Neurology* **64**(7), 1270–1272.

Cross, A. H., Stark, J. L., Lauber, J., Ramsbottom, M. J., and Lyons, J.-A. (2006). Rituximab reduces B cells and T cells in cerebrospinal fluid of multiple sclerosis patients. *J. Neuroimmunol.* **180**(1–2), 63–70.

Cua, D. J., Groux, H., Hinton, D. R., Stohlman, S. A., and Coffman, R. L. (1999). Transgenic interleukin 10 prevents induction of experimental autoimmune encephalomyelitis. *J. Exp. Med.* **189**(6), 1005–1010.

Czlonkowska, A., Ciesielska, A., Gromadzka, G., and Kurkowska-Jastrzebska, I. (2005). Estrogen and cytokines production - the possible cause of gender differences in neurological diseases. *Curr. Pharm. Des.* **11**(8), 1017–1030.

Dale, R. C., de Sousa, C., Chong, W. K., Cox, T. C., Harding, B., and Neville, B. G. (2000). Acute disseminated encephalomyelitis, multiphasic disseminated encephalomyelitis and multiple sclerosis in children. *Brain* **123**(12), 2407–2422.

Davidson, A., and Aranow, C. (2006). Pathogenesis and treatment of systemic lupus erythematosus nephritis. *Curr. Opin. Rheumatol.* **18**(5), 468–475.

Dorner, T., Egerer, K., Feist, E., and Burmester, G. R. (2004). Rheumatoid factor revisited. *Curr. Opin. Rheumatol.* **16**(3), 246–253.

Drake, L., McGovern-Brindisi, E. M., and Drake, J. R. (2006). BCR ubiquitination controls BCR-mediated antigen processing and presentation. *Blood* **108**(13), 4086–4093.

Duddy, M E., Alter, A., and Bar-Or, A. (2004). Distinct profiles of human B cell effector cytokines: A role in immune regulation? *J. Immunol.* **172**(6), 3422–3427.

Duddy, M., Niino, M., Adatia, F., Hebert, S., Freedman, M., Atkins, H., Kim, H. J., and Bar-Or, A. (2007). Distinct effector cytokine profiles of memory and naive human B cell subsets and implication in multiple sclerosis. *J. Immunol.* **178**(10), 6092–6099.

Edwards, J. C. W., and Cambridge, G. (2006). B-cell targeting in rheumatoid arthritis and other autoimmune diseases. *Nat. Rev. Immunol.* **6**(5), 394–403.

Edwards, J. C. W., Szczepanski, L., Szechinski, J., Filipowicz-Sosnowska, A., Emery, P., Close, D. R., Stevens, R. M., and Shaw, T. (2004). Efficacy of B-cell-targeted therapy with rituximab in patients with rheumatoid arthritis. *N. Engl. J. Med.* **350**(25), 2572–2581.

Elson, C. J., and Barker, R. N. (2000). Helper T cells in antibody-mediated, organ-specific autoimmunity. *Curr. Opin. Immunol.* **12**(6), 664–669.

Endres, R., Alimzhanov, M. B., Plitz, T., Futterer, A., Kosco-Vilbois, M. H., Nedospasov, S. A., Rajewsky, K., and Pfeffer, K. (1999). Mature follicular dendritic cell networks depend on expression of lymphotoxin beta receptor by radioresistant stromal cells and of lymphotoxin beta and tumor necrosis factor by B cells. *J. Exp. Med.* **189**(1), 159–168.

Fillatreau, S., Sweenie, C. H., McGeachy, M. J., Gray, D., and Anderton, S. M. (2002). B cells regulate autoimmunity by provision of IL-10. *Nat. Immunol.* **3**(10), 944–950.

Franciotta, D., Zardini, E., Bergamaschi, R., Grimaldi, L. M., Andreoni, L., and Cosi, V. (2005). Analysis of Chlamydia pneumoniae-specific oligoclonal bands in multiple sclerosis and other neurologic diseases. *Acta Neurol. Scand.* **112**(4), 238–241.

Fu, Y.-X. Huang, G., Wang, Y., and Chaplin, D. D. (1998). B lymphocytes induce the formation of follicular dendritic cell clusters in a lymphotoxin alpha -dependent fashion. *J. Exp Med.* **187**(7), 1009–1018.

Genain, C. P., Cannella, B., Hauser, S. L., and Raine, C. S. (1999). Identification of autoantibodies associated with myelin damage in multiple sclerosis. *Nat. Med.* **5**(2), 170–175.

Genain, C. P., Nguyen, M. H., Letvin, N. L., Pearl, R., Davis, R. L., Adelman, M., Lees, M. B., Linington, C., and Hauser, S. L. (1995). Antibody facilitation of multiple sclerosis-like lesions in a nonhuman primate. *J. Clin. Invest.* **96**(6), 2966–2974.

Gommerman, J. L., and Browning, J. L. (2003). Lymphotoxin/LIGHT, lymphoid microenvironments and autoimmune disease. *Nat. Rev. Immunol.* **3**(8), 642–655.

Gonzalez, M., Mackay, F., Browning, J. L., Kosco-Vilbois, M. H., and Noelle, R. J. (1998). The Sequential role of lymphotoxin and B cells in the development of splenic follicles. *J. Exp. Med.* **187**(7), 997–1007.

Goodin, D. S. (2006). Magnetic resonance imaging as a surrogate outcome measure of disability in multiple sclerosis: Have we been overly harsh in our assessment? *Ann. Neurol.* **59**(4), 597–605.

Gordy, C., Mishra, S., and Rodgers, W. (2004). Visualization of antigen presentation by actin-mediated targeting of glycolipid-enriched membrane domains to the immune synapse of B cell APCs. *J. Immunol.* **172**(4), 2030–2038.

Hafler, D. A., Slavik, J. M., Anderson, D. E., O'Connor, K. C., De Jager, P., and Baecher-Allan, C. (2005). Multiple sclerosis. *Immunol. Rev.* **204**(1), 208–231.

Haines, J. L., Ter-Minassian, M., Bazyk, A., Gusella, J. F., Kim, D. J., Terwedow, H., PericakVance, M. A., Rimmler, J. B., Haynes, C. S., Roses, A. D., Lee, A., Shaner, B., *et al.* (1996). A complete genomic screen for multiple sclerosis underscores a role for the major histocompatability complex. *Nat. Genet.* **13**(4), 469–471.

Harp, C., Lee, J., Lambracht-Washington, D., Cameron, E., Olsen, G., Frohman, E., Racke, M., and Monson, N. (2007). Cerebrospinal fluid B cells from multiple sclerosis patients are subject to normal germinal center selection. *J. Neuroimmunol.* **183**(1–2), 189–199.

Harris, D. P., Haynes, L., Sayles, P. C., Duso, D. K., Eaton, S. M., Lepak, N. M., LJohnson, A. L., Swain, S. L., and Lund, F. E. (2000). Reciprocal regulation of polarized cytokine production by effector B and T cells. *Nat. Immunol.* **1**(6), 475–482.

Harvey, B. P., Gee, R. J., Haberman, A. M., Shlomchik, M. J., and Mamula, M. J. (2007). Antigen presentation and transfer between B cells and macrophages. *Eur. J. Immunol.* **37**(7), 1739–1751.

Haubold, K., Owens, G. P., Kaur, P., Ritchie, A. M., Gilden, D. H., and Bennett, J. L. (2004). B-lymphocyte and plasma cell clonal expansion in monosymptomatic optic neuritis cerebrospinal fluid. *Ann. Neurol.* **56**(1), 97–107.

Hauser, S. L., and Oksenberg, J. R. (2006). The neurobiology of multiple sclerosis: Genes, inflammation, and neurodegeneration. *Neuron* **52**(1), 61–76.

Hauser, S. L., Waubant, E., Arnold, D. L., Vollmer, T., Antel, J., Fox, R. J., Bar-Or, A., Panzara, M., Sarkar, N., Agarwal, S., Langer-Gould, A., Smith, C. H., and the HERMES Trial Group (2008). B-Cell depletion with rituximab in relapsing-remitting multiple sclerosis. *N. Engl. J. Med.* **358**(7), 676–688.

Hickey, W. F. (2001). Basic principles of immunological surveillance of the normal central nervous system. *Glia* **36**(2), 118–124.

Hinson, S. R., Pittock, S. J., Lucchinetti, C. F., Roemer, S. F., Fryer, J. P., Kryzer, T. J., and Lennon, V. A. (2007). Pathogenic potential of IgG binding to water channel extracellular domain in neuromyelitis optica. *Neurology* **69**(24), 2221–2231.

Hjelmstrom, P., Juedes, A. E., Fjell, J., and Ruddle, N. H. (1998). Cutting edge: B cell-deficient mice develop experimental allergic encephalomyelitis with demyelination after myelin oligodendrocyte glycoprotein sensitization. *J. Immunol.* **161**(9), 4480–4483.

Huntington, N. D., Tomioka, R., Clavarino, C., Chow, A. M., Linares, D., Mana, P., Rossjohn, J., Cachero, T. G., Qian, F., Kalled, S. L., Bernard, C. C. A., and Reid, H. H. (2006). A BAFF antagonist suppresses experimental autoimmune encephalomyelitis by targeting cell-mediated and humoral immune responses. *Int. Immunol.* **18**(10), 1473–1485.

Kanter, J. L., Narayana, S., Ho, P. P., Catz, I., Warren, K. G., Sobel, R. A., Steinman, L., and Robinson, W. H. (2006). Lipid microarrays identify key mediators of autoimmune brain inflammation. *Nat. Med.* **12**(1), 138–143.

Kappos, L., Freedman, M. S., Polman, C. H., Edan, G., Hartung, H.-P., Miller, D. H., Montalban, X., Barkhof, F., Radu, E.-W., Bauer, L., Dahms, S., Lanius, V., *et al.* (2007).

Effect of early versus delayed interferon beta-1b treatment on disability after a first clinical event suggestive of multiple sclerosis: A 3-year follow-up analysis of the BENEFIT study. *Lancet* **370**(9585), 389–397.

Keegan, M., Konig, F., McClelland, R., Bruck, W., Morales, Y., Bitsch, A., Panitch, H., Lassmann, H., Weinshenker, B., Rodriguez, M., Parisi, J., and Lucchinetti, C. F. (2005). Relation between humoral pathological changes in multiple sclerosis and response to therapeutic plasma exchange. *Lancet* **366**(9485), 579–582.

Kennel De March, A., De Bouwerie, M., Kolopp-Sarda, M. N., Faure, G. C., Bene, M. C., and Bernard, C. C. A. (2003). Anti-myelin oligodendrocyte glycoprotein B-cell responses in multiple sclerosis. *J. Neuroimmunol.* **135**(1–2), 117–125.

Kent, S. C., Chen, Y., Bregoli, L., Clemmings, S. M., Kenyon, N. S., Ricordi, C., Hering, B. J., and Hafler, D. A. (2005). Expanded T cells from pancreatic lymph nodes of type 1 diabetic subjects recognize an insulin epitope. *Nature* **435**(7039), 224–228.

Keystone, E., Fleischmann, R., Emery, P., Furst, D. E., van Vollenhoven, R., Bathon, J., Dougados, M., Baldassare, A., Ferraccioli, G., Chubick, A., Udell, J., Cravets, M. W., *et al.* (2007). Safety and efficacy of additional courses of rituximab in patients with active rheumatoid arthritis: An open-label extension analysis. *Arthritis Rheum.* **56**(12), 3896–3908.

Klein, U., and Dalla-Favera, R. (2008). Germinal centres: Role in B-cell physiology and malignancy. *Nat. Rev. Immunol.* **8**(1), 22–33.

Kohn, L. D , and Harii, N. (2003). Thyrotropin receptor autoantibodies (TSHRAbs): Epitopes, origins and clinical significance. *Autoimmunity* **36**(6/7), 331–337.

Krishnamoorthy, G., Lassmann, H., Wekerle, H., and Holz, A. (2006). Spontaneous opticospinal encephalomyelitis in a double-transgenic mouse model of autoimmune T cell/B cell cooperation. *J. Clin. Invest.* **116**(9), 2385–2392.

Krumbholz, M., Theil, D., Derfuss, T., Rosenwald, A., Schrader, F., Monoranu, C.-M., Kalled, S. L., Hess, D. M., Serafini, B., Aloisi, F., Wekerle, H., Hohlfeld, R., *et al.* (2005). BAFF is produced by astrocytes and up-regulated in multiple sclerosis lesions and primary central nervous system lymphoma. *J. Exp. Med.* **201**(2), 195–200.

Krupp, L. B., Banwell, B., Tenembaum, S., and the International Pediatric MS Study Group (2007). Consensus definitions proposed for pediatric multiple sclerosis and related disorders. *Neurology* **68**(16, Suppl. 2), S7–S12.

Kuhle, J., Lindberg, R., Regeniter, A., Mehling, M., Hoffmann, F., Reindl, M., Berger, T., Radue, E., Leppert, D., and Kappos, L. (2007). Antimyelin antibodies in clinically isolated syndromes correlate with inflammation in MRI and CSF. *J. Neurol.* **254**(2), 160–168.

Kumar, P., Wu, H., McBride, J. L., Jung, K.-E., Hee Kim, M., Davidson, B. L., Kyung Lee, S., Shankar, P., and Manjunath, N. (2007). Transvascular delivery of small interfering RNA to the central nervous system. *Nature* **448**(7149), 39–43.

Kunkel, E. J., and Butcher, E. C. (2003). Plasma-cell homing. *Nat. Rev. Immunol.* **3**(10), 822–829.

Lalive, P. H., Menge, T., Delarasse, C., Della Gaspera, B., Pham-Dinh, D., Villoslada, P., von Büdingen, H.-C., and Genain, C. P. (2006). Antibodies to native myelin oligodendrocyte glycoprotein are serologic markers of early inflammation in multiple sclerosis. *Proc. Natl. Acad. Sci.* **103**(7), 2280–2285.

Lambracht-Washington, D., O'Connor, K. C., Cameron, E. M., Jowdry, A., Ward, E. S., Frohman, E., Racke, M. K., and Monson, N. L. (2007). Antigen specificity of clonally expanded and receptor edited cerebrospinal fluid B cells from patients with relapsing remitting MS. *J. Neuroimmunol.* **186**(1–2), 164–176.

Lampasona, V., Franciotta, D., Furlan, R., Zanaboni, S., Fazio, R., Bonifacio, E., Comi, G., and Martino, G. (2004). Similar low frequency of anti-MOG IgG and IgM in MS patients and healthy subjects. *Neurology* **62**(11), 2092–2094.

Lankar, D., Vincent-Schneider, H., Briken, V., Yokozeki, T., Raposo, G., and Bonnerot, C. (2002). Dynamics of major histocompatibility complex class II compartments during B cell receptor-mediated cell activation. *J. Exp. Med.* **195**(4), 461–472.

Lanzavecchia, A. (1985). Antigen-specific interaction between T and B cells. *Nature* **314**(6011), 537–539.

Lassmann, H., Bruck, W., and Lucchinetti, C. (2001). Heterogeneity of multiple sclerosis pathogenesis: Implications for diagnosis and therapy. *Trends Mol. Med.* **7**(3), 115–121.

Leake, J., Albani, S., Kao, A., Senac, M., Billman, G., Nespeca, M., Paulino, A., Quintela, E., Sawyer, M., and Bradley, J. (2004). Acute disseminated encephalomyelitis in childhood: Epidemiologic, clinical and laboratory features. *Pediatr. Infect. Dis. J.* **23**(8), 756–764.

Ledeen, R. W., and Chakraborty, G. (1998). Cytokines, signal transduction, and inflammatory demyelination: Review and hypothesis. *Neurochem. Res.* **23**(3), 277–289.

Lennon, V. A., Kryzer, T. J., Pittock, S. J., Verkman, A. S., and Hinson, S. R. (2005). IgG marker of optic-spinal multiple sclerosis binds to the aquaporin-4 water channel. *J. Exp. Med.* **202**(4), 473–477.

Lennon, V. A., Wingerchuk, D. M., Kryzer, T. J., Pittock, S. J., Lucchinetti, C. F., Fujihara, K., Nakashima, I., and Weinshenker, B. G. (2004). A serum autoantibody marker of neuromyelitis optica: Distinction from multiple sclerosis. *Lancet* **364**(9451), 2106–2112.

Lesley, R., Xu, Y., Kalled, S. L., Hess, D. M., Schwab, S. R., Shu, H.-B., and Cyster, J. G. (2004). Reduced competitiveness of autoantigen-engaged B cells due to increased dependence on BAFF. *Immunity* **20**(4), 441–453.

Lim, E. T., Berger, T., Reindl, M., Dalton, C. M., Fernando, K., Keir, G., Thompson, E. J., Miller, D. H., and Giovannoni, G. (2005). Anti-myelin antibodies do not allow earlier diagnosis of multiple sclerosis. *Mult. Scler.* **11**(4), 492–494.

Linington, C., Bradl, M., Lassmann, H., Brunner, C., and Vass, K. (1988). Augmentation of demyelination in rat acute allergic encephalomyelitis by circulating mouse monoclonal antibodies directed against a myelin/oligodendrocyte glycoprotein. *Am. J. Pathol.* **130**(3), 443–454.

Linington, C., Webb, M., and Woodhams, P. L. (1984). A novel myelin-associated glycoprotein defined by a mouse monoclonal antibody. *J. Neuroimmunol.* **6**(6), 387–396.

Link, H., and Huang, Y.-M. (2006). Oligoclonal bands in multiple sclerosis cerebrospinal fluid: An update on methodology and clinical usefulness. *J. Neuroimmunol.* **180**(1–2), 17–28.

Lorenz, R. G., Chaplin, D. D., McDonald, K. G., McDonough, J. S., and Newberry, R. D. (2003). Isolated lymphoid follicle formation is inducible and dependent upon lymphotoxin-sufficient B lymphocytes, lymphotoxin-beta receptor, and TNF receptor I function. *J. Immunol.* **170**(11), 5475–5482.

Lucchinetti, C., Brück, W., Parisi, J., Scheithauer, B., Rodriguez, M., and Lassmann, H. (2000). Heterogeneity of multiple sclerosis lesions: Implications for the pathogenesis of demyelination. *Ann. Neurol.* **47**(6), 707–717.

Lyons, J.-A., Ramsbottom, M. J., and Cross, A. H. (2002). Critical role of antigen-specific antibody in experimental autoimmune encephalomyelitis induced by recombinant myelin oligodendrocyte glycoprotein. *Eur. J. Immunol.* **32**(7), 1905–1913.

Lyons, J.-A., San, M., Happ, M. P., and Cross, A. H. (1999). B cells are critical to induction of experimental allergic encephalomyelitis by protein but not by a short encephalitogenic peptide. *Eur. J. Immunol.* **29**(11), 3432–3439.

MacLennan, I. C. M. (1994). Germinal centers. *Annu. Rev. Immunol.* **12**(1), 117–139.

Magliozzi, R., Columba-Cabezas, S., Serafini, B., and Aloisi, F. (2004). Intracerebral expression of CXCL13 and BAFF is accompanied by formation of lymphoid follicle-like structures in the meninges of mice with relapsing experimental autoimmune encephalomyelitis. *J. Neuroimmunol.* **148**(1–2), 11–23.

Magliozzi, R., Howell, O., Vora, A., Serafini, B., Nicholas, R., Puopolo, M., Reynolds, R., and Aloisi, F. (2007). Meningeal B-cell follicles in secondary progressive multiple sclerosis associate with early onset of disease and severe cortical pathology. *Brain* **130**(4), 1089–1104.

Mantegazza, R., Cristaldini, P., Bernasconi, P., Baggi, F., Pedotti, R., Piccini, I., Mascoli, N., Mantia, L. L., Antozzi, C., Simoncini, O., Cornelio, F., and Milanese, C. (2004). Anti-MOG autoantibodies in Italian multiple sclerosis patients: Specificity, sensitivity and clinical association. *Int. Immunol.* **16**(4), 559–565.

Markovic, M., Trajkovic, V., Drulovic, J., Mesaros, S., Stojsavljevic, N., Dujmovic, I., and Stojkovic, M. M. (2003). Antibodies against myelin oligodendrocyte glycoprotein in the cerebrospinal fluid of multiple sclerosis patients. *J. Neurol. Sci.* **211**(1–2), 67–73.

Mathey, E., Breithaupt, C., Schubart, A. S., and Linington, C. (2004). Commentary: Sorting the wheat from the chaff: Identifying demyelinating components of the myelin oligodendrocyte glycoprotein (MOG)-specific autoantibody repertoire. *Eur. J. Immunol.* **34**(8), 2065–2071.

Mathey, E. K., Derfuss, T., Storch, M. K., Williams, K. R., Hales, K., Woolley, D. R., Al-Hayani, A., Davies, S. N., Rasband, M. N., Olsson, T., Moldenhauer, A., Velhin, S., *et al.* (2007). Neurofascin as a novel target for autoantibody-mediated axonal injury. *J. Exp. Med.* **204**(10), 2363–2372.

McDonald, W. I., Compston, A., Edan, G., Goodkin, D., Hartung, H.-P., Lublin, F. D., McFarland, H. F., Paty, D. W., Polman, C. H., Reingold, S. C., Sandberg-Wollheim, M., Sibley, W., *et al.* (2001). Recommended diagnostic criteria for multiple sclerosis: Guidelines from the international panel on the diagnosis of multiple sclerosis. *Ann. Neurol.* **50**(1), 121–127.

McHeyzer-Williams, L. J., Malherbe, L. P., and McHeyzer-Williams, M. G. (2006). Checkpoints in memory B-cell evolution. *Immunol. Rev.* **211**(1), 255–268.

Menard, L. C., Minns, L. A., Darche, S., Mielcarz, D. W., Foureau, D. M., Roos, D., Dzierszinski, F., Kasper, L. H., and Buzoni-Gatel, D. (2007). B cells amplify IFN-γ production by T cells via a TNF-α-mediated mechanism. *J. Immunol.* **179**(7), 4857–4866.

Mikaeloff, Y., Caridade, G., Husson, B., Suissa, S., and Tardieu, M. (2007). Acute disseminated encephalomyelitis cohort study: Prognostic factors for relapse. *Eur. J. Paediatr. Neurol.* **11**(2), 90–95.

Mikaeloff, Y., Suissa, S., Vallee, L., Lubetzki, C., Ponsot, G., Confavreux, C., Tardieu, M., and Group, K. S. (2004). First episode of acute CNS inflammatory demyelination in childhood: Prognostic factors for multiple sclerosis and disability. *J. Pediatr.* **144**(2), 246–252.

Miller, J. (2004). The importance of early diagnosis of multiple sclerosis. *J. Manag. Care Pharm.* **10**(3, Suppl. b), S4–S11.

Monson, N. L., Cravens, P. D., Frohman, E. M., Hawker, K., and Racke, M. K. (2005). Effect of rituximab on the peripheral blood and cerebrospinal fluid B cells in patients with primary progressive multiple sclerosis. *Arch. Neurol.* **62**(2), 258–264.

Nielsen, S., Arnulf Nagelhus, E., Amiry-Moghaddam, M., Bourque, C., Agre, P., and Petter Ottersen, O. (1997). Specialized membrane domains for water transport in glial cells: High-resolution immunogold cytochemistry of aquaporin-4 in rat brain. *J. Neurosci.* **17**(1), 171–180.

Noseworthy, J. H., Lucchinetti, C., Rodriguez, M., and Weinshenker, B. G. (2000). Multiple sclerosis. *N. Engl. J. Med.* **343**(13), 938–952.

O'Connor, K. C., Appel, H., Bregoli, L., Call, M. E., Catz, I., Chan, J. A., Moore, N. H., Warren, K. G., Wong, S. J., Hafler, D. A., and Wucherpfennig, K. W. (2005). Antibodies from inflamed central nervous system tissue recognize myelin oligodendrocyte glycoprotein. *J. Immunol.* **175**(3), 1974–1982.

O'Connor, K. C., McLaughlin, K. A., De Jager, P. L., Chitnis, T., Bettelli, E., Xu, C., Robinson, W. H., Cherry, S. V., Bar-Or, A., Banwell, B., Fukaura, H., Fukazawa, T., *et al.* (2007). Self-antigen tetramers discriminate between myelin autoantibodies to native or denatured protein. *Nat. Med.* **13**(2), 211–217.

Owens, G. P., Kannus, H., Burgoon, M. P., Smith-Jensen, T., Devlin, M. E., and Gilden, D. H. (1998). Restricted use of VH4 Germline segments in an acute multiple sclerosis brain. *Ann. Neurol.* **43**(2), 236–243.

Paul, F., Jarius, S., Aktas, O., Bluthner, M., Bauer, O., Appelhans, H., Franciotta, D., Bergamaschi, R., Littleton, E., Palace, J., Seelig, H.-P., Hohlfeld, R., *et al.* (2007). Antibody to aquaporin 4 in the diagnosis of neuromyelitis optica. *PLoS Med.* **4**(4), e113.

Pelayo, R., Tintore, M., Montalban, X., Rovira, A., Espejo, C., Reindl, M., and Berger, T. (2007). Antimyelin antibodies with no progression to multiple sclerosis. *N. Engl. J. Med.* **356**(4), 426–428.

Pistoia, V. (1997). Production of cytokines by human B cells in health and disease. *Immunol. Today* **18**(7), 343–350.

Pittock, S. J., Reindl, M., Achenbach, S., Berger, T., Bruck, W., Konig, F., Morales, Y., Lassmann, H., Bryant, S., Moore, S. B., Keegan, B. M., and Lucchinetti, C. F. (2007). Myelin oligodendrocyte glycoprotein antibodies in pathologically proven multiple sclerosis: Frequency, stability and clinicopathologic correlations. *Mult. Scler.* **13**(1), 7–16.

Pittock, S. J., Weinshenker, B. G., Lucchinetti, C. F., Wingerchuk, D. M., Corboy, J. R., and Lennon, V. A. (2006). Neuromyelitis optica brain lesions localized at sites of high aquaporin 4 expression. *Arch. Neurol.* **63**(7), 964–968.

Poser, C. M., Paty, D. W., Scheinberg, L., McDonald, W. I., Davis, F. A., Ebers, G. C., Johnson, K. P., Sibley, W. A., Silberberg, D. H., and Tourtellotte, W. W. (1983). New diagnostic criteria for multiple sclerosis: Guidelines for research protocols. *Ann. Neurol.* **13**(3), 227–231.

Prineas, J. W., and Graham, J. S. (1981). Multiple sclerosis: Capping of surface immunoglobulin G on macrophages engaged in myelin breakdown. *Ann. Neurol.* **10**(2), 149–158.

Rahman, A., and Isenberg, D. A. (2008). Systemic lupus erythematosus. *N. Engl. J. Med.* **358** (9), 929–939.

Rastetter, W., White, C. A., and Molina, A. (2004). RITUXIMAB: Expanding role in therapy for lymphomas and autoimmune diseases. *Annu. Rev. Med.* **55**(1), 477–503.

Reff, M., Carner, K., Chambers, K., Chinn, P., Leonard, J., Raab, R., Newman, R., Hanna, N., and Anderson, D. (1994). Depletion of B cells *in vivo* by a chimeric mouse human monoclonal antibody to CD20. *Blood* **83**(2), 435–445.

Reindl, M., Linington, C., Brehm, U., Egg, R., Dilitz, E., Deisenhammer, F., Poewe, W., and Berger, T. (1999). Antibodies against the myelin oligodendrocyte glycoprotein and the myelin basic protein in multiple sclerosis and other neurological diseases: A comparative study. *Brain* **122**(11), 2047–2056.

Renaud, S., Fuhr, P., Gregor, M., Schweikert, K., Lorenz, D., Daniels, C., Deuschl, G., Gratwohl, A., and Steck, A. J. (2006). High-dose rituximab and anti-MAG-associated polyneuropathy. *Neurology* **66**(5), 742–744.

Renaud, S., Gregor, M., Fuhr, P., Lorenz, D., Deuschl, G., Gratwohl, A., and Steck, A. J. (2003). Rituximab in the treatment of polyneuropathy associated with anti-MAG antibodies. *Muscle Nerve* **27**(5), 611–615.

Rivera, A., Chen, C.-C., Ron, N., Dougherty, J. P., and Ron, Y. (2001). Role of B cells as antigen-presenting cells *in vivo* revisited: Antigen-specific B cells are essential for T cell expansion in lymph nodes and for systemic T cell responses to low antigen concentrations. *Int. Immunol.* **13**(12), 1583–1593.

Roemer, S. F., Parisi, J. E., Lennon, V. A., Benarroch, E. E., Lassmann, H., Bruck, W., Mandler, R. N., Weinshenker, B. G., Pittock, S. J., Wingerchuk, D. M., and Lucchinetti, C. F. (2007). Pattern-specific loss of aquaporin-4 immunoreactivity distinguishes neuromyelitis optica from multiple sclerosis. *Brain* **130**(5), 1194–1205.

Roncarolo, M. G., Gregori, S., Battaglia, M., Bacchetta, R., Fleischhauer, K., and Levings, M. K. (2006). Interleukin-10 secreting type 1 regulatory T cells in rodents and humans. *Immunol. Rev.* **212**(1), 28–50.

Rosati, G. (2001). The prevalence of multiple sclerosis in the world: An update. *Neurol. Sci.* **22**(2), 117–139.

Sabahi, R., and Anolik, J. H. (2006). B-cell-targeted therapy for systemic lupus erythematosus. *Drugs* **66**(15), 1933–1948.

Sawcer, S., Jones, H. B., Feakes, R., Gray, J., Smaldon, N., Chataway, J., Robertson, N., Clayton, D., Goodfellow, P. N., and Compston, A. (1996). A genome screen in multiple sclerosis reveals susceptibility loci on chromosome 6p21 and 17q22. *Nat. Genet.* **13**(4), 464–468.

Selmaj, K., Raine, C. S., Cannella, B., and Brosnan, C. F. (1991). Identification of lymphotoxin and tumor necrosis factor in multiple sclerosis lesions. *J. Clin. Invest.* **87**(3), 949–954.

Serafini, B., Rosicarelli, B., Magliozzi, R., Stigliano, E., and Aloisi, F. (2004). Detection of ectopic B-cell follicles with germinal centers in the meninges of patients with secondary progressive multiple sclerosis. *Brain Pathol.* **14**(2), 164–174.

Serreze, D., Chapman, H., Varnum, D., Hanson, M., Reifsnyder, P., Richard, S., Fleming, S., Leiter, E., and Shultz, L. (1996). B lymphocytes are essential for the initiation of T cell-mediated autoimmune diabetes: Analysis of a new "speed congenic" stock of NOD.Ig mu null mice. *J. Exp. Med.* **184**(5), 2049–2053.

Shilling, R. A., Bandukwala, H. S., and Sperling, A. I. (2006). Regulation of T:B cell interactions by the inducible costimulator molecule: Does ICOS "induce" disease? *Clin. Immunol.* **121**(1), 13–18.

Singh, R. R. (2005). SLE: Translating lessons from model systems to human disease. *Trends Immunol.* **26**(11), 572–579.

Smith-Jensen, T., Burgoon, M. P., Anthony, J., Kraus, H., Gilden, D. H., and Owens, G. P. (2000). Comparison of immunoglobulin G heavy-chain sequences in MS and SSPE brains reveals an antigen-driven response. *Neurology* **54**(6), 1227–1232.

Song, E., Zhu, P., Lee, S.-K., Chowdhury, D., Kussman, S., Dykxhoorn, D. M., Feng, Y., Palliser, D., Weiner, D. B., Shankar, P., Marasco, W. A., and Lieberman, J. (2005). Antibody mediated *in vivo* delivery of small interfering RNAs via cell-surface receptors. *Nat. Biotechnol.* **23**(6), 709–717.

Sospedra, M., and Martin, R. (2005). Immunology of multiple sclerosis. *Annu. Rev. Immunol.* **23**(1), 683–747.

Storch, M. K., Piddlesden, S., Haltia, M., Iivanainen, M., Morgan, P., and Lassmann, H. (1998). Multiple sclerosis: *In situ* evidence for antibody- and complement-mediated demyelination. *Ann. Neurol.* **43**(4), 465–471.

Stromnes, I. M., and Goverman, J. M. (2006a). Active induction of experimental allergic encephalomyelitis. *Nat. Protoc.* **1**(4), 1810–1819.

Stromnes, I. M., and Goverman, J. M. (2006b). Passive induction of experimental allergic encephalomyelitis. *Nat. Protoc.* **1**(4), 1952–1960.

Stuve, O., Cepok, S., Elias, B., Saleh, A., Hartung, H.-P., Hemmer, B., and Kieseier, B. C. (2005). Clinical stabilization and effective B-lymphocyte depletion in the cerebrospinal fluid and peripheral blood of a patient with fulminant relapsing-remitting multiple sclerosis. *Arch. Neurol.* **62**(10), 1620–1623.

Svensson, L., Abdul-Majid, K.-B., Bauer, J., Lassmann, H., Harris, R. A., and Holmdahl, R. (2002). A comparative analysis of B cell-mediated myelin oligodendrocyte glycoprotein-experimental autoimmune encephalomyelitis pathogenesis in B cell-deficient mice reveals an effect on demyelination. *Eur. J. Immunol.* **32**(7), 1939–1946.

Takahashi, T., Fujihara, K., Nakashima, I., Misu, T., Miyazawa, I., Nakamura, M., Watanabe, S., Shiga, Y., Kanaoka, C., Fujimori, J., Sato, S., and Itoyama, Y. (2007). Anti-aquaporin-4 antibody is involved in the pathogenesis of NMO: A study on antibody titre. *Brain* **130**(5), 1235–1243.

Tanaka, Y., Yamamoto, K., Takeuchi, T., Nishimoto, N., Miyasaka, N., Sumida, T., Shima, Y., Takeda, K., Matsumoto, I., Saito, K., and Koike, T. (2007). A multicenter phase I/II trial of rituximab for refractory systemic lupus erythematosus. *Mod. Rheumatol.* **17**(3), 191–197.

Tenembaum, S., Chamoles, N., and Fejerman, N. (2002). Acute disseminated encephalomyelitis: A long-term follow-up study of 84 pediatric patients. *Neurology* **59**(8), 1224–1231.

Tenembaum, S., Chitnis, T., Ness, J., Hahn, J. S., and the International Pediatric MS Study Group (2007). Acute disseminated encephalomyelitis. *Neurology* **68**(16, Suppl. 2), S23–S36.

The International Multiple Sclerosis Genetics Consortium. (2007). Risk alleles for multiple sclerosis identified by a genomewide study. *N. Engl. J. Med.* **357**(9), 851–862.

The Lenercept Multiple Sclerosis Study Group and The University of British Columbia MS/MRI Analysis Group. (1999). TNF neutralization in MS: Results of a randomized, placebo-controlled multicenter study. *Neurology* **53**(3), 457–465.

Thompson, A., Polman, C., Miller, D., McDonald, W., Brochet, B., Filippi M Montalban, X., and De Sa, J. (1997). Primary progressive multiple sclerosis. *Brain* **120**(6), 1085–1096.

Thrower, B. W. (2007). Clinically isolated syndromes: Predicting and delaying multiple sclerosis. *Neurology* **68**(24, Suppl. 4), S12–S15.

Townsend, S. E., and Goodnow, C. C. (1998). Abortive proliferation of rare T cells induced by direct or indirect antigen presentation by rare B cells *in vivo*. *J. Exp. Med.* **187**(10), 1611–1621.

Trapp, B. D., Peterson, J., Ransohoff, R. M., Rudick, R., Mork, S., and Bo, L. (1998). Axonal transection in the lesions of multiple sclerosis. *N. Engl. J. Med.* **338**(5), 278–285.

Vascotto, F., Le Roux, D., Lankar, D., Faure-Andre, G., Vargas, P., Guermonprez, P., and Lennon-Dumenil, A.-M. (2007). Antigen presentation by B lymphocytes: How receptor signaling directs membrane trafficking. *Curr. Opin. Immunol.* **19**(1), 93–98.

Verge, C., Gianani, R., Kawasaki, E., Yu, L., Pietropaolo, M., Jackson, R., Chase, H., and Eisenbarth, G. (1996). Prediction of type I diabetes in first-degree relatives using a combination of insulin, GAD, and ICA512bdc/IA-2 autoantibodies. *Diabetes* **45**(7), 926–933.

Vincent, A. (2002). Unravelling the pathogenesis of myasthenia gravis. *Nat. Rev. Immunol.* **2**(10), 797–804.

von Budingen, H.-C., Hauser, S. L., Fuhrmann, A., Nabavi, C. B., Lee, J. I., and Genain, C. P. (2002). Molecular characterization of antibody specificities against myelin/oligodendrocyte glycoprotein in autoimmune demyelination. *Proc. Natl. Acad. Sci.* **99**(12), 8207–8212.

von Budingen, H. C., Hauser, S. L., Ouallet, J. C., Tanuma, N., Menge, T., and Genain, C. P. (2004). Epitope recognition on the myelin/oligodendrocyte glycoprotein differentially influences disease phenotype and antibody effector functions in autoimmune demyelination. *Eur. J. Immunol.* **34**(8), 2072–2083.

Warren, K. G., and Catz, I. (1993). Autoantibodies to myelin basic protein within multiple sclerosis central nervous system tissue. *J. Neurol. Sci.* **115**(2), 169–176.

Weinshenker, B. G., O'Brien, P. C., Petterson, T. M., Noseworthy, J. H., Lucchinetti, C. F., Dodick, D. W., Pineda, A. A., Stevens, L. N., and Rodriguez, M. (1999). A randomized trial of plasma exchange in acute central nervous system inflammatory demyelinating disease. *Ann. Neurol.* **46**(6), 878–886.

West, M. A., Lucocq, J. M., and Watts, C. (1994). Antigen processing and class II MHC peptide-loading compartments in human B-lymphoblastoid cells. *Nature* **369**(6476), 147–151.

Wingerchuk, D. M. (2003). Postinfectious encephalomyelitis. *Curr. Neurol. Neurosci. Rep.* **3**(3), 256–264.

Wingerchuk, D. M., Lennon, V. A., Lucchinetti, C. F., Pittock, S. J., and Weinshenker, B. G. (2007). The spectrum of neuromyelitis optica. *Lancet Neurol.* **6**(9), 805–815.

Wingerchuk, D. M., Lennon, V. A., Pittock, S. J., Lucchinetti, C. F., and Weinshenker, B. G. (2006). Revised diagnostic criteria for neuromyelitis optica. *Neurology* **66**(10), 1485–1489.

Wong, F. S., Wen, L., Tang, M., Ramanathan, M., Visintin, I., Daugherty, J., Hannum, L. G., Janeway, C. A., Jr., and Shlomchik, M. J. (2004). Investigation of the Role of B-Cells in Type 1 Diabetes in the NOD Mouse. *Diabetes* **53**(10), 2581–2587.

Wucherpfennig, K. W., Catz, I., Hausmann, S., Strominger, J. L., Steinman, L., and Warren, K. G. (1997). Recognition of the immunodominant myelin basic protein peptide by autoantibodies and HLA-DR2-restricted T cell clones from multiple sclerosis patients. Identity of key contact residues in the B-cell and T-cell epitopes. *J. Clin. Invest.* **100**(5), 1114–1122.

Zhang, Y., Da, R.-R., Hilgenberg, L. G., Tourtellotte, W. W., Sobel, R. A., Smith, M. A., Olek, M., Nagra, R., Sudhir, G., van den Noort, S., and Qin, Y. (2005). Clonal expansion of IgA-positive plasma cells and axon-reactive antibodies in MS lesions. *J. Neuroimmunol.* **167**(1–2), 120–130.

Zhou, D., Srivastava, R., Nessler, S., Grummel, V., Sommer, N., Bruck, W., Hartung, H.-P., Stadelmann, C., and Hemmer, B. (2006). Identification of a pathogenic antibody response to native myelin oligodendrocyte glycoprotein in multiple sclerosis. *Proc. Natl. Acad. Sci.* **103**(50), 19057–19062.

Ziegler, A., Hummel, M., Schenker, M., and Bonifacio, E. (1999). Autoantibody appearance and risk for development of childhood diabetes in offspring of parents with type 1 diabetes: The 2-year analysis of the German BABYDIAB Study. *Diabetes* **48**(3), 460–468.

CHAPTER 5

Human B Cell Subsets

Stephen M. Jackson,* Patrick C. Wilson,*
Judith A. James,* and J. Donald Capra*

Contents

* Oklahoma Medical Research Foundation, Oklahoma City, Oklahoma 73104

Advances in Immunology, Volume 98
ISSN 0065-2776, DOI: 10.1016/S0065-2776(08)00405-7

1. INTRODUCTION

The past decade has witnessed the discovery of a large number of B-lymphocyte populations (subpopulations, subsets). These discoveries have been facilitated by combining multicolor flow cytometric analyses with functional studies that identify other important differences between cells circulating in the peripheral blood and the lymphoid tissues. When used in conjunction, these two approaches have proven to be powerful tools for further subdividing populations of B cells that were once considered to be homogeneous. Early on, lymphocytes were identified as a separate lineage of blood cells (reviewed in Pruzanski and Keystone, 1977; Warner, 1976) and then in the early 1970s, the pivotal separation into B and T lymphocytes fundamentally changed the field (Kwan and Norman, 1974; Nowak, 1983). The next major development was subsetting T cells: early on, largely through the work of Schlossman, T cells were divided into T4 and T8 (later called CD4 and CD8) and major advances followed as several investigators performed experiments that provided powerful insights into the separate functions of these two types of T cells (Evans et al., 1977, 1978; Strelkauskas et al., 1978). The use of the newly discovered technique of monoclonal antibody production, which Schlossman and his colleagues used to separate CD4 from CD8 cells was one of the pivotal developments in the T cell field not only because it uncovered unexpected functions of the various subsets of T cells, but also because it had immediate application in the clinic. As HIV/AIDS was moving to center stage, the ability to quantitate CD4 and CD8 T cells in the clinical laboratory had immediate implications for AIDS diagnosis and treatment not to mention implications for many other diseases. For many years those interested in the relationship between normal and malignant B cells or "diseased states" of the B cell lineage were at a serious disadvantage for the lack of similar reagents that so cleanly differentiated the human T cell subsets. It is fair to say that those who worked in the field of human B cell immunology did not catch up with the T cell aficionados for nearly two decades. Several outstanding reviews are available, including some that are related to murine B cells which will not be emphasized in this chapter (Anolik et al., 2004; Boise and Thompson, 1996; Burrows et al., 1995; Hardy et al., 2007; Kelsoe, 1995, 1996; Liu et al., 1997; Nossal, 1994a,b).

Thus, it was much later that B cells (particularly human B cells) became commonly subdivided. CD5 was an early marker (Antin et al., 1986; Casali and Notkins, 1989; Dono et al., 2004; Youinou et al., 1999) which proved useful, especially to hematologists who were searching for ways to distinguish the various human leukemias and lymphomas. Yet, CD5 was more problematic when used to separate normal human B cells. The introduction of the Bm (B mature) system (use of anti-CD38 and anti-IgD) (Liu and Arpin, 1997; Pascual et al., 1994) changed the field substantially. (CD38 and IgD were also used by Billian et al. (1996) as well as by others.) The contributions of Jacques Banchereau to this field cannot be overstated. Early, Banchereau and Rousset published a paper in *Nature* describing a method to grow human B lymphocytes (the CD40 system) (Banchereau and Rousset, 1991). Later, they and their colleagues contributed the seminal paper for human B cell subsetting (Liu and Banchereau, 1996). They then followed this up with expansion of separation schemes as well as showed that cells could be stimulated to differentiate from naïve to germinal center (Denepoux et al., 1997; Razanajaona et al., 1997). Their work on the initial classification is shown in Fig. 5.1A. Use of anti-IgD and anti-CD38 monoclonal antibodies roughly divided tonsillar lymphocytes into four "quadrants." Subsequently, these were expanded to five Bm subsets: "naïve" B cells (IgD$^+$CD38$^-$) emerging from the marrow were separated by CD23 into the Bm1 (CD23$^-$) and Bm2 (CD23$^+$) subsets. Germinal center B cells (IgD$^-$CD38$^+$) were subdivided by CD77 into centrocytes and centroblasts (Bm3 and Bm4) and finally, the Bm5 subset, IgD$^-$CD38$^-$ (the "double negatives") were shown to be memory B cells. Gradually each of these five subsets became further subdivided (see below). This was the first of the divisions of human B cells that worked both on peripheral blood as well as tonsil (and other secondary lymphoid organs) and was widely used to quantitate human B cell subsets in a variety of diseases including autoimmune diseases. But, most of all, the Bm system was pivotal in assisting in the "reclassification" of human lymphomas (in particular) and leukemias (in many cases). Yet, for all its utility, there were (and still remain) problems.

For example, shortly after the introduction of the Bm system, CD27 expression became commonly used as an exclusive marker for human memory B cells (Agematsu et al., 2000; Maurer et al., 1990). While it has served us well in this regard, there are now several reports that show that the total human CD27$^+$ B cell pool is much more diverse than was previously appreciated and can be subdivided into functionally distinct microsubsets of memory B cells that differ in surface Ig expression, BCR affinity for cognate antigen, mutation frequency of the BCR, and the Ig isotype secreted following in vitro culture (Agematsu et al., 1997, 2000; Dono et al., 2000; Klein et al., 1998; Maurer et al., 1992; Ody et al., 2007; Shi et al., 2003; Steiniger et al., 2005; Tangye and Good, 2007; Weller et al., 2004; Werner-Favre et al., 2001; Yokoi et al., 2003). Furthermore, independent studies by Fecteau and Neron (2003) and Xiao et al. (2004) suggest that CD27 is also expressed by some nonmemory B cells (some naïve B cells and centroblasts). This highlights a potential pitfall in using a single biomarker to enrich for homogeneous lymphocyte populations. Other studies by Fecteau et al. suggested that CD27 does not identify all peripheral blood memory B cells (i.e., some would be omitted by pan-CD27 staining), thereby demonstrating that a single marker is unlikely to identify all cells belonging to a large lymphocyte subset (Fecteau et al., 2006). At first glance, these findings seem to stress the importance of using more markers to subdivide populations. However, more important than quantity is the discriminatory power of the markers that are used. Here we review how improved combinations of commonly used markers during multicolor flow cytometry, followed by in-depth functional assays, have been used to identify novel human B cell populations and distinguish between different populations with partially-overlapping biomarker

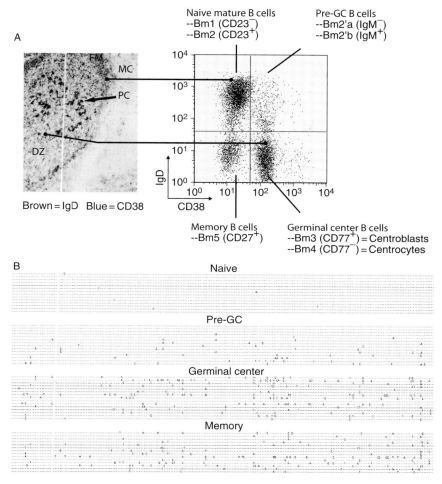

FIGURE 5.1   Bm classification of peripheral B cells. (A). IgD and CD38 were used to first subdivide peripheral B cells into the "four quadrants". CD23 separates IgD$^+$CD38$^-$ naïve B cells into Bm1 (CD23$^-$) and Bm2 (CD23$^+$) subsets. Differential IgM expression identifies IgD$^+$CD38$^+$ Bm2'a (IgM$^-$) and Bm2'b (IgM$^+$) Pre-GC B cells. CD77 distinguishes between Bm3 (CD77$^+$) centroblasts and Bm4 (CD77$^-$) centrocytes in the IgD$^-$CD38$^+$ quadrant. Finally, double negative IgD$^-$CD38$^-$ memory cells are identified as CD27$^+$. Immunohistochemical staining of tonsil cross-sections indicates where each subset commonly resides within lymphnodes. (B). Representative mutation frequencies within naïve, Pre-GC, GC, and memory B cell IgV$_H$ transcripts are indicated. VH4–34 gene segment transcripts are shown. All sequences except for those from GC and memory subpopulations were IgM$^+$. Otherwise, IgG was used. (panel B is derived from Kolar *et al.*). (See Plate 7 in Color Plate Section.)

profiles. Micro-subsetting larger populations has decreased the number of ambiguous, sometimes conflicting immunophenotypes that have perplexed investigators for years, such as the existence of mutations within the IgM^+IgD^+ population a population that was once considered to be entirely naïve B cells.

One cannot underestimate the impact that the complete nucleotide sequence of the human Ig heavy chain locus by the Honjo laboratory had on this field (Matsuda *et al.*, 1998). In 1998 two of the co-authors of this chapter (PW and JDC) wrote an opinion piece in the same issue of the Journal where the study was published stating that "the study of human VH genes has now entered the postgenomics era in which all human bioscience will be propelled in the near future" (Wilson and Capra, 1998). Indeed that has happened in B cell subsetting especially as related to autoimmune disease and lymphomas. In any patient, in any antibody variable region, in any clone of B cells, it is now possible to trace the expressed antibody back to the germline genes from which they arose. Thus, it became possible to finally understand the relationship between B cells that had been through a germinal center reaction from those that had not been subjected to that environment. As we shall see below, that information was critical to both prognostic and treatment issues.

2. HUMAN B CELL SUBSETS: CURRENT SEPARATION SCHEMES

2.1. Advancements in strategies used to subset lymphocyte populations

Cytometric and functional assays are typically combined in at least two ways. The first approach is to simultaneously stain with multiple markers whose expression patterns correlate (not necessarily exclusively) with lymphocyte development (i.e., IgD vs. CD38 as described above), the state of proliferation/cell cycle (e.g., Ki67, CFSE, Hoechst 33342 and pyronin Y), activation state (e.g., CD69 and CD71), predisposition for cell:cell interactions (e.g., CD40, CD25, CD70, CD80/86), and cell survival state (e.g., annexin V and propidium iodide) among others. The ability to design increasingly elaborate multicolor experiments has been facilitated by technological advances in fluorescent cytometry (and microscopy). For example, once previously limited to four colors not including forward scatter (FSC) and side scatter (SSC), it is now possible to simultaneously analyze upwards of 12 colors in a single experiment (or 14 parameters if one includes FSC and SSC) using a fluorescent flow cytometer such as the LSR II (BD Biosciences, San Diego, CA). The additional eight (or more) colors are therefore available to ask more probing questions about

B lymphocyte development and to identify physiological differences between corresponding B cell fractions isolated from normal subjects and patients with disease.

Cytometric and functional assays are uncoupled in the second approach to subdividing lymphocyte populations. Here, fluorescently stained cells are first sorted into one or more separate fractions, and sorted cells are then cultured *in vitro* under conditions designed to highlight functional (and other phenotypic) differences. The drawback to using intracellular markers during fluorometric analyses is they require that cells are prefixed, which limits their use in certain *in vitro* culture experiments. Nonetheless, the benefit of approaches that simultaneously define lymphoid lineage and function (e.g., $CD19^+Ki67$) solely using surface markers is that they are fast, efficient, reproducible, and can be easily applied to a large number of samples (Amu *et al.*, 2007; Jackson *et al.*, 2007a,b; Zanetti, 2007). On the other hand, *ex vivo* experimentation using presorted cells (followed by additional post-culture analyses) allows for a broader range of experiments that can better elucidate how cell immunophenotypes change over time under very specific environmental conditions (Avery *et al.*, 2005; Doucett *et al.*, 2005; Douglas *et al.*, 1998; Stein *et al.*, 1999). Both approaches are routinely used in basic and clinical laboratories for diagnostic and prognostic applications (Nedellec *et al.*, 2005; Ocana *et al.*, 2007; Vorob'ev *et al.*, 2006; Zucchetto *et al.*, 2006b). A recent explosion of information on B cell development is now available due to gene expression profiling, which analyzes potential changes in thousands of patterns in a single experiment (Hoffmann, 2005; Hystad *et al.*, 2007).

The mutational status of Ig transcripts is another approach that is particularly useful for separating B lymphocytes based on their developmental stage. Somatic hypermutation (SHM) alters the Ig V region nucleotide sequence which consequently changes BCR affinity (Liu and Arpin, 1997; MacLennan, 1994a; Wilson *et al.*, 1998). Beneficial changes that improve affinity for foreign antigen provide a better chance for continued development for that individual cell. In general, Ig V region mutation frequency (the number of mutations per Ig V region transcript) cumulatively increases over the course of B cell development such that early naïve B cells are typically unmutated (≤ 1 mutation) and nearly all terminal effector memory and plasma B cells contain mutated transcripts (many with more than 15 mutations per Ig V region transcript) (Jackson *et al.*, 2007b, 2008; Liu and Arpin, 1997; Pascual *et al.*, 1994) (see Fig. 5.1B; derived from Kolar *et al.* (2007)). SHM-associated BCR affinity maturation principally occurs during the germinal center reaction (Allen *et al.*, 2007; Han *et al.*, 1995; Liu and Arpin, 1997). As such, GC B cells collectively have a broader Ig mutation spectrum ranging between unmutated to highly mutated (≥ 20 mutations) (Jackson *et al.*, 2007b; Pascual *et al.*, 1994). Nonetheless, early stage GC B cells (i.e., Pre-GC) tend to have fewer mutations than later GC B cells that are on the verge of developing into the terminal B cell subsets (Jackson *et al.*, 2008). Though mutation status is a good indicator of B cell developmental stage, it is not a reliable marker for predicting B cell survival fate during the GC reaction. For example, Jackson and Capra (2005) showed that mutation frequencies were comparable between living and apoptotic GC B cell fractions that were first subdivided based on annexin V staining and/or residence within tingle body macrophages. Several investigators have also shown that SHM can occur independently of the GC reaction in the extrafollicular microenvironment (Fecteau and Neron, 2003; Kruetzmann *et al.*, 2003; Weill *et al.*, 2004; Weller *et al.*, 2004). Therefore, a fraction of mutated B cells may have never entered the GC stage. Instead, they may developmentally precede GC B cells, or alternatively may develop along a non-GC pathway as they differentiate into terminal antibody secreting and memory B cells. Further investigations are needed to better interpret how Ig mutation status correlates with peripheral B cell development, which in some cases may proceed along multiple pathways.

Another powerful application for the mutational status of Ig V region genes is to assess if the B cell has been through a germinal center reaction. In 1998–1999, a series of papers were published from several laboratories showing that the prognosis of patients with chronic lymphocytic leukemia (CLL) correlated with the extent of somatic hypermutation of V region gene segments (Fig. 5.2A; derived from Damle *et al.* (1999)). Specially, unmutated transcripts were far more likely to be derived from mature B cells of the "naïve" (Bm1, Bm2) subset, and had a poorer prognosis, while mutated transcripts were more likely to derive from more differentiated B cells, such as those derived from Bm3, Bm4, or Bm5 subsets and had a better prognosis. While other markers (like CD38, Fig. 5.2B) could be used to make the same distinction, the sequencing of Ig region transcripts quickly found its way into the clinical laboratory (Damle *et al.*, 1999; Hamblin *et al.*, 1999; Mockridge *et al.*, 2007; Naylor and Capra, 1999; van Zelm *et al.*, 2007).

2.2. A call to further subdivide B cell subsets based on functional differences

Though current binary approaches for subdividing B cells (i.e., the use of only two surface markers) are beneficial to enrich for certain populations, they less effectively distinguish between phenotypically different subsets that share the same surface profile. While this is true for any biomarker assay, overlapping profiles are much more likely to occur when fewer markers are used. As discussed above, the advancement of multicolor flow cytometric analysis will help to further distinguish populations based on surface markers, but unfortunately the ongoing trend is that as

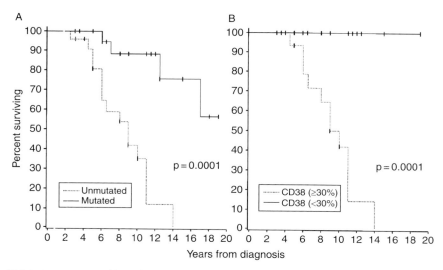

FIGURE 5.2   Survival based on V gene mutation status and CD38 expression. (A) Kaplan–Meier plot comparing survival based on the absence ("unmutated": · · ·) or presence ("mutated": ——) of significant numbers (H2%) of V gene mutations in 47 B-CLL cases (unmutated: 24 cases; mutated: 23). Median survival of unmutated group: 9 years; median survival of mutated group not reached. (B) Kaplan–Meier plot comparing survival based on the detection of CD38 expression.

the number of newly identified B cell populations grows, so does the occurrence of overlapping, nonexclusive immunophenotypes. For example, mutations were identified in the IgD^-CD38^+ B cell fraction over a decade ago (Pascual et al., 1994), but more recently mutations have also been detected in the IgD^+CD38^- B cell pool (Kolar et al., 2007). Therefore, a greater emphasis is now being placed on function-based approaches to subsetting B cells in addition to the traditional surface phenotype approach. More discriminating methods will help to identify and characterize the large number of cells that are not yet accounted. In addition, hopefully, this combined approach will allow a better determination of their niche on the developmental timeline.

Described earlier (Fig. 5.1), the Bm system subdivides peripheral B cells into separate pools based on IgD and CD38 expression. It has been applied in tonsillar (Jackson and Capra, 2005, 2007b; Kolar et al., 2006, 2007; Liu and Arpin, 1997; Pascual et al., 1994; Wilson et al., 1998, 2000) and peripheral blood studies (Bohnhorst et al., 2001a; d'Arbonneau et al., 2006; Pers et al., 2007) and can be used to identify cells during different stages of peripheral development (Fig. 5.3; derived from van Zelm et al. (2007)). Typically, cells within each IgD vs. CD38 quadrant have been named according to the predominating B cell subset that shares the same expression profile. For example, IgD^+CD38^- cells are collectively referred to as "naïve" since naïve B cells are known to express surface IgD and have unmutated Ig V region sequences (Pascual et al., 1994). Similarly, many studies respectively defined IgD^-CD38^-, IgD^-CD38^+, and IgD^+CD38^+ pools as memory, GC, and pre-GC B cells (see above).

However, additional B cell subpopulations have been identified within each IgD vs. CD38 fraction that are phenotypically and functionally distinct from the above subsets. For example, Liu et al. described a population of IgD^+CD38^+ B cells that contained up to 80 mutations in their immunoglobulin Ig VH region (Liu et al., 1996b). Termed IgD^{only} due to their lack of surface IgM, they were postulated to have completed or were in the mid/late stages of the germinal center reaction. Subsequently, they were shown to have greater autoreactive potential (Zheng et al., 2004). These properties are in sharp contrast to IgD^+CD38^+ pre-GC B cells, which average four or fewer mutations per IgV_H transcript (many are unmutated) and are among the earliest cells to initiate the GC reaction (Jackson et al., 2008; Zheng et al., 2004) (Fig. 5.1B). Unlike IgD^{only} B cells, in this subpopulation above-background levels of autoreactivity have not been conclusively demonstrated to date (Arce et al., 2001). Since IgD^{only} and pre-GC B cells have similar IgD vs. CD38 surface profiles, neither marker (whether used alone or simultaneously) accurately distinguishes between these two separate human B cell populations. However, the additional use of IgM effectively delineates IgM^-IgD^{only} and IgM^+ pre-GC B cells and has proven to be a useful marker to distinguish these key subsets of human B cells (see below).

2.3. Addition of IgM and CD27 increases discriminatory power

In this fashion, most studies now include IgM and/or CD27 expression to improve the discriminatory power of population enrichments (Fig. 5.3). This is especially true for naive ($IgM^+IgD^+CD38^-CD27^-$), memory ($IgD^-CD38^-CD27^+$), pre-GC ($IgM^+IgD^+CD38^+CD27^-$), and plasmablast ($IgD^-CD38^{++}CD27^+$) B cells. From the combined use of IgM, IgD, CD38, and CD27, it gradually became appreciated that each of the "original" IgD vs. CD38 fractions is more heterogeneous than was previously thought. For example $IgD^-CD38^-CD27^+$ memory B cells, which presumably represent the terminally differentiated post-GC memory pool, only account for $19\% \pm 15\%$ (average of three tonsils, 8–39% range) of the total IgD^-CD38^- tonsil B cell fraction (S.M. Jackson, unpublished data). This suggests that the majority of the IgD^-CD38^- B cells have yet to be fully

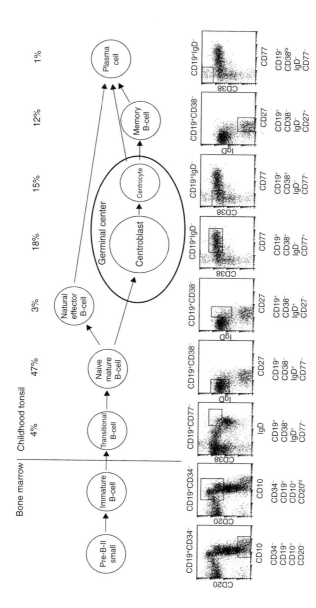

FIGURE 5.3 Scheme of the differentiation stages of mature B cell development. The B cell subsets that contain the intronRSS–Kde rearrangement were isolated from bone marrow and tonsil samples of young children using the indicated markers (derived from van Zelm *et al.*)

characterized. Naïve (65% ± 15%, range 48–74) and pre-GC B cells (61% ± 5%, range 55–65) accounted for larger percentages of their respective IgD vs. 38 fractions. GC, plasma cell, and plasmablast B cells collectively represented 98% ± 3% (range 96–99) of the IgD$^-$CD38$^+$ pool (S.M. Jackson, unpublished data). All together, probably more than a third of all tonsillar B cells are not specifically demarcated by the current antibody combinations used to identify the major B cell subsets, as we know them today. Several investigators such as Klein *et al.* (1993, 1998) and Weller *et al.* (2001, 2004), as well as others show that new populations can be identified simply by using new combinations of IgM, IgD, and especially CD27. Similar systems have been applied to peripheral blood B cells by staining for different combinations of IgD and CD27 with or without IgM (Agematsu, 2004; Klein *et al.*, 1998; Shi *et al.*, 2003). These findings are described below.

2.4. Further subdivision of known B cell subsets identifies distinct micro-populations with different functional properties

The general properties of the four major B cell populations (naïve, GC, plasma cell, and memory) have been well characterized (reviewed in Casola, 2007; Sagaert *et al.*, 2007; Thomas *et al.*, 2006; Youinou, 2007). Recent investigations continue to identify human B cell populations that are phenotypically and functionally similar to more than one major subset. Nonetheless, each newly identified population has been linked with the conventional subset whose overall immunophenotype "best fits" the majority of those exhibited by the new population. Though pragmatically permissible, it must be stressed that such associations may be temporary in some cases. The longevity of each assignment is contingent upon further studies that will presumably determine whether a newly identified population represents developmentally polarized (or transitioning) cells within a single major subset, or should instead be designated as an entirely new population classified by a new name (i.e., the Pro-GC population recently characterized by Kolar *et al.* (2007)).

2.5. Resolving the mysteries surrounding the IgM$^+$IgD$^+$CD27$^+$ population: Memory vs. naïve derivation

In an exemplary case, IgM$^+$IgD$^+$CD27$^+$ B cells have confounded investigators as to their source: whether they represent CD27-expressing naïve cells or IgD-expressing memory cells. On the one hand, their surface Ig expression profile is consistent with resting naïve B cells which are double positive for surface IgM and IgD and do not express CD27. On the other hand, CD27 is commonly thought to be a memory B cell marker

(Agematsu *et al.*, 2000). As such, this population is generally considered to belong to the memory B cell pool. Yet, it should be clear by now that it is impossible to assign $IgM^+IgD^+CD27^+$ B cells to either naïve or memory B cell subsets solely based on positive vs. negative expression of these three markers. Efforts to decipher this conundrum have recently been published and shed essential light on the issue (Klein *et al.*, 1998; Kruetzmann *et al.*, 2003; Tangye and Good, 2007; Weill *et al.*, 2004; Weller *et al.*, 2001, 2004). Most studies conclude that $IgM^+IgD^+CD27^+$ peripheral blood B cells represent memory cells.

This conclusion is based on several observations. First, these cells express CD27, a marker for memory. Pivotal evidence derived from sequence analysis of Ig V regions which conclusively showed that $IgM^+IgD^+CD27^+$ peripheral blood B cells Ig sequences were mutated with a frequency that was comparable to IgD^-CD27^+ memory B cells (Shi *et al.*, 2003; Werner-Favre *et al.*, 2001). Additionally, Kruetzmann *et al.* (2003) demonstrated that $IgM^+IgD^+CD27^+$ B cells participated in the T-independent (TI) response to *Streptococcus pneumoniae* and contained mutated Ig sequences. Mutations were also detected in $IgM^+IgD^+CD27^+$ in patients with X-linked Hyper-IgM syndrome, a disease that is partly characterized by the lack of germinal center formation (Facchetti *et al.*, 1995; Seyama *et al.*, 1998). This suggests that $IgM^+IgD^+CD27^+$ B cells do not necessarily represent post-GC B cells but raised the question as to their source of mutations since somatic hypermutation is thought to only occur during the germinal center reaction. Weller *et al.* presented intriguing data that may resolve this issue. They showed that mutations may be incorporated during early childhood (≤ 2 years of age) before cells mount an immune response to foreign antigen (Weill *et al.*, 2004; Weller *et al.*, 2004). In other words, they suggest the mutations introduced during the development of the preimmune Ig repertoire, as is the case in chickens. Nonetheless, Ig sequences expressed by naïve B cells are almost always unmutated, which suggests $IgM^+IgD^+CD27^+$ B cells are more closely related to memory cells than naïve cells.

Additionally, functional assays showed that sorted $IgM^+IgD^+CD27^+$ B cells could be driven to differentiate into antibody secreting cells when cultured with *Staphylococcus aureus* Cowan strain 1 (SAC) and IL-2 (Himmelmann *et al.*, 2001), producing large quantities of IgM antibodies shortly after stimulation (Himmelmann *et al.*, 2001; Shi *et al.*, 2003). IgG was minimally detected in culture supernatants, but concentrations increased significantly following culture with IL-4 (Werner-Favre *et al.*, 2001). Similar properties were observed among tonsillar IgD^+CD27^+ B cells (Maurer *et al.*, 1992). These findings are much more in line with these cells being memory cells.

Fecteau and Neron demonstrated that CD27 surface expression can be induced in peripheral blood naïve B cells following *in vitro* culture for 7 or more days with anti-CD40 and IL-4 (Fecteau and Neron, 2003). Like memory B cells, naïve cells began to differentiate in Ig-secreting cells producing both IgM and IgG antibodies. Proliferation was a prerequisite for differentiation. This raises the possibility that $IgM^+IgD^+CD27^+$ cells can be derived from naïve B cells responding to T-dependent (TD) antigen (CD27 upregulation required interactions with CD40L). Just as GC B cells can be derived from either naïve cells or recirculating memory cells (Bachmann *et al.*, 1994; Rasmussen *et al.*, 2004; Ridderstad and Tarlinton, 1998), the $IgM^+IgD^+CD27^+$ population may have multiple sources; namely naïve cells participating in a TD response, and memory cells participating in a TI response. Thus, while there remains some ambiguity, the combined use of multiple cell surface markers in addition to Ig V region sequencing, and functional assays provides important insights into this subpopulation of cells, and currently, they appear closest to what we have classically termed memory B cells.

2.6. Innovative combinations of multicolor cytometry, microscopy, and *in vitro* culture discern between candidate GC founder B cells

In two independent investigations, Arce *et al.* (2001) and Bohnhorst *et al.* (2001a) showed that IgD^+CD38^+ B cells (which the authors described as pre-GC B cells) were overrepresented in peripheral blood of patients with systemic lupus erythematosus (SLE) and Sjögren's syndrome (see below). It is unknown whether the postulated early GC B cells in this case played a direct role in disease pathogenesis. There is an increased interest in studying the germinal center reaction to be able to identify and isolate the earliest GC B cell population(s) (Kolar *et al.*, 2007). Once such cells are isolated, experiments can be performed to determine whether the atypically high number of $CD38^+$ peripheral blood B cells in some patients with autoreactive diseases is associated with deregulated CD38 expression (i.e., upregulation on an otherwise $CD38^-$ B cell) and/or some other factor (i.e., premature escape of an early GC B cell from lymphoid tissues). Such knowledge would be important in designing therapies (see below).

Using combined flow cytometric and *in vitro* functional analyses, Kolar *et al.* recently identified and characterized a novel B cell population whose immunophenotype was intermediate to naïve and GC B cells (Kolar *et al.*, 2007). It was hence termed Pro-GC and was postulated to be the earliest candidate naïve-like cell to populate and initiate the GC reaction (Fig. 5.4A and 4B; derived from Kolar *et al.* (2007)). The surface marker profile that identified Pro-GC B cells was identical to that of Bm1 naïve B cells such that they were $IgM^+IgD^+CD38^-CD27^-$. However, Pro-GC B cells differed from Bm1 cells in three physiologically relevant ways. First, forward scatter measurements were significantly higher than those

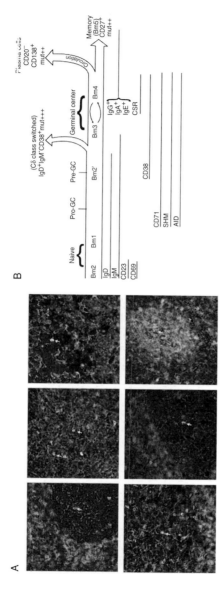

FIGURE 5.4 Pro-GC cells that are located in the germinal center. (A) Anti-IgD (red), anti-AID (green), and anti-CD38 (dark blue) were used to stain frozen tonsillar sections. Lymphoid follicles were identified as naive (red), germinal center (green), pre–germinal center (cyan appearance from combination of red, green, and dark blue), and IgD$^+$CD38$^-$AID$^+$ (yellow appearance from combination of red and green). IgD$^+$CD38$^-$AID$^+$ cells are labeled by a white arrow. Six representative sections are shown. (B) Pro-GC cells most likely exist as a transitional population between naive and germinal center cells because of their cell surface and molecular characteristics described here. We propose that pro-GC cells lie at a critical junction in which AID expression and somatic hypermutation (SHM) begin to take place but the cell has not yet matured with a full complement of germinal center surface markers. Because of cell size and the dynamics of CD69 and CD23 expression in the tonsil subpopulations, we place Bm2 prior to Bm1 in this diagram (derived from Kolar et al.). (See Plate 8 in Color Plate Section.)

for Bm1 cells. Morphological comparisons confirmed the larger Pro-GC size (>30 nm diameter vs. 10–20 for Bm1 cells). Second, Pro-GC B cells were predominantly in the G_1 stage of the cell cycle while Bm1 cells are in the resting G_0 stage (assessed by Hoechst and pyronin Y staining). This suggested Pro-GC cells were relatively more activated and were engaged in RNA/protein synthesis. A larger fraction of Pro-GC B cells were $CD71^+$, which is a commonly used marker of cellular proliferation. Nonetheless, the dearth of S, G_2, and M phase cells suggested they had not yet duplicated their DNA in preparation for cell division. Alternatively, they may have completed one or more rounds of division and were preferentially sustained in the G_1 stage poised to rapidly reenter S phase upon antigen and/or T cell help.

Pro-GC B cells immunophenotypically resembled early GC B cells in several regards. First, their Ig sequences were modestly mutated averaging one to two mutations per IgH transcript. In this regard, Pro-GC B cells were similar to $IgM^+IgD^+CD27^+$ B cells. However, Weller *et al.* (see above) has argued that mutations observed in the latter population could have been introduced early during B cell ontogeny before cells have initiated an immune response (Weller *et al.*, 2004). Thus, mutations within the Pro-GC Ig repertoire could have been introduced during a TD-antigen immune response. Additionally, Kolar *et al.* reported that Pro-GC B cells expressed AID protein (confirmed by immunoprecipitation and Western blot). They also began to express CD77, a marker for germinal center centroblasts.

Perhaps most convincingly, immunohistochemical staining of tonsil sections showed that half of the germinal centers analyzed contained only 1 or 2 centrally located B cells that fit the Pro-GC phenotype (see Fig. 5.4A). These germinal centers were typically smaller suggesting they were in the earlier stages of formation. Pro-GC B cells were not identified in the mantle zone, which is typically rich with resting naïve B cells (Frater and Hsi 2002; Kolar *et al.*, 2007; Shen *et al.*, 2004). Interestingly, Bm1 and Bm2 B cells begin to adopt the surface phenotype of Pro-GC cells when cultured with anti-IgM (BCR-crosslinking) and OKT3-stimulated irradiated T cells. A small percentage (<20%) of cells from both populations developed a $CD71^+CD23^-$ surface phenotype in a T-dependent fashion. Collectively, these findings strongly suggest that Pro-GC B cells are in transition from a naive-like cell towards a GC-like cell that requires B:T communication (Kolar *et al.*, 2007).

2.7. The germinal center reaction

Once naïve B cells enter the germinal center the process of antigen selection, affinity maturation, class switch recombination and further differentiation take place (Kosco-Vilbois *et al.*, 1997; Liu *et al.*, 1997). It has often been said, "life and death decisions are made in the germinal center."

Thus, selection and differentiation in germinal centers is a choice, and mature B cells make that "choice" during the germinal center reaction. There are really three possibilities: the B cell is selected for further rounds of division within the germinal center (presumably due to affinity of the BCR for antigen), and, if there is little or no affinity for antigen, they die, or the B cells progress to differentiation to memory B cells or to plasma cells. In the mouse, Rajewsky's group has tracked germinal center B cells expressing germline immunoglobulin gamma transcripts by conditional gene targeting and provided insights into the class switching process (Casola et al., 2006).

In many ways, an understanding of the fate of B cells as they go through the germinal center reaction is at the heart of subsetting human mature B cells. It was the ability to finally and unambiguously separate germinal center cells from naive and memory cells that allowed investigators in the mid 1990s to finally work with reasonably pure populations of these cells and to study the molecules involved in the process. Early, Banchereau's group pinpointed the inability to produce IL-6 as a functional feature of human germinal center B lymphocytes (Burdin et al., 1996). They postulated that a loss of IL-6 secretion was key, and that they swap from an autocrine to a paracrine IL-6 response that permitted a better control of B cell growth and differentiation during the germinal center reaction. This was one of the first papers to address the signals that resulted in the production of memory B cells or plasmablasts. They followed this up by a series of papers showing that negative selection of human germinal center B cells is accomplished by prolonged BCR cross-linking (Galibert et al., 1996a,b; Martinez-Valdez et al., 1996). In principle, the BCR can transmit either positive or negative signals to the cell leading to either death or activation/survival. The myriad of factors responsible for these "life or death" decisions depends on a variety of things, including (but not restricted to): (a) receptor density, (b) affinity of the receptor for antigen, (c) B cell differentiation stage, (d) duration of BCR occupancy, (e) antigenic valence, etc. This series of experiments pointed to specific roles for CD40, IL-4, IFN, Fas, c-myc, Bax, and CD19 and other factors in the pathway decision and as such provided a launching pad for a myriad of experiments to emerge from several laboratories that elucidated the molecules involved in these processes.

The hallmark of the T-dependent immune response in the germinal center reaction is the onset of somatic hypermutation and class switch recombination. The discovery of AID (activation-induced cytidine deaminase) by Honjo's group was central to studying its role in GC reactions and the ability to isolate this subset of human (and murine) B cells was pivotal for the studies that elucidated both AID's mechanism of action as well as its role in the GC reaction. We now know that the distribution of AID (nuclear vs. cytoplasmic) is important in its function (Cattoretti et al., 2006a).

A controversy that will be discussed in more detail later in this chapter concerns the role of reactivated memory B cells and the role they play in the GC reaction as well as in the eventual production of antibody secreting cells (ACS; plasmablasts, plasma cells, etc). A seminal paper recently described following the migration trails of antigen-responsive B cells in murine lymphoid tissue by isolating Ig transcripts. They concluded that recurring somatic hypermutation progressively drives the Ig repertoire of memory B cells to higher affinities and suggest that the high affinity clones do not arise during a single GC passage, but can be collected during successive recall responses. This is a point of view that has been discussed in detail for decades but is only recently amenable to experimental test (see also Wrammert *et al.* 2008).

Finally, there have been a series of outstanding papers that detail the dynamics of the germinal center reaction by multiphoton imaging. Each of the papers has links to movie clips that are visual masterpieces. Although these studies have largely been done in mice, the mechanisms are likely similar in humans (Cahalan and Parker, 2005; Carrasco and Batista, 2007; Hauser *et al.*, 2007a,b; Miller *et al.*, 2002).

2.8. Plasma cells

Germinal center cells differentiate into plasma cells, which then travel from the secondary lymphoid tissue to the bone marrow where they become the antibody secreting cells of long lasting humoral immunity. The majority of serum antibody is produced by terminally differentiated plasma cells. Apparently, these cells do not participate in any other function that involve other B cells are involved with (antigen processing, presentation, etc). Rather they have been likened to "antibody factories" that fundamentally secret antibodies until they die. It is still not clear how plasma cells are formed. One school argues that memory cells are constantly replenishing the plasma cell pool (that is, rather than deriving from germinal center cells, plasma cells can derive from the memory compartment). Others have shown that plasma cells are long lived and do not need such replenishment (Manz *et al.*, 1997; Slifka *et al.*, 1998). There is now abundant evidence that many plasma cells have a very long half-life (over 3 months). Additionally, there are memory cells that constantly replenish the plasma cell pool. Early work suggested that a subpopulation of human memory B cells generated during antigen responses recirculates to the bone marrow (Paramithiotis and Cooper, 1997). This has been dissected in recent work by our group (Wrammert *et al.*, 2008) (see below). There is also new evidence that only high affinity germinal center B cells form plasma cells (Phan *et al.*, 2006).

2.9. Implications of newly defined B cell subsets for human disease diagnosis and pathogenesis

A discussion of the full implications for the new subsetting of human B cells for human disease is outside the scope of this chapter, yet a few comments are worth mentioning briefly. Above we noted the remarkable association between V_H transcript sequencing and treatment/prognosis in chronic lymphocytic leukemia/lymphoma (Fig. 5.2). A number of papers have fully documented the utility of this approach, and is a remarkable unexpected outcome of understanding human B cell subsets, and the origin of lymphoma (Krober et al., 2006; Wu, 1995). A corollary of this is noted in Fig. 5.5. Here, the result of work in numerous laboratories has illustrated the "normal B cell of origin" of the common (and some uncommon) human B cell malignancies. Certainly there are some cells/malignancies that pose no surprise: plasma cells give rise to myeloma; but other malignancies were rather unexpected. Again, as noted above, the "Ig transcript" sequence ("has a cell been thru a germinal center reaction?") while not perfect, has been of considerable help in directing clinicians into appreciating that some CLLs are indeed more undifferentiated than others, that some needed more aggressive therapy than others, etc. Although there are certainly exceptions, this simple molecular test has been useful in deducing the stage at which malignancy develops and as such has helped to guide treatment.

Similarly, the identification of the B cell of origin of the Burkitt cell lymphoma (Cogne et al., 1988; Haluska et al., 1987; Lister et al., 1996; Riboldi et al., 1994), the mantel cell lymphoma, etc. has helped in determining key treatment decisions for hematologist/oncologists everywhere (Bomben et al., 2007; Cerhan et al., 2007; Dent, 2005; Gonzalez et al., 2007; Moskowitz, 2006; Muller-Hermelink and Rudiger, 2006; Natkunam et al., 2007; Rettig et al., 1996; Robillard et al., 2003; Szczepek et al., 1998). These associations are in considerable flux as with new insights into the molecular basis for some of the malignant lymphomas other, even more powerful means of differentiating them from each other have been developed (Bahler and Levy, 1992; Bentz et al., 1996; Matolcsy et al., 1996). Yet, the classification based on the original idea of IgD and CD38 expression, along with the myriad of additional cell surface markers, clearly led to an explosion in our understanding of the cell of origin of a number of the B cell malignancies. Good reviews on this subject are available (Blade et al., 1994; Blum et al., 2004; Caligaris-Cappio et al., 1997; Kuliszkiewicz-Janus et al., 2005; Kyle and Rajkumar, 2004). The impact of the newer information on lymphoma classification cannot be underestimated (Harris et al., 1994; Hiddemann et al., 1996; Pileri et al., 1995; Pritsch et al., 1997) (see Table 5.1, derived from Craig and Foon (2008)). (Later in this chapter we will address the perturbation of these B cell subsets in autoimmune disease (especially systemic lupus).)

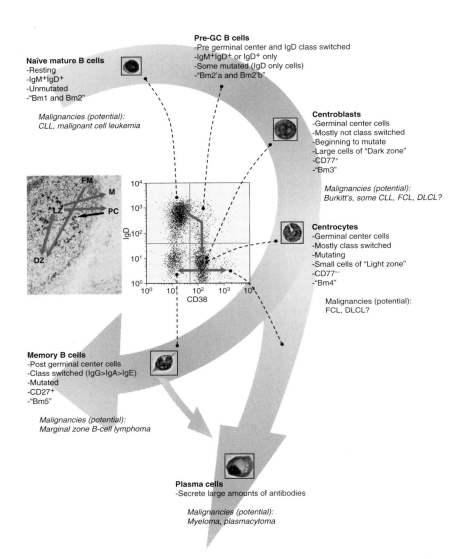

FIGURE 5.5 Hallmark phenotypes of peripheral B cell subsets and their associated malignancies. Properties that are commonly associated with the major peripheral B cell subsets are indicated including Bm classification, mutation status, and cell morphology. The sequential order of how surface IgD and CD38 is thought to change during peripheral development is indicated by red arrows on the IgD vs. CD38 dotplot. B cell migration pathways within the follicular microenvironment are also indicated by red arrows in the immunohistochemically stained tissue section to the left. When appropriate, malignancies that are predominantly associated with a given B cell subset are listed. (See Plate 9 in Color Plate Section.)

TABLE 5.1 Reagents of clinical utility in the evaluation of mature B-cell lymphoid neoplasms (derived from Craig et al.)

Reagent	Normal distribution of staining	Clinical utility in mature B-cell lymphoid malignancy	Comments
CD5	T cells and minor B-cell subset.	Expression on B cells: CLL, MCL.	—
CD10	Immature T cells and B cells, subset of mature T cells and B cells, and neutrophils.	Germinal center–like phenotype: FL, DLBCL, BL. Frequently present in ALL.	
CD19	All B cells, including lymphoblasts, mature B lymphoid cells, and most plasma cells.	Indicates B-cell lineage. May demonstrate abnormal intensity in B-cell neoplasms. Usually absent in plasma cell neoplasms.	Aberrant expression on myeloid cells in AML or MDS.
CD20	Acquired during maturation of precursor B cells (hematogones). Mature B-lymphoid cells positive. Absent on most BM plasma cells. Minor T-cellsubset.	Supports B-cell lineage. Intensity often differs between subtypes: CLL/SLL dim, FL brighter. Aberrant expression on ALL or PCN.	Present on T-cell lymphoid neoplasms.
CD45	All B cells (weaker intensity on precursors and plasma cells), all T cells (weaker intensity on precursors).	Useful in distinguishing mature lymphoid neoplasms (bright intensity) from ALL and PCN (weak intensity to negative).	—
Kappa and lambda, surface	Mature B cells.	Immunoglobulin light chain restriction.	—

(continued)

TABLE 5.1 (continued)

Reagent	Normal distribution of staining	Clinical utility in mature B-cell lymphoid malignancy	Comments
CD9*	Precursor B cells, activated T cells, platelets.	Precursor B-cell ALL.	—
CD11c*	Some B cells, some T cells.	Hairy cell leukemia CD11c (_br).	Frequent weaker expression on CLL, MCL and others.
CD15*	Myeloid and monocytic cells.	May be aberrantly expressed in B-cell neoplasia.	More frequently seen in ALL
CD22	Cytoplasmic expression in early B cells. Surface expression acquired during maturation of precursor B cells.	Indicates B-cell lineage in ALL and mature lymphoid neoplasms. Intensity often differs between subtypes of mature B-cell neoplasm: CLL/SLL dim.	Cross reactivity of some clones with monocytes and basophils.
CD23*	Weak intensity expression on resting B cells and increased with activation.	Distinguish CD5_B-cell lymphoid neoplasms: CLL/SLL (_br).	—
CD25*	Activated B cells and T cells.	Hairy cell leukemia in combination with CD11c and CD103.	—
CD13*	Myeloid and monocytic cells.	May be aberrantly expressed in B-cell neoplasia.	More frequently seen in ALL than in mature neoplasm.

Marker			
CD33*	Myeloid and monocytic cells.	May be aberrantly expressed in B-cell neoplasia.	More frequently seen in ALL than in mature neoplasm. Also AML.
CD34*	B-cell and T-cell precursors and myeloblasts.	ALL.	
CD38*	Precursor B cells (hematogones), normal follicle center B cells, immature and activated T cells, plasma cells (bright intensity), myeloid and monocytic cells, and erythroid precursors.	Bright intensity staining may indicate plasmacytic differentiation. Prognostic marker in CLL/SLL.	
CD43*	T cells, myeloid, monocytes, small B-cell subset.	Aberrant expression in CLL, MCL, some MZL.	—
CD58*	Leukocytes including bright intensity staining of precursors and decreased intensity with maturation.	Distinction of ALL from normal precursor B-cell (hematogones) including detection of MRD.	—
CD79a and b*	Cytoplasmic staining in precursor B cells, plasma cells positive, variable expression mature B cells.	Indicates B-cell lineage in ALL and mature lymphoid neoplasms. Intensity often differs between subtypes of mature B-cell neoplasm: CLL/SLL dim CD79b.	CD79a staining has been reported in some T-ALL and rare mature T-cell lymphoid neoplasms.

(continued)

TABLE 5.1 *(continued)*

Reagent	Normal distribution of staining	Clinical utility in mature B-cell lymphoid malignancy	Comments
CD103*	B-cell subset, intramucosal T cells.	Hairy cell leukemia and some MZL.	Also EATCL.
FMC-7*	B cells.	Distinguish CD5_ lymphoid neoplasm: CLL_, MCL often positive. Also HCL_.	
bcl-2*	T cells, some B cells; negative normal germinal center cells.	Distinguish CD10_ lymphoid neoplasms: FL_, BL_.	Variable staining in DLBCL.
Kappa and lambda, cytoplasmic*	Plasma cells.	Light chain restriction in cells with plasmacytic differentiation.	Most flow cytometric assays detect surface and cytoplasmic Ig.
Zap-70*	T cells, NK cells, precursor B cells.	Prognostic marker in CLL/SLL.	—
TdT*	B-cell and T-cell precursors.	ALL.	Also some AML.
cIgM*	First Ig component in precursor B cells. Expressed by subset of plasma cells and mature B cells.	IgM producing neoplasms that might be associated with Waldenstrom macroglobulinemia	—

Reagents included in this table were recommended in the consensus guidelines. _, indicates usually positive; _, usually negative; br, bright or strong intensity; Ig, immunoglobulin; TdT, terminal deoxynucleotidyl transferase; cIg, cytoplasmic immunoglobulin; and —, not applicable.
* These reagents may be considered for secondary evaluation, after other reagents listed have been used in the initial evaluation.

3. TRANSITIONS IN PERIPHERAL B CELL DEVELOPMENT

3.1. Introduction

Though we know many of the properties characteristic of different B cells that participate in autoreactive and neoplastic diseases (antibody secretion/proliferation), it has been difficult to accurately identify the specific cells responsible for the onset of disease. More and more attention has focused on the potential culpability of B cells that are in transition between developmental and functional states. The highest frequency of transformation from a normal to diseased condition appears to happen among peripheral B cells that developmentally lie between the resting naïve and resting memory stages, neither of which are excluded from implication (Tiller *et al.*, 2007; Tsuiji *et al.*, 2006). Investigators are now taking a closer look at "transitioning" B cells in humans and their pathophysiologic significance in autoreactive and neoplastic diseases. This is because many of the biomolecular processes that occur during cellular transitions (i.e., from a non-proliferating to rapidly dividing state) have been shown to cause disease when deregulated (Hipfner and Cohen, 2004; Truffinet *et al.*, 2007; Wang and Boxer, 2005). Subsetting B cell populations using some of the increasingly elaborate schemes described above has the potential to identify disease-causing cells that would shed light on the events leading to disease.

B cells undergo multiple types of transition in the periphery, including developmental (i.e., from one subset to the next proximal subset), activation-based (i.e., between resting and active states), and survival-based (i.e., from life to death). All three are usually interlinked since developmental transitions (i.e., GC→memory) require the prior receipt of activation-inducing stimuli under the proper conditions. Failed or inappropriately received signals can result in cell death or anergy. In B cells, the central factor affecting cellular transitions is the B cell receptor (BCR) and its related signal strength with which the BCR recognizes foreign and self antigen (Dorner and Lipsky, 2006; Healy and Goodnow, 1998). BCR signal strength is extracellularly influenced by the avidity and affinity for cognate antigen. The number of Ig molecules on the cell surface governs avidity. Therefore, surface Ig density is one parameter that can be used to identify cells undergoing BCR-induced transitions.

BCR affinity is determined by the antibody sequence. During the germinal center reaction, somatic hypermutation alters the nucleotide sequence, which consequently changes BCR affinity. Beneficial changes that improve affinity for foreign antigen provide a better chance for continued development for that individual cell. On the other hand, mutated GC B cells that no longer bind the eliciting foreign antigen typically undergo apoptosis. Therefore, Ig sequence and mutation status can also be used to subdivide B cells based on the outcome of BCR:antigen interactions.

Mutations can also change a previously benign BCR sequence to one that recognizes self (Bobeck *et al.*, 2007; Brard *et al.*, 1999; Portanova *et al.*, 1995; Putterman *et al.*, 2000; Richter *et al.*, 1995; Shlomchik *et al.*, 1987; Yu *et al.*, 2003). Human diseases with autoreactive features are nonetheless relatively rare considering the surprisingly high percentage of auto-specific BCRs within the GC (Koelsch *et al.*, 2007; Wilson *et al.*, 2000), memory (Tiller *et al.*, 2007; Tsuiji *et al.*, 2006), and naïve (Munakata *et al.*, 1998) (antigen-inexperienced) B cell Ig repertoires. The low incidence of disease is primarily due to counter-selection against such B cells in the form of cell death or clonal anergy. However, pathology likely occurs when B cells that bind self-antigen transition from a resting to an active state, begin to proliferate, and develop into autoantibody-secreting cells. Dysregulation at any point can lead to cell death in the best scenario, or disease in the worst. Therefore, BCR-mediated selection is another parameter that can distinguish between cells with different survival fates or cells with the potential for causing disease. There is a growing appreciation for the need to identify "transitioning" B cells. *Ex vivo* experimentation on pure populations will not only improve our biomechanistic understanding of the processes associated with each transition point but will hopefully lead toward therapeutics against specific autoreactive and neoplastic diseases. Understanding these transition states of B cells will be critical for an appreciation of the rest of this chapter, as these transitional cells are clearly implicated in many aspects of B cell function we will be covering below.

3.2. Markers for micro-subsetting B cell populations to identify cells in transition

Advances in fluorometric-based methods for micro-subsetting have been (and will be) accomplished by the use of new markers (Ody *et al.*, 2007) and by simply taking a closer look at conventional markers in unconventional ways (Jackson *et al.*, 2007a,b, 2008). Certain markers such as CD19 and CD4 generate histograms with clearly defined positive and negative fractions. However, others such as CD69 produce a more spectral or bell-curved distribution pattern due to the variable levels of activation. In this case, there has been a shift toward categorizing cells based on incremental increases or decreases in expression, each identified as negative (−), low (−/+), intermediate (+/−), positive (+), and bright (++). This approach to drawing cytometric gates has a greater potential for identifying cells in transition where surface expression gradually increases or decreases (Arce *et al.*, 2004).

3.3. CD38 as a marker for transition

Reviewed in Malavasi *et al.* (2006) and Morabito *et al.* (2006) CD38 is a type-II transmembrane glycoprotein. Its extracellular domain enzymatically functions as a glycohydrolase converting nicotinamide adenine dinucleotide (NAD^+) to cyclic ADP-ribose. cADPR is a secondary messenger for Ca^{2+} mobilization. Within the peripheral B cell pool, CD38 is primarily expressed on germinal center and antibody-secreting cells. CD38 is upregulated on the surface of $CD38^-$ B cells (including naïve and memory) after stimulation with IL-10, IL-2, and/or IFN type I and II T cell factors (Avery *et al.*, 2005; Bauvois *et al.*, 1999). CD38 does not physically associate with CD19 (Kitanaka *et al.*, 1997) or the BCR (Lund *et al.*, 1996). Nonetheless, Lund *et al.* demonstrated that the BCR was required for efficient CD38 signaling (Lund *et al.*, 1996). CD38-crosslinking led to increased kinase activity (especially Fyn and Lyn) (Kitanaka *et al.*, 1997; Yasue *et al.*, 1997) and an early influx of calcium (Lund *et al.*, 1996). Ca^{2+} mobilization is an early response to receptor signaling and potentiates several downstream events. It is therefore not surprising that CD38 is multifunctional and influences cellular processes including activation, proliferation, apoptosis, and protein tyrosine phosphorylation (Funaro *et al.*, 1997; Malavasi *et al.*, 2006; Mehta *et al.*, 1996; Silvennoinen *et al.*, 1996; Yokoyama, 1999). Anti-CD38 antibodies also suppressed lymphopoeisis (Kitanaka *et al.*, 1996; Kumagai *et al.*, 1995).

These observations (many of which were made after the widespread use of CD38 in the Bm system) help interpret, and in some cases, predict functional differences between cells that differentially express CD38. For example, in an important paper, Arce *et al.* (2004), showed that $CD38^{LO}$ IgG-secreting B cells resembled early plasma cells, were primarily found in tonsils, and increased their surface CD38 expression during transition towards a mature plasma cell phenotype. In this case, differential CD38 expression could segregate cell populations based on environmental location and the degree of maturation. Similar findings were reported by Avery *et al.* showing that a subset of $CD27^+$ memory B cells upmodulated surface CD38 after extended culture with anti-CD40, Il-2, and Il-10 (Avery *et al.*, 2005). Interestingly, those cells that retained high surface CD27 were immunophenotypically similar to antibody secreting cells; an observation supported by higher Blimp-1/Xbp1 and reduced Pax5 transcription factors. This pattern was less apparent in the $CD27^{LO}$ fraction which has more of a germinal center-like phenotype (i.e., increased Bcl-6) (Allman *et al.*, 1996). Therefore, the relative expression levels of CD38 and CD27 could be used to identify B cells that are at different stages of transition from a memory cell towards either a high affinity antibody-secreting cell, or a recirculating in or very early mutated founder GC B cell (Pro-GC cell). Ten years ago, both were hypothetical populations. Now we know that discriminatory power is greater when both markers are used in conjunction.

As in normal B cells, CD38 functions as a signaling molecule in B cell chronic lymphocytic leukemia (CLL) (Deaglio *et al.*, 2003) and has been linked with disease pathogenesis (Deaglio *et al.* 2006) (See above). This may stem from its ability to enhance BCR responses by lowering its signaling threshold (Lund *et al.* 1996). CD38 is commonly used as a prognostic marker (Boonstra *et al.*, 2006; D'Arena *et al.*, 2001; Ibrahim *et al.*, 2001; Mhes *et al.*, 2003) in patients with newly diagnosed CLL and generally correlates with increased cell proliferation and shorter patient survival. Its effectiveness as a biomarker increases when combined with additional factors including CD20 (Hsi *et al.*, 2003), CD49a (Zucchetto *et al.*, 2006a), ZAP-70 (Hus *et al.*, 2006; Schroers *et al.*, 2005), and Ig mutation frequency (Bagli *et al.*, 2006; Damle *et al.*, 1999; Del I *et al.*, 2006; Krober *et al.*, 2002; Matrai *et al.*, 2001).

3.4. A new role for CD45 in B cell subsetting

CD45 is expressed by all lymphoid lineage cells and is commonly used to subset T cell populations. To date, there are surprisingly few reports that utilize CD45 expression to subdivide B cells. This is in part due to CD45 being considered as a pan-B cell marker, with presumably inconsequential variations in expression level. We recently published findings that differential CD45 expression can be used as a marker for separating B cells based on their state of activation and proliferation, survival fate, and stage of BCR-mediated selection (Jackson *et al.*, 2007a,b, 2008). Additionally, using CD45 GC B cells can be subdivided into separate fractions enriched for cells with significantly different mutation frequencies. This ability has been one of the long sought goals of those interested in the GC reaction (Jackson *et al.*, 2007a,b, 2008; Kolar *et al.*, 2007).

3.5. CD45 exists as multiple isoforms with concerted functions

CD45 is a protein tyrosine phosphatase and regulates the activity of multiple kinases (especially Src family kinases). It exists as several different isoforms that are generated by alternative splicing of the long ~220 kDA CD45RA (RA) pre-mRNA (Heyd *et al.*, 2006; Lemaire *et al.*, 1999; Sarkissian *et al.*, 1996). Isoform size is determined by the combination of exons (A, B, and C) that are excluded from the final transcript. CD45RO (RO) is the shortest and excludes all three exons. CD45RB (RB) excludes exon A and may exclude exon C in some cases. To date, isoform-specific functions have yet to be clearly distinguished. Dawes *et al.* (2006) and Tchilian and Beverley (2006) recently published findings that lymphocyte immunophenotypes better correlate with the ratio of CD45 isoforms that are expressed. In other words, different isoforms appear to work in concert to produce a given phenotype that can change with respect to the relative surface density of each isoform (Fukuhara *et al.*, 2002; Hathcock *et al.*, 1993; Matto *et al.*, 2005; McNeill *et al.*, 2004; Tchilian *et al.*, 2004). The influence of a given isoform can vary between B and T cells. For example, mature peripheral T cells develop in CD45RO^{only} transgenic mice, but the peripheral B cell pool does not (Fleming *et al.*, 2004; Ogilvy *et al.*, 2003; Tchilian *et al.*, 2004). Findings by Novak *et al.* (1994) and Hermiston *et al.* (2003) suggested that CD45RO plays a role in T cell deactivation and influences signaling thresholds.

3.6. CD45 and T cells

As noted above, relative CD45 isoform expression has been extensively investigated in T cells where its been shown to influence cell development, activation and differentiation (Alexander, 2000; Holmes, 2006; Huntington and Tarlinton, 2004; McNeill *et al.*, 2004). Simultaneous staining with RA and RO (henceforth, CD45RA will be designated RA; CD45RO will be designated RO and CD45RB will be designated RB) showed that resting naïve T cells are predominantly RA$^+$RO$^-$. In contrast, the RA$^-$RO$^+$ T cell fraction is generally activated and enriched for memory and other effector cells (Young *et al.*, 1997). Interestingly, an intermediate RA$^{+/-}$RO$^{+/-}$ T cell fraction was identified that was shown to be in a state of bidirectional transition toward either an RO$^-$ resting or an RO$^+$ activated stage (Deans *et al.*, 1992; Hamann *et al.*, 1996; Kristensson *et al.*, 1992). Collectively, these studies showed that differential CD45 isoform expression profiles could reproducibly subdivide the total peripheral T cell pool into separate subsets that differed in their developmental stage (naïve vs. memory) and activation state (RO$^-$ vs. RO$^+$). While a few investigators have made correlations with CD45RA in human B cells in patients with leukemia or autoimmune diseases (Dawes *et al.*, 2006; Rodig *et al.*, 2005; Yu *et al.*, 2002) there has been no comprehensive attempt to use CD45 as a tool to subdivide human B cells. What follows is a synopsis of our studies in this regard which parallel the studies that have been done in T cells. Most RO-associated phenotypes observed among T cells were similarly observed among B cells including increased activation and transitional development towards alternative subsets.

3.7. CD45 and B cells

Described above, BCR-crosslinking (plus T cell stimulatory factors) can induce cell activation that promotes cell proliferation, diversification, and differentiation towards downstream subsets. The outcome of BCR-mediated positive and negative selection is governed in part by BCR signal strength. As in T cells, CD45 plays a critical role in each of these processes in B cells and affects BCR signaling thresholds (Cyster, 1997;

Cyster *et al.*, 1996; Hermiston *et al.*, 2003; Huntington *et al.*, 2006; Rodig *et al.*, 2005). A better understanding of the how activation and diversification correlate with differential CD45 isoform expression would allow us to identify and sort B cells engaged in processes with high pathogenic potential (i.e., auto-specific BCRs generated via SHM). Once accomplished, physiologically relevant B cell fractions can be sorted and cultured *in vitro* under conditions that could potentially counteract anomalous cellular responses that lead to the onset of B cell malignancies and autoimmunity (Dahlke *et al.*, 2004; Tchilian and Beverley, 2006). We recently reported that a small, but significant percentage of GC B cells (<5%) expressed high levels of surface RO. Surface RO expression increased sevenfold within 3 h of treating sorted GC B cells with anti-IgM to crosslink their BCR receptors. This suggested that at least a fraction of RO-expressing GC B cells successfully bound antigen presented by accessory cells *in vivo*. If valid, surface RO levels could be used to identify the subset of GC B cells that are positively selected to differentiate into terminal effector memory and plasma cells.

3.8. CD45RO enriches for highly mutated GC B cells with increased signs of selection

The ability to sort live populations of GC B cells would be a resource for investigating how mutations potentially lead to disease. In an early study, Pascual *et al.* (1994) showed that Ig mutation frequencies varied according to B cell developmental stage such that naïve B cells contained unmutated IgV_H transcripts and essentially all memory B cells contained mutated IgV_H transcripts. GC B cell mutation frequencies spanned the complete spectrum ranging from unmutated to highly mutated. Mutated GC B cells have been implicated in the pathogenesis of several autoreactive and neoplastic diseases prompting attempts to specifically identify and isolate the live mutating fraction. Initial strategies used CD77 (globotriaosylceramide glycolipid, or Gb3) as a distinguishing marker to separate human GC B cells into two subsets, $CD77^+$ centroblasts and $CD77^-$ centrocytes (Bailey *et al.*, 2005; Hogerkorp and Borrebaeck, 2006; Liu *et al.*, 1996a; Pascual *et al.*, 1994). SHM is primarily associated with centroblasts while selection principally occurs during the centrocyte developmental stage (Cattoretti *et al.*, 2006b; Liu *et al.*, 1996a,c; MacLennan, 1994b). Remarkably, a number of mutation analyses showed that mutation frequencies were indistinguishable between $CD77^+$ centroblasts and $CD77^-$ centrocytes (Jackson and Capra, 2005; Kimoto *et al.*, 1997; Pascual *et al.*, 1994).

Studies by Zhou *et al.* (2003) demonstrated that CD45 negatively regulates AID expression. AID regulates SHM and plays a critical role in initiating and perpetuating the mutation process (Muramatsu *et al.*, 1999, 2000). We therefore hypothesized that GC B cells that expressed different surface levels of one or more CD45 isoforms would also differentially express AID that would correlate with differences in somatic mutation frequencies. Thus, to investigate, we examined IgV_H mutation frequencies varied among GC B cells that differentially expressed surface CD45 (see Fig. 5.6A–D; derived from Jackson *et al.* (2007a,b, 2008)). Where CD77 was of limited use in separating B cells containing Ig transcripts with different mutation levels, CD45RO separated the total GC B cell subset into smaller RO^- and RO^+ fractions with distinctly different mutation frequencies. For example, mutation frequency increased concordantly with RO upregulation (Jackson *et al.*, 2007a,b). This pattern was consistently observed between both IgM and IgG sequences (Fig. 5.6A). The discriminatory power was greatest when RO was used in conjunction with CD69, the well-known activation marker. These findings suggested strongly that extracellular surface markers could effectively and reproducibly separate normal human GC B cells into pools enriched for low or higher mutation averages.

The majority of RO^+ GC B cells were centrocytes demonstrating that RO also correlated with differences in B cell developmental stage (Jackson *et al.*, 2007a,b). This however made it difficult to interpret whether higher mutation frequencies were due to differences in SHM activity (i.e., RO^+ cells mutate more frequently) or due to phenotypic differences between centroblasts and centrocytes (i.e., positive selection of centrocytes whose higher mutation frequency confers selective advantages during interactions with T cells and/or follicular dendritic cells). We observed that RO^+ centrocytes were significantly more mutated than their RO^- counterpart was (Jackson *et al.*, 2007a) (greater percentage of highly mutated IgV_H transcript sequences; Fig. 5.6B). In contrast, RO^- and RO-expressing centroblasts were comparably mutated. This suggested that the RO-associated increase in mutations was linked with BCR-mediated selection. Consistent with this hypothesis, IgM surface density and the ratio of replacement to silent mutations (R:S ratio) were higher among RO^+ centrocytes. Surface IgM also increased among RO^+ centroblasts, but their R:S ratio pattern was inverted (higher in the RO^- fraction).

3.9. Somatic hypermutation most-likely occurs during the RO^- GC B cell stage

The RO^- GC B cell fraction displayed several characteristics that were consistent with ongoing somatic hypermutation (summarized in Fig. 5.7). Prerequisites for the initiation and persistence of SHM included that cells expressed AID and were proliferating (indicated by positive Ki67 expression) (Muramatsu *et al.*, 1999, 2000; Muramatsu, 2006; Okazaki *et al.*, 2003). Both AID (mRNA transcripts) and Ki67 (intracellular protein) expression were higher among RO^- GC B cells than their RO^+ counterpart, suggesting

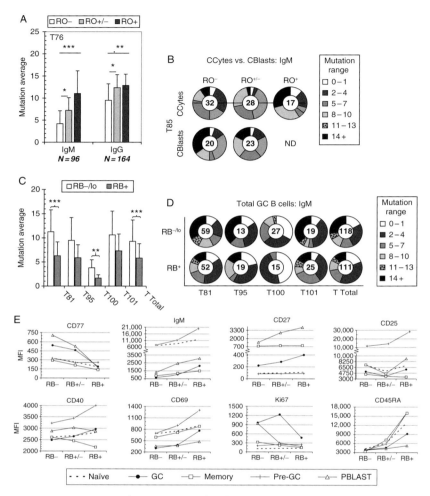

FIGURE 5.6 CD45RO and RB expression level subdivides peripheral B cells based on differences in mutation frequency, activation, signs of selection, and terminal differentiation. (A) GC B cells were sorted into RO⁻, RO⁺ᐟ⁻, and RO⁺ fractions based on their surface RO expression. The average number of mutations per IgV$_H$4 IgM and IgG transcript was calculated for each fraction (data from a representative tonsil is shown; T76). Asterisks indicate the level of statistical significance between inter-sample comparisons, such that *, **, and *** represent $p \leq 0.05$, ≤ 0.01, and 0.001, respectively. N-values for RO⁻ (white bar), RO⁺ᐟ⁻ (grey bar), and RO⁺ (black bar) fractions are: IgM (31, 24, 41) and IgG (76, 30, 58), respectively. (B) IgM Mutation frequencies were similarly assessed for RO-subdivided centrocyte (CCytes) and centroblast (CBlasts) fractions. The mutation spectrum (0–14⁺ mutations) was subdivided into six groups (3 bp increments). Each clone was assigned to a group based on its mutation frequency. Data depicted are from one representative tonsil, T85. ND means RO⁺ CB B cells were not sorted in this analysis. (C) Mutation frequencies were calculated for RB-subdivided GC B cells as in

SHM was more likely to occur during the RO⁻ stage (Jackson *et al.*, 2007b). However, it was not possible to rule out SHM activity among RO-expressing GC B cells since AID and Ki67 were still detected, though at a reduced level. It was possible that proliferating RO^+AID^+ cells were involved in other AID-associated processes, including class switch

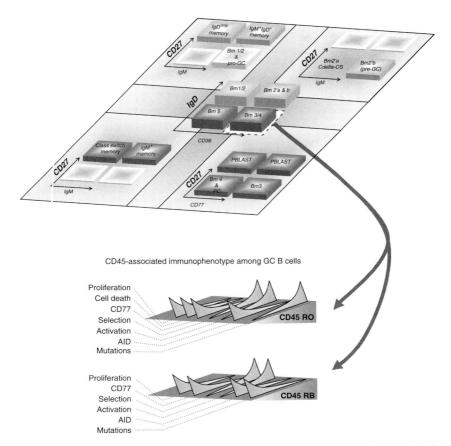

FIGURE 5.7 Summary Slide: Surface marker combinations used to identify peripheral B cell subsets and CD45-associated immunophenotype of GC B cells. (See Plate 10 in Color Plate Section.)

panel A. Note the inverse correlation between SHM frequency and RB expression. (D) The mutation spectrum was determined for RB-subdivided GC B cells using the same approach described in panel B. (E) Indicated B cell subsets were subdivided based on RB surface expression. Mean fluorescence intensities (MFI) were determined for markers that were partially indicative of selection (CD77, IgM), late stage development or terminal differentiation (CD27), potential interactions with accessory cells (CD25, CD40), and activation/proliferation (CD69, Ki67).

recombination (CSR). Investigations by Liu *et al.* (1996a), suggested that SHM precedes CSR, but our recent findings are difficult to reconcile with this view. Though rare, we repeatedly isolated unmutated IgG sequences suggesting that CSR can precede SHM. The RO-associated increase in mutations also applied to IgG transcripts (some of which were clonally related) suggesting that SHM can reinitiate in class-switched B cells.

RA$^+$RO$^-$ resting naïve T cells upregulate surface RO (RA$^{+/-}$RO$^{+/-}$) in response to activation-inducing stimuli during "transition" towards memory and effector cells. We therefore hypothesized that GC B cells that upregulated surface RO were in the midst of transition, whether from one physiological state to the next (i.e., proliferating to nonproliferating), or differentiation to other B cell subsets (Jackson *et al.*, 2007a). Interestingly, CD27 expression was higher among RO$^+$ GC B cells suggesting they may be differentiating into early CD27$^+$ plasmablasts or CD27$^+$ memory B cells. Further studies are needed to address this issue more directly. Nonetheless, our findings showed that RO-expressing GC B cells, especially RO$^+$, exhibited characteristics that were consistent with transitioning cells, including increased activation (CD69), enrichment for CD77$^-$ centrocytes, and a decreased propensity for cell death (Annexin V$^-$).

3.10. CD45RB provides new insights into B cell subsetting approaches

Expression levels of different CD45 isoforms typically change during lymphocyte activation and developmental processes. Though isoform-specific functions have yet to be conclusively determined, consistent changes in isoform surface expression have been reproducibly associated with changes in lymphocyte immunophenotypes (described above). In some cases, the strength of association is greatest when it corresponds with simultaneous changes in multiple surface CD45 isoforms (Dawes *et al.*, 2006; Horgan *et al.*, 1994). For example, experiments by Horgan *et al.* demonstrated that differential CD45RB expression could be used to separate the CD4$^+$CD45RO$^+$ memory T cell pool into two Mem1 (RO$^+$RB^{bright}) and Mem2 (RO$^+$RB^{intermediate}) fractions (Horgan *et al.*, 1994). If used alone as a distinguishing marker, high RB expression did not exclusively identify memory T cells. Anti-RB antibodies also stained a fraction of naïve RB^{bright} T cells. Interestingly, most naïve T cells are also RO$^-$ (Young *et al.*, 1997) which substantiates that in some cases, CD45RO and CD45RB can be independently expressed on a given cell since. Taken together, these data show that CD45RO and RB relative surface expression levels can be used to enrich for lymphocyte fractions with different immunological properties.

We recently reported that like T cells, a fraction of peripheral B cells also independently expressed CD45RB and CD45RO on their surfaces

(Jackson *et al.*, 2008). To date, there are relatively few reports that shed light on the physiological significance of differential RB expression among B cells. Potential influences were inferred from reports in T cells where high RB expression correlated with T cell selection in the thymus (Lucas *et al.*, 1994; Wallace *et al.*, 1992). Described above, we showed that high CD45RO expression directly correlated with activated GC B cells that had most likely undergone BCR-mediated selection (Jackson *et al.*, 2007a,b). We therefore investigated whether GC B cell selection also differentially correlated with RB expression, as it does in T cells. We also evaluated potential RB correlations with various other immunophenotypes including somatic hypermutation, cell activation, and proliferation (Fig. 5.6C–E), each of which varied with respect to relative RO surface expression.

The majority of RO-expressing GC B cells were also positive for RB (88%) (Jackson *et al.*, 2008), which led to the hypothesis that some of the properties associated with RO expression would also be observed within the RB$^+$ fraction. Like CD45RO, high CD45RB expression reproducibly correlated with GC B cell selection and differentiation towards terminal effector B cell stages (Fig. 5.6E). For example, compared with their RB$^-$ counterpart, RB$^+$ GC B cells had elevated surface Ig levels, were enriched for CD77$^-$ centrocytes, had a higher percentage of CD27$^+$ cells, and had elevated CD25 and CD40 surface expression (as measured by mean fluorescence intensity). Collectively, these data suggested that RB$^+$ GC B cells were either better positioned to, or had already received T cell/follicular dendritic cell secondary help during the GC reaction. Accordingly, RB$^+$ GC B cells (or a micropopulation within this fraction) may represent cells that are differentiating towards post-GC B cell stages including CD27$^+$ memory B cells.

3.11. Summary

In summation, by adding CD45RO and CD45RB analyses to the current strategies for identifying human GC B cells (see Fig. 5.7), one introduces a "temporal-gauge" whereby cells can be separated into pools enriched for early modestly mutated cells and highly mutated cells (later stage, i.e., more rounds of division) that showed increased signs of having undergone selection, or at least were better prepared to undergo selection (increased surface Ig expression). Surface RO and RB isoforms are also independently expressed by other non-GC B cell fractions (Jackson *et al.*, 2008) and may correlate with other physiologically relevant phenotypes that may be nonexistent in GC B cells (and thus may have been missed in these tonsillar GC studies). Collectively, investigations by Jackson *et al.* (2007a,b, 2008) document that the addition of surface RO expression to the panel of reagents currently used to discriminate between human B cell subsets provides new insights into the relationship between SHM and B cell selection, and how they lead to terminal B cell differentiation. These findings should ultimately lead to new therapeutic approaches to counteract defects that may occur during transitional development, activation, proliferation, and diversification.

4. MONOCLONAL ANTIBODIES FROM HUMAN B CELL SUBSETS: INSIGHTS INTO AUTOREACTIVITY AND FUNCTION

4.1. Introduction

The growing divisions of mature human B cell subsets have begged the question: what are the functions of these many B cell types? Is a particular B cell type serving a distinctive role in immunity or is it autoreactive, and as such, is the phenotype related to tolerance? Early on it became apparent that the specificity of the B cell receptor must influence cell fate and provide the most direct evidence of function. However, until very recently, analysis of B cell specificity was hampered by the need to immortalize B cells to determine their specificity. B cells were immortalized through fusion with neoplastic partners to produce hybridomas (Kohler and Milstein, 1975) or by using oncogenic viral infection (Steinitz *et al.*, 1977). Although hybridoma technology revolutionized biology and medicine, few questions involving B cell function were answered because the technological processes are inefficient and in some cases create a bias, since some B cell subpopulations cannot be easily transformed. Thus, while monoclonal antibodies can be generated from many purified B cell subpopulations using fusion/hybridoma technology, they are very difficult—if not impossible—to generate from some rare B cell types. This hindrance was particularly true for human B cells that are much more resistant to immortalization than mouse B cells.

4.2. It's in the specificity: Advances from analyses of recombinant monoclonal antibodies from humans

A major advance came with the realization that both the heavy and light chain variable region genes could be identified from a single B cell using a cocktail of PCR primers. Simultaneously, flow cytometry became capable of isolating single B cells of any phenotype into PCR plates. By combining these technologies, the variable region genes could be cloned from single human B cells of any type and then expressed in mammalian expression systems. This enabled investigators to learn of the specificity of each single B cell, regardless of transformation potential or the ability to survive manipulation. Since the procedure captured over 90% of individual B cells, bias was eliminated. Briefly, single B cells are sorted by flow cytometry from which the variable region genes are then isolated by RT-PCR and cloned into expression vectors. The expression vectors are then transfected into 293 human embryonic kidney cells generating monoclonal antibodies (Fig. 5.8). The associated (boxed) anecdote describing the first antibodies that we generated by this approach in Michel Nussenzweig's laboratory provides some background for this discovery.

This powerful technology has now opened the door to a plethora of new information and the prospects for many fully human therapeutic antibodies. While many contributed to the various steps along the way, the champion of this "complete" technology was Michel Nussenzweig who, with his fellows Hedda Wardemann and Eric Meffre, first used the approach to illustrate that most new immature B cells bore autoreactive receptors (Wardemann *et al.*, 2003). These autoreactive B cells were progressively reduced during maturation until the naïve B cell stage. Surprisingly, 20% of mature naïve B cells (Bm1, Bm2) in humans maintain a significant level of autoreactivity, including a subset of autoreactive B cells recently subjected to receptor editing that persisted into the naïve pool (Meffre *et al.*, 2000, 2004). Conversely, analysis of single-cell antibodies demonstrated that mature but naïve (Bm1 and Bm2) B cells of systemic lupus erythematosus (SLE) patients—either with active disease (Yurasov *et al.*, 2005) or when asymptomatic (Yurasov *et al.*, 2006)—maintain high frequencies of autoreactive BCR expression similar to the preselected B cell repertoire. Similarly, Meffre and colleagues found that rheumatoid arthritis patients had a surprising abundance of antigen-inexperienced naïve B cells that expressed autoantibodies (Samuels *et al.*, 2005). These observations

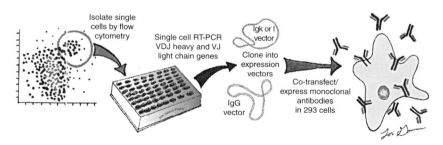

FIGURE 5.8 Strategy for generating recombinant mAbs from the variable genes of sorted single B cells Single B cells are sorted by flow cytometry and the VH and Vκ or Vλ genes amplified by reverse transcription and PCR with primers to the variable gene leader region and constant regions. The identified light and heavy chain genes are then cloned into an expression vector and cotransfected into 293 cells. Artwork by Lori Garman, OMRF. (See Plate 11 in Color Plate Section.)

illustrate the important concept that lupus and rheumatoid arthritis patients have an apparent global disruption of immune tolerance evident even at the earliest stages of B cell selection.

In our studies, we found that normal memory B cells that are not class switched (are IgM$^+$) have variable gene repertoires reminiscent of naïve B cells, whereas IgG$^+$ cells differ markedly by several criteria, including VH and JH genes used and CDR3 length, suggesting that IgG but not IgM memory cells have been particularly selected to avoid production of autoantibodies (Zheng *et al.*, 2004). Therefore, it was quite surprising when Tsuji and colleagues in the Nussenzweig group demonstrated that IgM memory B cells rarely express anti-nuclear antigen (ANA)-reactive or polyreactive BCRs (Tsuiji *et al.*, 2006), less so than naïve B cells. In contrast, Tiller, *et al.*, also from the Nussenzweig group (Tiller *et al.*, 2007) as well as our own laboratory (Koelsch *et al.*, 2007) independently found that IgG$^+$ memory B cells are more often autoreactive than naïve B cells, though it should be noted that the affinity for autoantigens was quite low. Tiller and colleagues found that these surprising observations likely result in large part because of accumulated somatic mutations that alter the BCR specificity and thus, to some degree, possibly negate the selective processes during primary B cell development that remove autoreactive BCR from the repertoire (Tiller *et al.*, 2007). These results also suggest that there are different selective criteria dictating allowable levels of autoreactivity for newly formed B cells versus those that are recruited into immune responses. It seems obvious that the IgG antibodies that are often ANA-reactive in healthy people are not pathological (because healthy people are not ill). Thus, it will be interesting to learn what level of autoantigen binding or what specificity of an autoantibody from a lupus patient, for example, causes autoimmune pathology. There is also the persistent dilemma of how does a B cell "know" that it is autoreactive versus binding a foreign antigen, particularly if the cell is contributing to an immune response and acquires it's autoreactivity due to a somatic mutation. The textbook model is that B cells binding external antigens will present foreign peptides to T cells and receive costimulation, whereas those that bind self-antigens will not receive T cell help and so die by attrition. However, presumably some B cells with receptors that become autoreactive by accumulating somatic mutations will maintain binding for a foreign antigen, thus allowing presentation of peptides and T cell help. Why do these cells not generate autoantibodies in healthy people? Maybe autoreactive B cells also present self-peptides to regulatory T cells whose negative regulatory signals can trump any activating stimulation or cause active deletion of the autoreactive B cell. These topics are of central importance to our understanding of B cell selection.

A number of other interesting observations have been made using recombinant monoclonal antibodies. For example, Wardemann and colleagues found that most autoreactive antibodies in the naïve human B cell repertoire could be eliminated by exchanging their light chains, particularly if lambda light chains are expressed (Wardemann *et al.*, 2004). Thus, lambda light chains appear to be the "better editors." Also, by expressing recombinant monoclonal antibodies, the Meffre laboratory found that expression of Bruton's tyrosine kinase (Btk) (Ng *et al.*, 2004) is essential for proper human B cell tolerance. Meffre's laboratory also recently demonstrated that CD40 ligand and class II HLA (Herve *et al.*, 2007) are each essential for proper human B cell tolerance. This important study illustrated that like generation of secondary B cell immunity B:T cell interactions and peptide presentation to T cells are also important processes for tolerizing B cells. Interestingly, using recombinant antibodies, Meffre, in collaboration with the Chiorazzi laboratory, found that most human B cell CLL specimens are autoreactive (Herve *et al.*, 2005). This surprising observation for the first time linked autoreactive specificity to malignant transformation and led to fascinating new models for how B cells could become cancerous (Chiorazzi *et al.*, 2005). Further work by the Nussenzweig group, by his now independent trainees and by their collaborators regarding B cell selection is detailed in an excellent recent review article (Yurasov and Nussenzweig, 2007).

4.3. Mature B cell populations involved in immune tolerance

In recent years our laboratory has focused on characterizing mature B cells in healthy people that are autoreactive. Because they are autoreactive in healthy people, we surmise that these B cells must be selected or differentiate due to a mechanism of immune tolerance. For example, analyses in mice have revealed various autoreactive specificities that lead to inactivation of B cells, or anergy (reviewed in Cambier *et al.* (2007)), the classic model being that of the anti-HEL/HEL mouse from the Goodnow laboratory (Goodnow *et al.*, 1989). These anergic B cells have a unique cell surface phenotype that allows their isolation by flow cytometry. We predicted that a similar B cell population could be isolated from humans, linking the past 25 years of studies in transgenic mice to human biology. Thus, we could finally determine the specific role of immune tolerance mechanisms in the cause or avoidance of autoreactivity directly in humans. For example, we can ask whether particular autoreactive B cell populations become activated more often and produce the autoantibodies that cause lupus pathology. These cells can then be targeted to treat SLE. Another possibility is to determine how autoreactive B cells are induced to obtain phenotypes such as those associated with anergy and therefore could limit their role in immune responses. Treatments that enhance or delete these populations (depending on the circumstances) could be developed to treat autoimmune diseases.

The hurdle to this point has been that there was no simple way to learn the fate of autoreactive B cells because we did not know which cells were autoreactive. For murine models, this obstacle was removed years ago by the generation of transgenic mice expressing only self-reactive BCRs. Now we can carefully dissect the BCR specificity from single cells of a given human B cell type, leading to the discovery of two new types of B cell that express autoreactive BCRs. Thus, a central focus of our laboratory has been to identify and characterize B cell populations in healthy people that are autoreactive and therefore likely to be involved in mechanisms of immune tolerance. The first is the human counterpart to anergic B cells. Secondly, we have characterized the specificity of the enigmatic IgD (Cδ) class switched (or Cδ–CS) B cell population and learned that it too is autoreactive and likely generated through a mechanism of immune tolerance (Koelsch *et al.*, 2007; Zheng *et al.*, 2004).

To date, we have used two approaches in successfully identifying these two autoreactive B cell populations in humans. First, there is some predictive value to simply sequencing the immunoglobulin variable region genes of various B cell types. Over the years our laboratories have sequenced over 7,000 variable region transcripts from some 20 unique, mature human B cell populations for a variety of studies (based on cell surface phenotype) (Jackson and Capra, 2005; Jackson *et al.*, 2007a, b; Koelsch *et al.*, 2007; Kolar and Capra, 2004a,b; Kolar *et al.*, 2004, 2006, 2007; Pascual *et al.*, 1992a, 1994, 1997; Wilson *et al.*, 1998, 2000; Zheng *et al.*, 2004, 2005). There are certain variable region genes or characteristics of the antibody repertoire that are either counter-selected during B cell development in conjunction with removal of autoreactive cells or associated with lupus and other autoimmune diseases. One example is the utilization of the VH4–34 gene segment that encodes a natural autoreactivity associated with pathological cold-agglutinin disease (Li *et al.*, 1996; Pascual *et al.*, 1991, 1992b; Potter *et al.*, 1993, 1999; Thompson *et al.*, 1991). Subsequently, we and others demonstrated that VH4–34 expressing B cells have a limited role in secondary immune responses in healthy people (Pugh-Bernard *et al.*, 2001; Zheng *et al.*, 2004) and others have shown that VH4–34[+] cells appear to be unregulated in SLE patients (Cappione III *et al.*, 2005). Other characteristics of the variable gene repertoire associated with autoreactivity include (i) increased use of the JH6 gene segment that is found more often in the preselected (pre-B cell) repertoire (Elagib *et al.*, 1999; Meffre *et al.*, 2000; Wardemann *et al.*, 2003); (ii) increased average CDR3 length and the frequency of B cells with long third-hypervariable regions in mouse and human SLE and also found in the preselected repertoire (Klonowski *et al.*, 1999; Link and Schroeder, Jr., 2002; Meffre *et al.*, 2000; Wardemann *et al.*, 2003); (iii) increased frequency of CDR-localized charged amino acids (Barbas *et al.*, 1995; Link and Schroeder, 2002; Meffre *et al.*, 2004; Radic and Weigert, 2004; Shlomchik *et al.*, 1990, 2004; Wardemann *et al.*, 2003);

and (iv) bias for distal Jκ gene segment utilization suggestive of receptor editing (Meffre et al., 2000; Prak and Weigert, 1995). All of these characteristics of the sequence of a variable region give clues as to its origin and function, although no single antibody sequence can be certain to have a particular origin or function based on the sequence these are best seen as population phenomena.

Using these criteria, we reported that antibodies from B cells of healthy people that are class switched to IgD via genetic recombination (Cδ-CS) display each of these unusual variable gene repertoire characteristics such as predominant use of the JH6 gene segment and long CDR3 elements (Fig. 5.9A and B), or frequent charged amino acid residues within the antigen-binding surface (Zheng et al., 2004). During these analyses, we also noted that an atypical subset of naïve B cells that have down-regulated expression of surface IgM while retaining IgD ("B_{ND}" cells for B Naïve IgD) have a similar variable gene repertoire with increased use of the JH6 gene and particularly long CDR3 elements (Fig. 5.9A and B).

As noted above, though suggestive, analysis of the variable region repertoire alone does not conclusively prove that a B cell population is autoreactive. To directly test for autoreactivity, we produced recombinant monoclonal antibodies from the variable region genes of single cells sorted by using flow cytometry. As indicated in Fig. 5.9C, a commercial immunosorbant assay for ANA binding in wide use in the clinical laboratory to assist in the diagnosis of lupus found that over half of the Cδ-CS and B_{ND} antibodies were more strongly ANA reactive than antibodies from naïve B cells (based on average absorbencies to the ANA assay). Analysis of sera or antibodies for binding to HEp-2 cells adhered to microscope slides and detected by immunofluorescence is a classic test of autoreactivity and ANA binding that is commonly used to diagnose autoimmune diseases such as SLE (Bradwell et al., 1995). Over half (56%) of the antibodies derived from B_{ND} cells and 60% of Cδ-CS B cells were reactive with human HEp-2 cells by using immunofluorescent staining and microscopy (Fig. 5.9D). The HEp-2 binding patterns for both the B_{ND} and Cδ-CS cells were varied, indicating an assortment of autoreactivities (examples of B_{ND} antibody reactivities are presented in Fig. 5.9D). Another hallmark of autoreactivity in lupus is the production of antibodies that react to DNA (Koffler et al., 1969; Gay et al., 1993; Radic and Weigert, 1995; Stollar et al., 1986). Antibodies reactive with single-stranded DNA (ssDNA) are more common in autoimmune conditions than in healthy people but are not diagnostic of a pathological state. However, antibodies to double-stranded DNA (dsDNA) are generally indicative of pathology. We found that both the B_{ND} and Cδ-CS populations were often anti-DNA reactive (Fig. 5.9E). In total over 60% of both B_{ND} and Cδ-CS cells were ANA$^+$ or bound DNA, and thus unlike IgM$^+$IgD$^+$ naïve or IgG memory B cells, most B_{ND} and Cδ-CS B cells are

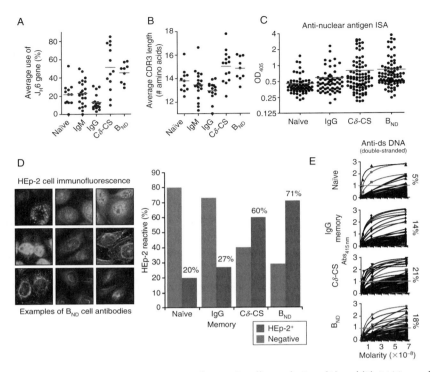

FIGURE 5.9 Detecting autoreactivity in human B cell populations (A) and (B). JH6 is used more often, and the CDR3 elements are longer in autoreactive B cells. Each point represents a single donor from which an average of 40 VH genes was sequenced. Red lines = mean. (C) Monoclonal antibodies were expressed from the variable region genes of sorted single B cells that were amplified by single cell RT-PCR. Antibodies included 100 from Cδ-CS GC B cells (IgD$^+$IgM$^-$CD38$^+$) isolated from three donors ($n = 45, 34$, and 21 antibodies by donor) or 94 from B$_{ND}$ B cells (IgD$^+$IgM$^{-/low}$CD27$^-$) from four donors ($n = 27, 21, 15$, and 30 by donor), and were compared to 78 antibodies from naïve B cells from three donors (IgD$^+$IgM$^+$CD38$^-$, $n = 37, 27$, and 14 antibodies by donor), and 57 antibodies from IgG memory cells from three donors (IgD$^-$IgM$^-$CD27$^+$, by donor $n = 25, 12, 20$). Antinuclear-antigen reactivity was measured by immunosorbant assay (Bion, inc.). (D) Micrographs of 9 HEp-2+ autoantibodies from B$_{ND}$ cells (left). Frequency of HEp-2 binding antibodies (right). The increased frequencies of both B$_{ND}$ and Cδ-CS cells that were HEp-2 reactive was highly significant when considering all antibodies as a whole (φ^2, $p<0.0001$) and when considered by average frequency between donors (t test $p<0.05$ for either B$_{ND}$ or Cδ-CS versus naïve cell antibodies). (E) Saturation curves plus histograms of anti-DNA antibodies. Red lines in anti-DNA antibodies 3H9, squares, and HN241. (See Plate 12 in Color Plate Section.)

autoreactive. Importantly, the reactivity of the B$_{ND}$ cells to autoantigens was *natural* as the variable genes harbor no somatic mutations (since these derive from a "naïve" pool, by definition they have not entered a germinal center reaction). In contrast, the Cδ-CS cells have excessively mutated variable region genes (Arpin *et al.*, 1998; Liu *et al.*, 1996b; Zheng *et al.*, 2004) and we demonstrated that for some antibodies the autoreactivity could be removed by expressing the variable region genes without any mutations (Koelsch *et al.*, 2007). These two populations of B cells, B_{ND} and Cδ-CS are each similarly surface IgD^{+} but IgM^{-} and express autoreactive BCRs, linking sole expression of surface IgD with autoreactivity. However, the similarity between these B cell subsets ends at that. Cδ-CS B cells apparently derive from immune responses, have highly mutated variable region genes (Liu *et al.*, 1996b) and are immunoglobulin class switched at the genetic level (White *et al.*, 1990; Yasui *et al.*, 1989). They also have quite distinct phenotypes. Cδ-CS B cells are more accurately classified as a B cell lineage than as a subtype because these cells can be found as germinal center (Liu *et al.*, 1996b), memory (Klein *et al.*, 1998), plasma cells (Arpin *et al.*, 1998; Zheng *et al.*, 2004) and even as a distinct form of myeloma (Arpin *et al.*, 1998). Conversely, B_{ND} cells are a subset of naïve, antigen inexperienced B cells by every measure. As such, we postulate that B_{ND} cells may be anergic B cells that have been limited in their role for participating in immune responses and so we are currently characterizing their signaling capacity. The autoreactive Cδ-CS B population has been the subject of much additional speculation by many laboratories.

4.4. The enigmatic Cδ-CS B cell population

The "Cδ-CS" for "Cδ class switched" have also been called the "Bm2′b" (Liu *et al.*, 1996b) population or "IgD-only" (Zheng *et al.*, 2004) B cells. For a recent review of the class switch process see Chaudhuri and Jasin (2007). These cells arise in humans at approximately the same frequency as IgE or IgG4 class switched populations (about 1% of B cells) (Koelsch *et al.*, 2007; Zheng *et al.*, 2004). Unlike normal Cμ versus Cδ usage involving differential splicing of a single VDJ-Cμ-Cδ transcript, Cδ-CS B cells actually class switch from Cμ to Cδ at the genetic level using cryptic switch regions between the Cμ and Cδ exons (Fig. 5.10A). Interestingly, Cδ-CS appears to occur via two mechanisms involving either a standard class switch between the μ switch region and a cryptic δ switch region, or by what appears to be a homologous recombination mechanism involving identical 443-bp sequence motifs (σμ) located 5′ of the μSR and 3′ of the cryptic δ switch region (Fig. 5.10A) (White *et al.*, 1990; Yasui *et al.*, 1989). Cδ-CS GC cells can differentiate into IgD-secreting PCs (Arpin *et al.*, 1998; Zheng *et al.*, 2004) and memory cells (Klein *et al.*, 1998), and have the unusual characteristics that they use >90% lambda light chains (Arpin *et al.*, 1998) and accumulate more somatic mutations than any other B cell population (Liu *et al.*, 1996b). Also, we have shown that this population has the highest frequency of receptor edited heavy (Wilson *et al.*, 2000) and light chain (Zheng *et al.*, 2004) V regions. As described above, we have found

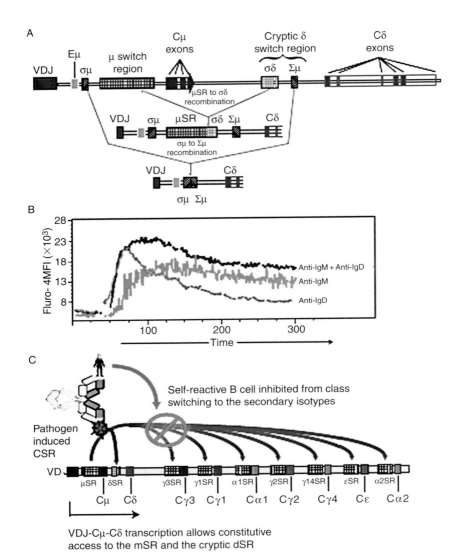

FIGURE 5.10   Cδ-CS B cells class switch to IgD. A. Immunoglobulin CSR from Cμ to Cδ can occur in a minor population of GC B cells and has been detected in PCs and in a class of IgD plasmacytomas. μ-to-δ CSR occurs via recombination between the μSR or a nearby cryptic μSR sequence (σμ) to cryptic SRs in the Cμ-Cδ intron (σμ and σδ), resulting in permanent loss of Cμ. B. Flux of calcium in CD27⁻IgM⁺IgD⁺ B cells differs both qualitatively and quantitatively when anti-IgM versus anti-IgD alone is used to crosslink the BCR. Thus, Cδ-CS B cells may have limited capacity to signal in a fashion analogous to anergy. C. *Switch-inhibition hypothesis*: B cells previously or currently tolerized due to expression of autoreactive IgM/D receptors (panel C, man icon) are inhibited from class switch specifically involving the secondary switch regions

that Cδ-CS B cells use Ig variable regions associated with autoreactive or counter-selected B cell populations (Zheng *et al.*, 2004) and indeed express predominantly autoreactive B cell receptors (Koelsch *et al.*, 2007). These links to autoreactivity and tolerance lead to the intriguing possibility that the Cδ-CS lineage arises due to a mechanism of tolerance causing class switch from Cμ to Cδ for B cells with autospecific surface Ig. Alternatively, there might be selective survival of Cδ-CS cells that are autoreactive. Each of these possibilities is explored below.

It is intriguing to consider the possibility that the Cδ class switch results from an active process. We have devised two models that could cause autoreactive B cells to class switch to IgD. Both models involve molecular mechanisms that can be tested. The first is simple: there may be a directed mechanism whereby the cryptic Cδ switch region is targeted in activated human B cells that are self-reactive. Such a mechanism may have evolved to rescue autoreactive B cells for some specialized or reduced role in immunity involving the IgD isotype. There has been speculation for many years that the elusive function of IgD may be uniquely involved in self tolerance (reviewed in Preud'homme *et al.* (2000)). In support of this possibility, we have noted that when compared to IgM, IgD crosslinkage of naïve human B cells results in greatly reduced calcium flux sustainability (Data unpublished, Fig. 5.10B). However, analyses of transgenic mice expressing IgD or IgM alone as autoantibodies suggested normal B cell tolerance (Brink *et al.*, 1992) although these studies did not specifically involve autoreactive B cells. In addition, a number of studies suggest that the function of IgD is to augment IgM signaling with a role instead in B cell positive selection into the functional repertoire (reviewed in Preud'homme *et al.* (2000) and Geisberger *et al.* (2006)).

A second more complex model is that autoreactive B cells may be inhibited from class switching to the downstream Ig isotypes (IgG, IgA, IgE), leaving only the cryptic Cδ switch region available for switch in cells that previously bound self antigen (Fig. 5.10C). Class switch recombination is believed to be controlled by regulated sterile transcription of the various switch regions (e.g., transcription of the γ1-SR rather than α-SR will induce IgG1 rather than IgA class switch). In our second model, class switch to the downstream switch regions may be inhibited for autoreactive human B cells upon activation by antigen, possibly as a manifestation

---

(red stop symbol). Therefore when autoreactive B cells are activated by foreign antigen the only switch partner available to the CμSR and recombination machinery is the cryptic switch regions (cδSR) between the Cμ and Cδ exons (detailed in panel A) leading to the predominantly autoreactive Cδ-CS B cell lineage as indicated by the small blue arrow. (See Plate 13 in Color Plate Section.)

of anergy to avoid the more pathogenic IgG autoantibodies. However, Cμ-SR to Cδ-SR recombination may be unaffected because these signal sequences are constitutively accessible due to transcription via the variable region gene promoter to produce the VDJ-Cμ-Cδ transcript. Thus, the only class switch option for autoreactive B cells may be to IgD. The cryptic nature of the Cδ-SR may bias most cells that are not autoreactive to switch in a normal fashion to IgG, leaving the Cδ-CS lineage predominantly autoreactive and as rare as it is observed. This model is attractive because unlike direct class switch to IgD for B cells, "class switch inhibition" suggests a global process that would exclude most autoreactive B cells from the IgG pool with Cδ-CS cells resulting only rarely.

Both of the above models are interesting because they each suggest that tolerizing mechanisms can directly control immunoglobulin class switch. Clearly, such a mechanism would be beneficial because most pathological autoantibodies in diseases such as SLE and rheumatoid arthritis are of the IgG class. However, as suggested above, it is not clear what *level or manner* of autoreactivity would elicit this kind of mechanism. In other words, at what affinity and to what particular autoantigen is one versus another mechanism of peripheral immune tolerance induced. Although we have observed selective changes in the B cell repertoire comparing IgG to IgM B cells that suggest avoidance of IgG-autoreactivity (Zheng *et al.*, 2004), the work of Tiller and colleagues (Tiller *et al.*, 2007) and by our laboratory (Koelsch *et al.*, 2007) has demonstrated that IgG$^+$ memory B cells can frequently bind self antigens, although with quite low affinity.

An alternative model is that class switch to IgD is random such that any cell can switch to IgD at some low frequency. Subsequently, there is selection of the Cδ-CS B cells after class switch, causing preferential survival of the autoreactive clones. It has been suggested that compared to IgM receptors alone, expression of IgD provides preferential survival (Brink *et al.*, 1994) and recruitment of B cells into immune responses. Thus, expression of IgD alone may allow rescue by chronic stimulation with self-antigen (Roes and Rajewsky, 1993). This tendency could be exaggerated because IgD expression alone on the B cell surface may not signal in a fashion that causes B cell death if stimulated by self-antigens. For example, it has been suggested that IgD internalization differs from IgM internalization such that IgD-bound antigens may not be processed in a similar fashion for presentation to T cells (Geisberger *et al.*, 2006; Guarini *et al.*, 1989; Snider *et al.*, 1991). Thus, autoreactive Cδ-CS B cells may differ in their uptake of self-antigens bound for presentation of self-peptides to regulatory T cells, resulting in an escape from the tolerizing influence of T cells. The latter could also limit the role of Cδ-CS B cells in normal immune responses due to inefficient uptake of pathogenic antigens. In any event, because there are also Cδ-CS plasma cells that secrete autoantibodies, we suspect that this population serves some function in immunity, though in a controlled fashion because IgD antibodies have fewer effector functions than IgG, and the role of Cδ-CS B cells in immunity and the chance for autoimmune pathology is reduced. Characterization of the mechanism causing the majority of Cδ-CS B cells to be autoreactive may provide important advances to our understanding of peripheral mechanisms of immune tolerance and perhaps a better understanding of the elusive function of the IgD isotype.

5. VACCINE-INDUCED PLASMA CELLS AND RECOMBINANT ANTIBODIES: A RAPID SOURCE OF FULLY HUMAN MABS

Memory B lymphocytes are capable of rapidly generating antibody-secreting plasma cells (ASCs) that can provide critical early defense against infection. We have recently characterized the human ASC response to influenza vaccination (Wrammert et al., 2008). We and others (Bernasconi et al., 2002; Brokstad et al., 1995a,b, 2001, 2002; Odendahl et al., 2005; Sasaki et al., 2007) have found that this ASC response occurs as a rapid burst of antigen-specific IgG$^+$ ASCs that peaks at day 7 after vaccination and precedes the generation of memory cells and high titers of anti-antigen antibodies in the serum. During the response to influenza vaccination, we found that the ASC response consistently accounts for an impressive 6% of all peripheral blood B cells (Wrammert et al., 2008). These ASCs could be distinguished from influenza specific IgG$^+$ memory B cells that peak 14–21 days after vaccination and average (a similarly impressive) 1% of all B cells. Importantly, when assayed by ELISPOT, as many as 80% of ASCs purified at the peak of the response (day 7) are influenza specific. We used multiplex RT-PCR to clone the immunoglobulin variable regions from sorted single ASCs. We used these to produce over sixty recombinant human monoclonal antibodies (as per Fig. 5.8) that bound to the three influenza vaccine strains with high affinity (Fig. 5.11). Thus, we now demonstrate that numerous completely human mAbs can be readily produced from ASCs. These antibodies have been selected at all stages of B cell development in healthy people and bind clearly relevant epitopes, making them likely to produce safe pharmaceutical reagents. This illustrates a powerful application of "B cell subsetting" in conjunction with recombinant monoclonal antibody technology to gain critical insights into the immune response.

This study also helps resolve a major obstacle in the field of medicine since Christmas night, 1891. It is said that on this night, based on the work of Behring and Kitasato, doctors in Berlin used animal sera to passively immunize and cure a young boy of otherwise fatal diphtheria (Llewelyn et al., 1992). Since then, antibody or serum therapy has been demonstrated

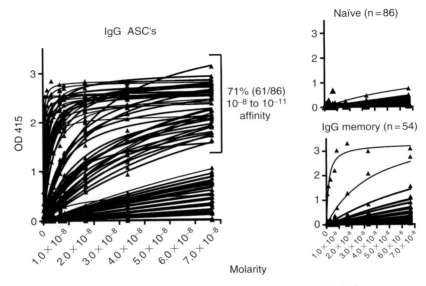

FIGURE 5.11 Post-vaccination plasmablasts are a ready source of fully human monoclonal antibodies. Recombinant monoclonal antibodies from day 7 IgG anti-influenza
ASCs bind to a mixture of the three influenza vaccine strain virions with high affinity. In
total, 71% of the ASC antibodies bound either influenza virus strains freshly grown in
eggs, indicating reactivity with physiologically relevant epitopes (62% or 53/86), or to
antigens within the vaccine only (9% or 8/86, not shown). Each of the five donors was
influenza specific (by donor: 34, 13, 11, 15, and 21 antibodies were generated of which
45–85% were influenza specific). No naïve and only rare IgG memory B cell antibodies
were influenza specific.

to be effective to treat a plethora of diseases but is not widely used
because fatal anaphylactic responses and serum sickness are common.
These obstacles can only be overcome by using fully human mAbs.
However, current technologies (such as hybridomas or viral transformation) are so inefficient at generating mAbs that not a single fully human
antibody exists that is currently used therapeutically. Today, there is a
major resurgence in interest in monoclonal antibody therapy against
many of the most dangerous threats to world health, such as influenza,
HIV, the hepatitis viruses, and many other pathogens. This approach
appears broadly applicable because a similar ASC response has been
observed following various vaccinations, including Tetanus Toxin
(Odendahl *et al.*, 2005) (from which mAbs have even been generated in
E. coli!) and pneumococcal pneumoniae vaccination ("Pneumovax")
(Nieminen *et al.*, 1998). We also have found the response occurs in donors
vaccinated for Hepatitis B, Anthrax toxin, and for vaccinia (unpublished
observation). In addition, the approach may be expanded to isolate ASCs

during primary infections that could lead to therapeutic antibodies against various pathogens for which there are currently no effective vaccines, including HIV, avian influenza and hepatitis C. Thus, we anticipate that antibodies produced recombinantly from activated ASCs will generate substantial advances for the passive immunization against or treatment of a variety of infectious diseases.

6. AUTOIMMUNE DISEASE AND HUMAN B CELL SUBSETS

6.1. Introduction

Although significant advances are occurring in our ability to subset human B cell populations and to understand the roles of these various subsets in aspects of normal immunity, of how these subsets contribute to onset, pathogenesis and disease activity within various autoimmune diseases is still evolving. B cells are known to have important pathogenic roles in humans and in animal models of systemic lupus erythematosus (SLE), rheumatoid arthritis, Sjögren's syndrome and many other autoimmune diseases (Dorner, 2006). However, the role of B cell subset abnormalities in disease pathogenesis as well as associations with disease activity and treatment are best understood in SLE. Therefore, this chapter will focus on the advances made in understanding the role of B cells subsets in human SLE.

6.2. B cell abnormalities in SLE

SLE patients consistently have hyperactive B cells, and the presence of serum autoantibodies is nearly universal in these patients. Evidence suggests that B cells from SLE patients display increased activation and proliferation (Klapper et al., 2004). Additionally, a larger number of B cells spontaneously release immunoglobulin, a phenomenon that correlates with disease activity (Balow and Tsokos, 1984; Tsokos and Balow, 1981). Lupus-specific autoantibodies are found in the sera of individuals long before the onset of clinical SLE (Arbuckle et al., 2003) and autoantibodies regularly occur before their associated clinical manifestations (Heinlen et al., 2007). Thus, autoreactive B cells are likely to be present before the onset of clinical disease.

B cell tolerance checkpoints appear abnormal in SLE patients. Both the immature B cell stage in the bone marrow and in the periphery before maturation appears to be compromised in pediatric SLE patients (Jacobi and Diamond, 2005; Yurasov et al., 2005). Adult SLE patients have been shown to have unchecked progression of VH4-34 cells to final plasma-cell stage (see below) without normal regulation checkpoints (Pugh-Bernard *et al.*, 2001). The VH4.34 gene segment alone has been associated with cold agglutinin disease, as well as a number of other autoimmune diseases (Pascual and Capra, 1992). In addition, germinal center exclusion has been shown in autoreactive B cells from SLE patient tonsils (Cappione *et al.*, 2005).

Overall, lymphopenia is a consistent finding in a large majority of SLE patients; this has led to its inclusion as an ACR diagnostic criterion for SLE (Hochberg, 1997; Tan *et al.*, 1982). As one example, peripheral B and T cell numbers were lower in a cohort of 71 SLE patients, but activated B and T cell levels were elevated despite disease activity measures (Erkeller-Yusel *et al.*, 1993). When only patients without immunosuppressive modulation were studied, SLE patients, as well as pure discoid lupus patients, had slightly elevated numbers of $CD5^+CD19^+$ peripheral B cells (Wouters *et al.*, 2004).

6.3. Perturbations of B cell subsets in adult SLE

Recent work has focused on detailed evaluation of select peripheral B cell subpopulations in adult and pediatric lupus. A study examining 13 SLE patients (6 with disease flare and 7 with quiescent disease), 9 patients with other autoimmune diseases, and 14 healthy controls found consistent and significant abnormalities in various B cell compartments in only the SLE patients (Odendahl *et al.*, 2000). SLE patients regardless of disease activity were found to have lower numbers of $CD19^+$ B cells and higher numbers of CD27-expressing cells. The higher frequency of $CD27^+$ cells was found to be generated by a greater reduction in $CD19^+/CD27^-$ naïve cells compared to the lesser decline in $CD19^+/CD27^+$ memory B cells in both SLE groups. Interestingly, those SLE patients with high expression of CD27 were more likely to have active disease. These $CD27^{high}$ cells were also $CD38^+$, $CD95^+$, $CD138^+$, $CD20^-$ and had intracellular but not soluble Ig (sIG), suggesting these cells may be a subset of plasma cells. These cells were more likely to utilize the VH4–34 gene segment. This study suggests that a relative expansion of this $CD27^{high}$ B cell subset occurs in SLE patients during disease flare and that a decrease in numbers of $CD27^{high}$ cells occurs with treatment and disease suppression. This may demonstrate a key group of cells in SLE disease pathogenesis, as well as other dysregulated B cell subsets (Odendahl *et al.*, 2000). Also a higher number of $CD20^+CD38^+$ peripheral B cells is correlated with greater disease activity and higher soluble levels of Fas (Bijl *et al.* 1998) in SLE patients.

Additional work studying 19 SLE patients and healthy controls shows that lupus patients have a relatively higher representation of peripheral transitional B cells (Sims *et al.*, 2005) as defined by IgD^+CD38^+. Using a cohort of 20 SLE patients, patients with various autoimmune diseases, and healthy controls, another group demonstrated higher numbers of transitional peripheral B cells in patients with either SLE or other autoimmune diseases (rheumatoid arthritis, Sjögren's syndrome, Wegener's granulomatosus, systemic sclerosis, anti-phospholipid antibody syndrome and others) (Wehr *et al.*, 2004). SLE patients with active disease also had higher numbers of plasmablasts (CD27^{high}CD38$^+$) in this study.

In addition to having relatively higher numbers of CD27^{high} cells in the periphery, adult SLE patients also had an expanded population of a novel subset of memory B cells that lacked expression of CD27 and IgD (DN population) (Anolik *et al.*, 2004; Huang *et al.*, 2002; Wei *et al.*, 2007). In a detailed study of 36 SLE patients, 45 rheumatoid arthritis [RA] patients, 7 hepatitis C patients and 29 healthy controls, the Sanz laboratory characterized this DN human B cell subset. The frequency of DN cells was higher in 72% of SLE patients compared to healthy controls whose DN cells were in the normal range. Both RA and hepatitis patients had numbers of DN cells similar to the healthy controls. These SLE CD27/IgD DN B cells have phenotypic, functional and genetic characteristics that are consistent with memory B cells. It is of interest that the number of DN cells is elevated in SLE patients with nephritis as well as in SLE patients with higher levels of anti-dsDNA, anti-Sm and/or anti-nRNP (Wei *et al.*, 2007).

6.4. The B cell repertoire in SLE

A number of investigators have characterized the B cells and their antibodies in the serum, lymph nodes, tonsils, and bone marrow, from normal individuals. Thus, it was natural that these studies would gravitate to a study of both the naïve and the pre and post treatment repertoire in SLE. Zouali (Zouali, 1992) published an excellent review on this issue, that is still quite up to date. A more recent review by Grimaldi *et al.* (2005) details the role of the BCR in both positive and negative regulation in SLE as well as reviews the heavy chain V gene segment repertoire in these patients. One of the implications of the overrepresentation of a variety of V gene segments is that these can be targeted at virtually any stage of B cell development including the plasma cell stage (Tarlinton and Hodgkin, 2004). Monoclonal antibody therapy directed to specific V gene segments could potentially be useful in deleting specific clones of B cells producing autoantibodies, as there is a wealth of information available that many of the autoantibodies seen in SLE utilize a restricted subset of V gene segments. Additionally, plasma cells could be targeted with agents that interfere with the interaction of plasma cells and stromal elements such as cytokines and adhesion molecules. Grammer *et al.* (2003) used an anti-CD154 mAb to treat patients with SLE and noted the disappearance of a CD38^{high} subset of Ig-secreting cells (plasma cells or plasmablasts) in the periphery in the post treatment period. This is a rich area for future investigation.

6.5. Perturbations of B cell subsets in pediatric SLE

Pediatric SLE patients also have dysregulated B cell subsets in the periphery. Using samples from 68 pediatric SLE patients, 10 juvenile dermatomyositis, 35 healthy children and 17 health adults, Pascual and colleagues studied four peripheral B cell subpopulations. SLE pediatric patients had a 75% to 80% reduction in peripheral T and B cells, irrespective of disease activity or therapeutic interventions. Pediatric SLE patients were also found to have lower numbers of naïve and memory B cell subsets. These patients also had an approximate 3-fold enrichment in plasma cell precursors ($CD20^-CD19^+CD27^{high}CD38^+CD138^-$) and a relative enrichment in a novel B cell subset with a pre-germinal center phenotype ($IgD^+CD38^+centerin^+$). No marked differences were found between pediatric SLE patients with high disease activity (SLEDAI>10) compared to lower disease activity (Arce *et al.*, 2001). A small study of 11 SLE pediatric patients and 14 healthy children also showed a significant increase in $CD27^{high}$ cells; many of these cells expressed CD138, which suggests that plasmablasts/plasma cells circulate in patients with active disease. However, immunosuppressive therapy did not appear to decrease the number of $CD27^{high}$ cells in pediatric SLE patients compared to number in adult patients (Odendahl *et al.*, 2003).

6.6. B cell subsets predict SLE disease activity

Several groups have reported that $CD27^{high}$ cells are enriched in the peripheral blood of SLE patients. Preliminary data also show that $CD27^{high}$ levels may correlate with disease activity. A recent study examined the correlation between $CD27^{high}$ peripheral blood cells and different measures of clinical and serologic disease activity (Jacobi *et al.*, 2003). This cross-sectional study of 36 SLE patients showed that the number and frequency of $CD27^{high}$ cells significantly correlated with SLE disease activity when measured with two standard approaches (SLE Disease Activity Index and European Consensus Lupus Activity Measure) and with anti-dsDNA titer. Increased levels of $CD27^{high}$ cells also correlated with increased disease and serologic activity; the predictive value for active versus nonactive disease was nearly 80%. B cell subsets did not appear to be significantly influenced by age or sex; however, $CD27^{high}$ cell levels appeared to increase with disease duration. Additional information regarding the pathogenic impact and potential causes for these dysregulated cell populations in SLE are presented in more detail by Dorner and Lipsky (2004).

While these results are supportive of the potential for measurements of the CD27^high B cell subset helping to serve as a biomarker for SLE disease activity, several potential limitations are present. Longitudinal assessments of CD27^high levels with concurrent measures of disease activity in a cohort of patients over time are needed to assess the temporal importance of these findings. CD27^high cell numbers dramatically increase during acute infections (Harada *et al.*, 1996). Therefore, this potential biomarker will probably not be useful to determine the difference between an SLE disease flare and concurrent infection. CD27^high cells peak 2–5 days after infection and rapidly return to baseline levels in healthy individuals (Ten *et al.*, 2007). CD27^high cells also dramatically increase after vaccination in healthy individuals (Jego *et al.*, 1999). Other confounders have not yet been studied. Interestingly, CD27^high levels are not increased in healthy individuals with antinuclear antibodies (Ten *et al.*, 2007) or in individuals with Sjögren's syndrome (Bohnhorst *et al.*, 2001b; Hansen *et al.*, 2002).

6.7. Recovery of B cell subsets after depletion therapy

Since key roles for B cells in autoimmunity have been found, recent therapeutic advances have focused on the effectiveness of B cell depletion therapy in RA, SLE, and other autoimmune disorders. These interventional trials have provided unique opportunities to examine the influence of B cell depletion of CD20 positive peripheral B cell subsets and to establish the manner of repopulation of peripheral B cell populations after ablative therapy.

Rituximab, a mouse/human chimeric monoclonal antibody against CD20, depletes B cells and is effective in treating RA and SLE (Anolik *et al.*, 2003; Chambers and Isenberg, 2005; Cohen *et al.*, 2006; Edwards *et al.*, 2004; Looney *et al.*, 2004a, b; Looney, 2005; Silverman, 2006). In one of the initial Phase I/II trials, SLE patients had serial measurements of B cell phenotypes before and after anti-CD20 therapy. Again, SLE patients had naïve B cell lymphopenia and relative increased memory B cell pools, expansion of CD27^high/CD38^high/CD20^− plasmablasts and a greater population of CD27/IgD^− DN B cells in their peripheral blood. Although B cell depletion was highly variable, after one year of treatment an increase in naïve B cells was found in the both RA and SLE patients. This was more significant in the subset of patients with effective B cell depletion (to <1%). A significant decrease in CD27^high plasmablasts was found in both SLE patients with and without significant depletion and a significant decrease in the DN cells was found only in patients with successful B cell depletion. These findings contrast the lack of change in serum autoantibodies in a significant fraction of patients with successful B cell depletion but good clinical responses (Anolik *et al.*, 2004).

This obviously provides clinicians with yet another tool to measure early changes in disease activity, as well as potential response to therapy.

Fifteen SLE patients have now undergone long-term follow-up after anti-CD20 treatment to assess the reconstitution of the peripheral B cell subsets after B cell depletion therapy. Reconstitution of CD27$^+$ memory B cells appears to be delayed for several years in patients with autoantibody normalization and prolonged clinical responses, whereas patients with shorter-term or no clinical response had a faster memory B cell recovery compared with long-term responders. Long-term responders also had an increased reconstitution with transitional B cells (defined with CD38/CD24 designation scheme), albeit with a significant degree of patient heterogeneity. These transitional B cells appear to have signs of functional immaturity, such as reduced proliferation, increased apoptosis and inefficient Rhodamine 123 extrusion (Anolik *et al.*, 2007a,b). Additional long-term evaluation, especially in individuals with repeated treatment after flare or with additional disease activity measurements, will be helpful in further defining the reoccurrence of select B cell subsets and their correlation with disease pathogenesis and/or activity.

6.8. Other cell surface phenotypic changes in B cells from SLE patients

Peripheral B cells from SLE patients have a variety of other cell surface changes (see Table 5.2). CD19 expression levels appear decreased on naïve B cells and increased up to two-fold in memory B cells. These CD19 overexpressing cells in SLE appear to have a distinct autoantibody profile with high levels of antinuclear ribonucleoprotein specificities (Culton *et al.*, 2007). SLE patients have a 2.5–7-fold higher percentage of CD86$^+$ cells in their peripheral blood on resting and activated B cells, respectively. CD80-expressing cells were also modestly increased in SLE patients compared to controls, but were only found on activated cells. Even thought only adult patients with quiescent disease were studied some results also suggest that at least some SLE patients overexpress molecules of the B7 family (Folzenlogen *et al.*, 1997). Active SLE patients compared to healthy controls also have a dramatic increase, perhaps as much as 20-fold, in CD40L$^+$ B cells. B cells from SLE patients compared to controls also increased surface expression of CD40L after activation by 2 twofold. Neutralizing anti-CD40L monoclonal antibodies also decreased anti-dsDNA production *in vitro* (Datta and Kalled, 1997). The few human patients that were treated with anti-CD40L showed dramatic decreases in anti-dsDNA-producing B cells (Huang *et al.*, 2002). Some studies suggest that CD95 levels may be differentially expressed in SLE patients (Huck *et al.*, 1998) and that ICOS-L is downregulated in SLE memory B cells (Higuchi *et al.*, 2002). In addition,

TABLE 5.2 B cell subset abnormalities in SLE patient peripheral blood

Subset name	Surface marker phenotype	Effect (i.e., increased or decreased frequency)	Reference
No subset given (peripheral B cells)	$CD20^+CD38^+$	High numbers in quiescent disease	Bijl et al. (1998)
Naive	$CD27^-$	Number and frequency decreased with time in SLE	Jacobi et al. (2003)
Naive	IgM^+CD27^-	No change in SLE	Wehr et al. (2004)
Memory	IgM^+CD27^+	Low numbers in SLE	Wehr et al. (2004)
Uncharacterized previously	$CD19^{hi}CD21^{lo}CD38^{lo}CD86^{int}$	Expanded in autoimmune diseases	Wehr et al. (2004)
Transitional B cell	$CD19^+IgM^{hi}IgD^+CD24^{hi}CD38^{hi}$	Expanded in SLE and other autoimmune disease	Wehr et al. (2004)
Plasma cell	$CD27^{hi}$	Number and frequency increased with time in SLE	Jacobi et al. (2003)
Plasma cell	$CD27^{hi}CD38^+CD19^{dim}$ surface $Ig^{lo}CD20^-CD138^+$	High frequency in active SLE	Odendahl et al. (2003)
Plasmablast	$CD19^{lo}CD21^{lo}CD27^{++}CD38^{++}$	High in active SLE	Wehr et al. (2004)
Plasmablast	$CD27^{hi}CD38^+HLA\text{-}DR^{dim}$, surface Ig^{lo}	Increased in active adults and pediatric SLE	Odendahl et al. (2003)

RP105-negative B cells are enriched in SLE patients and when cultured *in vitro* with IL-6 were able to secrete anti-dsDNA (Kikuchi *et al.*, 2002).

7. SUMMARY

B cell subsetting (especially human B cell subsetting) has now reached the same level of sophistication as T cell subsetting. We are now able to define the various subsets, relate them to autoimmune disease, understand their function, correlate them as the cell of origin of the B cell lymphomas and leukemias, and exploit them as diagnostics and therapeutics. What is left is the use of mAbs to target specific human B cell subsets in disease, a feature our T cell brethren have used for over 25 years. We are confident that that breakthrough is close.

REFERENCES

Agematsu, K. (2004). [Molecules involved in characteristics of naive/memory B cells]. *Nihon Rinsho Meneki. Gakkai Kaishi* **27**, 309–314.

Agematsu, K., Hokibara, S., Nagumo, H., and Komiyama, A. (2000). CD27: A memory B-cell marker. *Immunol. Today* **21**, 204–206.

Agematsu, K., Nagumo, H., Yang, F. C., Nakazawa, T., Fukushima, K., Ito, S., Sugita, K., Mori, T., Kobata, T., Morimoto, C., and Komiyama, A. (1997). B cell subpopulations separated by CD27 and crucial collaboration of CD27+ B cells and helper T cells in immunoglobulin production. *Eur. J. Immunol.* **27**, 2073–2079.

Alexander, D. R. (2000). The CD45 tyrosine phosphatase: A positive and negative regulator of immune cell function. *Semin. Immunol.* **12**, 349–359.

Allen, C. D., Okada, T., and Cyster, J. G. (2007). Germinal-center organization and cellular dynamics. *Immunity.* **27**, 190–202.

Allman, D., Jain, A., Dent, A., Maile, R. R., Selvaggi, T., Kehry, M. R., and Staudt, L. M. (1996). BCL-6 expression during B-cell activation. *Blood* **87**, 5257–5268.

Amu, S., Tarkowski, A., Dorner, T., Bokarewa, M., and Brisslert, M. (2007). The human immunomodulatory CD25+ B cell population belongs to the memory B cell pool. *Scand. J. Immunol.* **66**, 77–86.

Anolik, J., Sanz, I., and Looney, R. J. (2003). B cell depletion therapy in systemic lupus erythematosus. *Curr. Rheumatol. Rep.* **5**, 350–356.

Anolik, J. H., Barnard, J., Cappione, A., Pugh-Bernard, A. E., Felgar, R. E., Looney, R. J., and Sanz, I. (2004). Rituximab improves peripheral B cell abnormalities in human systemic lupus erythematosus. *Arthritis Rheum.* **50**, 3580–3590.

Anolik, J. H., Barnard, J., Owen, T., Zheng, B., Kemshetti, S., Looney, R. J., and Sanz, I. (2007a). Delayed memory B cell recovery in peripheral blood and lymphoid tissue in systemic lupus erythematosus after B cell depletion therapy. *Arthritis Rheum.* **56**, 3044–3056.

Anolik, J. H., Friedberg, J. W., Zheng, B., Barnard, J., Owen, T., Cushing, E., Kelly, J., Milner, E. C., Fisher, R. I., and Sanz, I. (2007b). B cell reconstitution after rituximab treatment of lymphoma recapitulates B cell ontogeny. *Clin. Immunol.* **122**, 139–145.

Antin, J. H., Emerson, S. G., Martin, P., Gadol, N., and Ault, K. A. (1986). Leu-1+ (CD5+) B cells. A major lymphoid subpopulation in human fetal spleen: Phenotypic and functional studies. *J. Immunol.* **136,** 505–510.

Arbuckle, M. R., McClain, M. T., Rubertone, M. V., Scofield, R. H., Dennis, G. J., James, J. A., and Harley, J. B. (2003). Development of autoantibodies before the clinical onset of systemic lupus erythematosus. *N. Engl. J. Med.* **349,** 1526–1533.

Arce, E., Jackson, D. G., Gill, M. A., Bennett, L. B., Banchereau, J., and Pascual, V. (2001). Increased frequency of pre-germinal center B cells and plasma cell precursors in the blood of children with systemic lupus erythematosus. *J. Immunol.* **167,** 2361–2369.

Arce, S., Luger, E., Muehlinghaus, G., Cassese, G., Hauser, A., Horst, A., Lehnert, K., Odendahl, M., Honemann, D., Heller, K. D., Kleinschmidt, H., Berek, C., *et al.* (2004). CD38 low IgG-secreting cells are precursors of various CD38 high-expressing plasma cell populations. *J. Leukoc. Biol.* **75,** 1022–1028.

Arpin, C., de, B. O., Razanajaona, D., Fugier-Vivier, I., Briere, F., Banchereau, J., Lebecque, S., and Liu, Y. J. (1998). The normal counterpart of IgD myeloma cells in germinal center displays extensively mutated IgVH gene, Cmu-Cdelta switch, and lambda light chain expression. *J. Exp. Med.* **187,** 1169–1178.

Avery, D. T., Ellyard, J. I., Mackay, F., Corcoran, L. M., Hodgkin, P. D., and Tangye, S. G. (2005). Increased expression of CD27 on activated human memory B cells correlates with their commitment to the plasma cell lineage. *J. Immunol.* **174,** 4034–4042.

Bachmann, M. F., Kundig, T. M., Odermatt, B., Hengartner, H., and Zinkernagel, R. M. (1994) Free recirculation of memory B cells versus antigen-dependent differentiation to antibody-forming cells. *J. Immunol.* **153,** 3386–3397.

Bagli, L., Zucchini, A., Innoceta, A. M., Zaccaria, A., Cipriani, R., Fattori, P. P., and Ravaioli, A. (2006). Immunoglobulin V(H) genes and CD38 expression analysis in B-cell chronic lymphocytic leukemia. *Acta Haematol.* **116,** 72–74.

Bahler, D. W., and Levy, R. (1992). Clonal evolution of a follicular lymphoma: evidence for antigen selection. *Proc. Natl. Acad. Sci. USA* **89,** 6770–6774.

Bailey, S., Mardell, C., Wheatland, L., Zola, H., and Macardle, P. J. (2005). A comparison of Verotoxin B-subunit (Stx1B) and CD77 antibody to define germinal centre populations. *Cell Immunol.* **236,** 167–170.

Balow, J. E., and Tsokos, G. C. (1984). T and B lymphocyte function in patients with lupus nephritis: Correlation with renal pathology. *Clin. Nephrol.* **21,** 93–97.

Banchereau, J., and Rousset, F. (1991). Growing human B lymphocytes in the CD40 system. *Nature* **353,** 678–679.

Barbas, S. M., Ditzel, H. J., Salonen, E. M., Yang, W. P., Silverman, G. J., and Burton, D. R. (1995). Human autoantibody recognition of DNA. *Proc. Natl. Acad. Sci. USA* **92,** 2529–2533.

Bauvois, B., Durant, L., Laboureau, J., Barthelemy, E., Rouillard, D., Boulla, G., and Deterre, P. (1999). Upregulation of CD38 gene expression in leukemic B cells by interferon types I and II. *J. Interferon Cytokine Res.* **19,** 1059–1066.

Bentz, M., Werner, C. A., Dohner, H., Joos, S., Barth, T. F., Siebert, R., Schroder, M., Stilgenbauer, S., Fischer, K., Moller, P., and Lichter, P. (1996). High incidence of chromosomal imbalances and gene amplifications in the classical follicular variant of follicle center lymphoma. *Blood* **88,** 1437–1444.

Bernasconi, N. L., Traggiai, E., and Lanzavecchia, A. (2002). Maintenance of serological memory by polyclonal activation of human memory B cells. *Science* **298,** 2199–2202.

Bijl, M. van, L. T., Limburg, P. C., Spronk, P. E., Jaegers, S. M., Aarden, L. A., Smeenk, R. J., and Kallenberg, G. G. (1998). Do elevated levels of serum-soluble fas contribute to the persistence of activated lymphocytes in systemic lupus erythematosus? *J. Autoimmun.* **11,** 457–463.

Billian, G., Bella, C., Mondiere, P., and Defrance, T. (1996). Identification of a tonsil IgD+ B cell subset with phenotypical and functional characteristics of germinal center B cells. *Eur. J. Immunol.* **26,** 1712–1719.

Blade, J., Lust, J. A., and Kyle, R. A. (1994). Immunoglobulin D multiple myeloma: Presenting features, response to therapy, and survival in a series of 53 cases. *J. Clin. Oncol.* **12,** 2398–2404.

Blum, K. A., Lozanski, G., and Byrd, J. C. (2004). Adult Burkitt leukemia and lymphoma. *Blood* **104,** 3009–3020.

Bobeck, M. J., Cleary, J., Beckingham, J. A., Ackroyd, P. C., and Glick, G. D. (2007). Effect of somatic mutation on DNA binding properties of anti-DNA autoantibodies. *Biopolymers* **85,** 471–480.

Bohnhorst, J. O., Bjorgan, M. B., Thoen, J. E., Natvig, J. B., and Thompson, K. M. (2001a). Bm1-Bm5 classification of peripheral blood B cells reveals circulating germinal center founder cells in healthy individuals and disturbance in the B cell subpopulations in patients with primary Sjogren's syndrome. *J. Immunol.* **167,** 3610–3618.

Bohnhorst, J. O., Thoen, J. E., Natvig, J. B., and Thompson, K. M. (2001b). Significantly depressed percentage of CD27+ (memory) B cells among peripheral blood B cells in patients with primary Sjogren's syndrome. *Scand. J. Immunol.* **54,** 421–427.

Boise, L. H., and Thompson, C. B. (1996). Hierarchical control of lymphocyte survival. *Science* **274,** 67–68.

Bomben, R., Dal, B. M., Capello, D., Benedetti, D., Marconi, D., Zucchetto, A., Forconi, F., Maffei, R., Ghia, E. M., Laurenti, L., Bulian, P., Del Principe, M. I., *et al.* (2007). Comprehensive characterization of IGHV3-21-expressing B-cell chronic lymphocytic leukemia: An Italian multicenter study. *Blood* **109,** 2989–2998.

Boonstra, J. G., van, L. K., Langerak, A. W., Graveland, W. J., Valk, P. J., Kraan, J., van, V, and Gratama, J. W. (2006). CD38 as a prognostic factor in B cell chronic lymphocytic leukaemia (B-CLL): Comparison of three approaches to analyze its expression. *Cytometry B Clin. Cytom.* **70,** 136–141.

Bradwell, A. R., Stokes, R. P., and Johnson, G. D. (1995). Atlas of HEp-2 patterns The Binding Site LTD. Birmingham, England.

Brard, F., Shannon, M., Prak, E. L., Litwin, S., and Weigert, M. (1999). Somatic mutation and light chain rearrangement generate autoimmunity in anti-single-stranded DNA transgenic MRL/lpr mice. *J. Exp. Med.* **190,** 691–704.

Brink, R., Fulcher, D. A., Goodnow, C. C., and Basten, A. (1994). Differential regulation of early and late stages of B lymphocyte development by the mu and delta membrane heavy chains of Ig. *Int. Immunol.* **6,** 1905–1916.

Brink, R., Goodnow, C. C., Crosbie, J., Adams, E., Eris, J., Mason, D. Y., Hartley, S. B., and Basten, A. (1992). Immunoglobulin M and D antigen receptors are both capable of mediating B lymphocyte activation, deletion, or anergy after interaction with specific antigen. *J. Exp. Med.* **176,** 991–1005.

Brokstad, K. A., Cox, R. J., Eriksson, J. C., Olofsson, J., Jonsson, R., and Davidsson, A. (2001). High prevalence of influenza specific antibody secreting cells in nasal mucosa. *Scand. J. Immunol.* **54,** 243–247.

Brokstad, K. A., Cox, R. J., Major, D., Wood, J. M., and Haaheim, L. R. (1995a). Cross-reaction but no avidity change of the serum antibody response after influenza vaccination. *Vaccine* **13,** 1522–1528.

Brokstad, K. A., Cox, R. J., Olofsson, J., Jonsson, R., and Haaheim, L. R. (1995b). Parenteral influenza vaccination induces a rapid systemic and local immune response. *J. Infect. Dis.* **171,** 198–203.

Brokstad, K. A., Eriksson, J. C., Cox, R. J., Tynning, T., Olofsson, J., Jonsson, R., and Davidsson, A. (2002). Parenteral vaccination against influenza does not induce a local antigen-specific immune response in the nasal mucosa. *J. Infect. Dis.* **185,** 878–884.

Burdin, N., Galibert, L., Garrone, P., Durand, I., Banchereau, J., and Rousset, F. (1996). Inability to produce IL-6 is a functional feature of human germinal center B lymphocytes. *J. Immunol.* **156,** 4107–4113.

Burrows, P. D., Schroeder, H. W., and Cooper, M. D. (1995). B-cell differentiation in humans. *In* Immunoglobulin Genes, (T. Honjo, and F. W. Alt, Eds.), pp. 3–32. Academic Press, San Diego, CA.

Cahalan, M. D., and Parker, I. (2005). Close encounters of the first and second kind: T-DC and T-B interactions in the lymph node. *Semin. Immunol.* **17,** 442–451.

Caligaris-Cappio, F., Gregoretti, M. G., and Nilsson, K. (1997). B cell populations: The multiple myeloma model. *Chem. Immunol.* **67,** 102–113.

Cambier, J. C., Gauld, S. B., Merrell, K. T., and Vilen, B. J. (2007). B-cell anergy: From transgenic models to naturally occurring anergic B cells? *Nat. Rev. Immunol.* **7,** 633–643.

Cappione, A., III, Anolik, J. H., Pugh-Bernard, A., Barnard, J., Dutcher, P., Silverman, G., and Sanz, I. (2005). Germinal center exclusion of autoreactive B cells is defective in human systemic lupus erythematosus. *J. Clin. Invest* **115,** 3205–3216.

Carrasco, Y. R., and Batista, F. D. (2007). B cells acquire particulate antigen in a macrophage-rich area at the boundary between the follicle and the subcapsular sinus of the lymph node. *Immunity.* **27,** 160–171.

Casali, P., and Notkins, A. L. (1989). CD5+ B lymphocytes, polyreactive antibodies and the human B-cell repertoire. *Immunol. Today* **10,** 364–368.

Casola, S. (2007). Control of peripheral B-cell development. *Curr. Opin. Immunol.* **19,** 143–149.

Casola, S., Cattoretti, G., Uyttersprot, N., Koralov, S. B., Seagal, J., Hao, Z., Waisman, A., Egert, A., Ghitza, D., and Rajewsky, K. (2006). Tracking germinal center B cells expressing germ-line immunoglobulin gamma1 transcripts by conditional gene targeting. *Proc. Natl. Acad. Sci. USA* **103,** 7396–7401.

Cattoretti, G., Buttner, M., Shaknovich, R., Kremmer, E., Alobeid, B., and Niedobitek, G. (2006a). Nuclear and cytoplasmic AID in extrafollicular and germinal center B cells. *Blood* **107,** 3967–3975.

Cattoretti, G., Shaknovich, R., Smith, P. M., Jack, H. M., Murty, V. V., and Alobeid, B. (2006b). Stages of germinal center transit are defined by B cell transcription factor coexpression and relative abundance. *J. Immunol.* **177,** 6930–6939.

Cerhan, J. R., Wang, S., Maurer, M. J., Ansell, S. M., Geyer, S. M., Cozen, W., Morton, L. M., Davis, S., Severson, R. K., Rothman, N., Lynch, C. F., Wacholder, S., *et al.* (2007). Prognostic significance of host immune gene polymorphisms in follicular lymphoma survival. *Blood* **109,** 5439–5446.

Chambers, S. A., and Isenberg, D. (2005). Anti-B cell therapy (rituximab) in the treatment of autoimmune diseases. *Lupus* **14,** 210–214.

Chaudhuri, J., and Jasin, M. (2007). Immunology. Antibodies get a break. Science **315,** 335–336.

Chiorazzi, N., Rai, K. R., and Ferrarini, M. (2005). Chronic lymphocytic leukemia. *N. Engl. J. Med.* **352,** 804–815.

Cogne, M., Mounir, S., Preud'homme, J. L., Nau, F., and Guglielmi, P. (1988). Burkitt's lymphoma cell lines producing truncated mu immunoglobulin heavy chains lacking part of the variable region. *Eur. J. Immunol.* **18,** 1485–1489.

Cohen, S. B., Emery, P., Greenwald, M. W., Dougados, M., Furie, R. A., Genovese, M. C., Keystone, E. C., Loveless, J. E., Burmester, G. R., Cravets, M. W., Hessey, E. W., Shaw, T., *et al.* (2006). Rituximab for rheumatoid arthritis refractory to anti-tumor necrosis factor therapy: Results of a multicenter, randomized, double-blind, placebo-controlled, phase III trial evaluating primary efficacy and safety at twenty-four weeks. *Arthritis Rheum.* **54,** 2793–2806.

Craig, F. E., and Foon, K. A. (2008). Flow cytometric immunophenotyping for hematologic neoplasms. *Blood* **111,** 3941–3967.

Culton, D. A., Nicholas, M. W., Bunch, D. O., Zhen, Q. L., Kepler, T. B., Dooley, M. A., Mohan, C., Nachman, P. H., and Clarke, S. H. (2007). Similar CD19 dysregulation in two autoantibody-associated autoimmune diseases suggests a shared mechanism of B-cell tolerance loss. *J. Clin. Immunol.* **27,** 53–68.

Cyster, J. G. (1997). Signaling thresholds and interclonal competition in preimmune B-cell selection. *Immunol. Rev.* **156,** 87–101.

Cyster, J. G., Healy, J. I., Kishihara, K., Mak, T. W., Thomas, M. L., and Goodnow, C. C. (1996). Regulation of B-lymphocyte negative and positive selection by tyrosine phosphatase CD45. *Nature* **381,** 325–328.

d'Arbonneau, F., Pers, J. O., Devauchelle, V., Pennec, Y., Saraux, A., and Youinou, P. (2006). BAFF-induced changes in B cell antigen receptor-containing lipid rafts in Sjogren's syndrome. *Arthritis Rheum.* **54,** 115–126.

D'Arena, G., Musto, P., Cascavilla, N., Dell'Olio, M., Di, R. N., Perla, G., Savino, L., and Carotenuto, M. (2001). CD38 expression correlates with adverse biological features and predicts poor clinical outcome in B-cell chronic lymphocytic leukemia. *Leuk. Lymphoma* **42,** 109–114.

Dahlke, M. H., Larsen, S. R., Rasko, J. E., and Schlitt, H. J. (2004). The biology of CD45 and its use as a therapeutic target. *Leuk. Lymphoma* **45,** 229–236.

Damle, R. N., Wasil, T., Fais, F., Ghiotto, F., Valetto, A., Allen, S. L., Buchbinder, A., Budman, D., Dittmar, K., Kolitz, J., Lichtman, S. M., Schulman, P., *et al.* (1999). Ig V gene mutation status and CD38 expression as novel prognostic indicators in chronic lymphocytic leukemia. *Blood* **94,** 1840–1847.

Datta, S. K., and Kalled, S. L. (1997). CD40-CD40 ligand interaction in autoimmune disease. *Arthritis Rheum.* **40,** 1735–1745.

Dawes, R., Petrova, S., Liu, Z., Wraith, D., Beverley, P. C., and Tchilian, E. Z. (2006). Combinations of CD45 isoforms are crucial for immune function and disease. *J. Immunol.* **176,** 3417–3425.

Deaglio, S., Capobianco, A., Bergui, L., Durig, J., Morabito, F., Duhrsen, U., and Malavasi, F. (2003). CD38 is a signaling molecule in B-cell chronic lymphocytic leukemia cells. *Blood* **102,** 2146–2155.

Deaglio, S., Vaisitti, T., Aydin, S., Ferrero, E., and Malavasi, F. (2006). In-tandem insight from basic science combined with clinical research: CD38 as both marker and key component of the pathogenetic network underlying chronic lymphocytic leukemia. *Blood* **108,** 1135–1144.

Deans, J. P., Serra, H. M., Shaw, J., Shen, Y. J., Torres, R. M., and Pilarski, L. M. (1992). Transient accumulation and subsequent rapid loss of messenger RNA encoding high molecular mass CD45 isoforms after T cell activation. *J. Immunol.* **148,** 1898–1905.

Del, G., I, Davis, Z., Matutes, E., Osuji, N., Parry-Jones, N., Morilla, A., Brito-Babapulle, V., Oscier, D., and Catovsky, D. (2006). IgVH genes mutation and usage, ZAP-70 and CD38 expression provide new insights on B-cell prolymphocytic leukemia (B-PLL). *Leukemia* **20,** 1231–1237.

Denepoux, S., Razanajaona, D., Blanchard, D., Meffre, G., Capra, J. D., Banchereau, J., and Lebecque, S. (1997). Induction of somatic mutation in a human B cell line *in vitro*. *Immunity.* **6,** 35–46.

Dent, A. (2005). B-cell lymphoma: Suppressing a tumor suppressor. *Nat. Med.* **11,** 22.

Dono, M., Cerruti, G., and Zupo, S. (2004). The CD5+ B-cell. *Int. J. Biochem. Cell Biol.* **36,** 2105–2111.

Dono, M., Zupo, S., Leanza, N., Melioli, G., Fogli, M., Melagrana, A., Chiorazzi, N., and Ferrarini, M. (2000). Heterogeneity of tonsillar subepithelial B lymphocytes, the splenic marginal zone equivalents. *J. Immunol.* **164,** 5596–5604.

Dorner, T. (2006). Crossroads of B cell activation in autoimmunity: Rationale of targeting B cells. *J. Rheumatol. Suppl* **77,** 3–11.

Dorner, T., and Lipsky, P. E. (2004). Correlation of circulating CD27high plasma cells and disease activity in systemic lupus erythematosus. *Lupus* **13,** 283–289.

Dorner, T., and Lipsky, P. E. (2006). Signalling pathways in B cells: implications for autoimmunity. *Curr. Top. Microbiol. Immunol.* **305,** 213–240.

Doucett, V. P., Gerhard, W., Owler, K., Curry, D., Brown, L., and Baumgarth, N. (2005). Enumeration and characterization of virus-specific B cells by multicolor flow cytometry. *J. Immunol. Methods* **303,** 40–52.

Douglas, R. S., Pletcher, C. H., Jr., Nowell, P. C., and Moore, J. S. (1998). Novel approach for simultaneous evaluation of cell phenotype, apoptosis, and cell cycle using multiparameter flow cytometry. *Cytometry* **32,** 57–65.

Edwards, J. C., Szczepanski, L., Szechinski, J., Filipowicz-Sosnowska, A., Emery, P., Close, D. R., Stevens, R. M., and Shaw, T. (2004). Efficacy of B-cell-targeted therapy with rituximab in patients with rheumatoid arthritis. *N. Engl. J. Med.* **350,** 2572–2581.

Elagib, K. E., Borretzen, M., Jonsson, R., Haga, H. J., Thoen, J., Thompson, K. M., and Natvig, J. B. (1999). Rheumatoid factors in primary Sjogren's syndrome (pSS) use diverse VH region genes, the majority of which show no evidence of somatic hypermutation. *Clin. Exp. Immunol.* **117,** 388–394.

Erkeller-Yusel, F., Hulstaart, F., Hannet, I., Isenberg, D., and Lydyard, P. (1993). Lymphocyte subsets in a large cohort of patients with systemic lupus erythematosus. *Lupus* **2,** 227–231.

Evans, R. L., Breard, J. M., Lazarus, H., Schlossman, S. F., and Chess, L. (1977). Detection, isolation, and functional characterization of two human T-cell subclasses bearing unique differentiation antigens. *J. Exp. Med.* **145,** 221–233.

Evans, R. L., Lazarus, H., Penta, A. C., and Schlossman, S. F. (1978). Two functionally distinct subpopulations of human T cells that collaborate in the generation of cytotoxic cells responsible for cell-mediated lympholysis. *J. Immunol.* **120,** 1423–1428.

Facchetti, F., Appiani, C., Salvi, L., Levy, J., and Notarangelo, L. D. (1995). Immunohistologic analysis of ineffective CD40-CD40 ligand interaction in lymphoid tissues from patients with X-linked immunodeficiency with hyper-IgM. Abortive germinal center cell reaction and severe depletion of follicular dendritic cells. *J. Immunol.* **154,** 6624–6633.

Fecteau, J. F., Cote, G., and Neron, S. (2006). A new memory CD27-IgG+ B cell population in peripheral blood expressing VH genes with low frequency of somatic mutation. *J. Immunol.* **177,** 3728–3736.

Fecteau, J F., and Neron, S. (2003). CD40 stimulation of human peripheral B lymphocytes: distinct response from naive and memory cells. *J. Immunol.* **171,** 4621–4629.

Fleming, H. E., Milne, C. D., and Paige, C. J. (2004). CD45-deficient mice accumulate Pro-B cells both *in vivo* and *in vitro*. *J. Immunol.* **173,** 2542–2551.

Folzenlogen, D., Hofer, M. F., Leung, D. Y., Freed, J. H., and Newell, M. K. (1997). Analysis of CD80 and CD86 expression on peripheral blood B lymphocytes reveals increased expression of CD86 in lupus patients. *Clin. Immunol. Immunopathol.* **83,** 199–204.

Frater, J. L., and Hsi, E. D. (2002). Properties of the mantle cell and mantle cell lymphoma. *Curr. Opin. Hematol.* **9,** 56–62.

Frazer, J. K., Jackson, D. G., Gaillard, J. P., Lutter, M., Liu, Y. J., Banchereau, J., Capra, J. D., and Pascual, V. (2000). Identification of centerin: A novel human germinal center B cell-restricted serpin. *Eur. J. Immunol.* **30,** 3039–3048.

Frazer, J. K., LeGros, J., de, B. O., Liu, Y. J., Banchereau, J., Pascual, V., and Capra, J. D. (1997). Identification and cloning of genes expressed by human tonsillar B lymphocyte subsets. *Ann. N. Y. Acad. Sci.* **815,** 316–318.

Fukuhara, K., Okumura, M., Shiono, H., Inoue, M., Kadota, Y., Miyoshi, S., and Matsuda, H. (2002). A study on CD45 isoform expression during T-cell development and selection events in the human thymus. *Hum. Immunol.* **63,** 394–404.

Funaro, A., Morra, M., Calosso, L., Zini, M. G., Ausiello, C. M., and Malavasi, F. (1997). Role of the human CD38 molecule in B cell activation and proliferation. *Tissue Antigens* **49,** 7–15.

Galibert, L., Burdin, N., Barthelemy, C., Meffre, G., Durand, I., Garcia, E., Garrone, P., Rousset, F., Banchereau, J., and Liu, Y. J. (1996a). Negative selection of human germinal center B cells by prolonged BCR cross-linking. *J. Exp. Med.* **183,** 2075–2085.

Galibert, L., Burdin, N., de Saint-Vis, B., Garrone, P., Van, K. C., Banchereau, J., and Rousset, F. (1996b). CD40 and B cell antigen receptor dual triggering of resting B lymphocytes turns on a partial germinal center phenotype. *J. Exp. Med.* **183,** 77–85.

Gay, D., Saunders, T., Camper, S., and Weigert, M. (1993). Receptor editing: An approach by autoreactive B cells to escape tolerance. *J. Exp. Med.* **177,** 999–1008.

Geisberger, R., Lamers, M., and Achatz, G. (2006). The riddle of the dual expression of IgM and IgD. *Immunology* **118,** 429–437.

Gonzalez, D., van der, B. M., Garcia-Sanz, R., Fenton, J. A., Langerak, A. W., Gonzalez, M., van Dongen, J. J., San Miguel, J. F., and Morgan, G. J. (2007). Immunoglobulin gene rearrangements and the pathogenesis of multiple myeloma. *Blood* **110,** 3112–3121.

Goodnow, C. C., Crosbie, J., Jorgensen, H., Brink, R. A., and Basten, A. (1989). Induction of self-tolerance in mature peripheral B lymphocytes. *Nature* **342,** 385–391.

Grammer, A. C., Slota, R., Fischer, R., Gur, H., Girschick, H., Yarboro, C., Illei, G. G., and Lipsky, P. E. (2003). Abnormal germinal center reactions in systemic lupus erythematosus demonstrated by blockade of CD154-CD40 interactions. *J. Clin. Invest* **112,** 1506–1520.

Grimaldi, C. M., Hicks, R., and Diamond, B. (2005). B cell selection and susceptibility to autoimmunity. *J. Immunol.* **174,** 1775–1781.

Guarini, L., Weber, D. A., and Pernis, B. (1989). Differential endocytosis of IgM and IgD in murine B cell lines. *Cell Immunol.* **123,** 456–461.

Haluska, F. G., Tsujimoto, Y., and Croce, C. M. (1987). The t(8;14) chromosome translocation of the Burkitt lymphoma cell line Daudi occurred during immunoglobulin gene rearrangement and involved the heavy chain diversity region. *Proc. Natl. Acad. Sci. USA* **84,** 6835–6839.

Hamann, D., Baars, P. A., Hooibrink, B., and van Lier, R. W. (1996). Heterogeneity of the human CD4+ T-cell population: Two distinct CD4+ T-cell subsets characterized by coexpression of CD45RA and CD45RO isoforms. *Blood* **88,** 3513–3521.

Hamblin, T. J., Davis, Z., Gardiner, A., Oscier, D. G., and Stevenson, F. K. (1999). Unmutated Ig V(H) genes are associated with a more aggressive form of chronic lymphocytic leukemia. *Blood* **94,** 1848–1854.

Han, S., Hathcock, K., Zheng, B., Kepler, T. B., Hodes, R., and Kelsoe, G. (1995). Cellular interaction in germinal centers. Roles of CD40 ligand and B7-2 in established germinal centers. *J. Immunol.* **155,** 556–567.

Hansen, A., Odendahl, M., Reiter, K., Jacobi, A. M., Feist, E., Scholze, J., Burmester, G. R., Lipsky, P. E., and Dorner, T. (2002). Diminished peripheral blood memory B cells and accumulation of memory B cells in the salivary glands of patients with Sjogren's syndrome. *Arthritis Rheum.* **46,** 2160–2171.

Harada, Y., Kawano, M. M., Huang, N., Mahmoud, M. S., Lisukov, I. A., Mihara, K., Tsujimoto, T., and Kuramoto, A. (1996). Identification of early plasma cells in peripheral blood and their clinical significance. *Br. J. Haematol.* **92,** 184–191.

Hardy, R. R., Kincade, P. W., and Dorshkind, K. (2007). The protean nature of cells in the B lymphocyte lineage. *Immunity.* **26,** 703–714.

Harris, N. L., Jaffe, E. S., Stein, H., Banks, P. M., Chan, J. K., Cleary, M. L., Delsol, G., De Wolf-Peeters, C., Falini, B., and Gatter, K. C. (1994). A revised European-American classification of lymphoid neoplasms: A proposal from the International Lymphoma Study Group. *Blood* **84,** 1361–1392.

Hathcock, K. S., Hirano, H., and Hodes, R. J. (1993). CD45 expression by murine B cells and T cells: Alteration of CD45 isoforms in subpopulations of activated B cells. *Immunol. Res.* **12,** 21–36.

Hauser, A. E., Junt, T., Mempel, T. R., Sneddon, M. W., Kleinstein, S. H., Henrickson, S. E., von Andrian, U. H., Shlomchik, M. J., and Haberman, A. M. (2007). Definition of germinal-center B cell migration *in vivo* reveals predominant intrazonal circulation patterns. *Immunity.* **26,** 655–667.

Hauser, A. E., Shlomchik, M. J., and Haberman, A. M. (2007). *In vivo* imaging studies shed light on germinal-centre development. *Nat. Rev. Immunol.* **7**, 499–504.

Healy, J. I., and Goodnow, C. C. (1998). Positive versus negative signaling by lymphocyte antigen receptors. *Annu. Rev. Immunol.* **16**, 645–670.

Heinlen, L. D., McClain, M. T., Merrill, J., Akbarali, Y. W., Edgerton, C. C., Harley, J. B., and James, J. A. (2007). Clinical criteria for systemic lupus erythematosus precede diagnosis, and associated autoantibodies are present before clinical symptoms. *Arthritis Rheum.* **56**, 2344–2351.

Hermiston, M. L., Xu, Z., and Weiss, A. (2003). CD45: A critical regulator of signaling thresholds in immune cells. *Annu. Rev. Immunol.* **21**, 107–137.

Herve, M., Isnardi, I., Ng, Y. S., Bussel, J. B., Ochs, H. D., Cunningham-Rundles, C., and Meffre, E. (2007). CD40 ligand and MHC class II expression are essential for human peripheral B cell tolerance. *J. Exp. Med.* **204**, 1583–1593.

Herve, M., Xu, K., Ng, Y. S., Wardemann, H., Albesiano, E., Messmer, B. T., Chiorazzi, N., and Meffre, E. (2005). Unmutated and mutated chronic lymphocytic leukemias derive from self-reactive B cell precursors despite expressing different antibody reactivity. *J. Clin. Invest* **115**, 1636–1643.

Heyd, F., ten, D. G., and Moroy, T. (2006). Auxiliary splice factor U2AF26 and transcription factor Gfi1 cooperate directly in regulating CD45 alternative splicing. *Nat. Immunol.* **7**, 859–867.

Hiddemann, W., Longo, D. L., Coiffier, B., Fisher, R. I., Cabanillas, F., Cavalli, F., Nadler, L. M., De, V., V, Lister, T. A., and Armitage, J. O. (1996). Lymphoma classification–the gap between biology and clinical management is closing. *Blood* **88**, 4085–4089.

Higuchi, T., Aiba, Y., Nomura, T., Matsuda, J., Mochida, K., Suzuki, M., Kikutani, H., Honjo, T., Nishioka, K., and Tsubata, T. (2002). Cutting Edge: Ectopic expression of CD40 ligand on B cells induces lupus-like autoimmune disease. *J. Immunol.* **168**, 9–12.

Himmelmann, A., Gautschi, O., Nawrath, M., Bolliger, U., Fehr, J., and Stahel, R. A. (2001). Persistent polyclonal B-cell lymphocytosis is an expansion of functional IgD(+)CD27(+) memory B cells. *Br. J. Haematol.* **114**, 400–405.

Hipfner, D. R., and Cohen, S. M. (2004). Connecting proliferation and apoptosis in development and disease. *Nat. Rev. Mol. Cell Biol.* **5**, 805–815.

Hochberg, M. C. (1997). Updating the American College of Rheumatology revised criteria for the classification of systemic lupus erythematosus. *Arthritis Rheum.* **40**, 1725.

Hoffmann, R. (2005). Gene expression patterns in human and mouse B cell development. *Curr. Top. Microbiol. Immunol.* **294**, 19–29.

Hogerkorp, C. M., and Borrebaeck, C. A. (2006). The human. *J. Immunol.* **177**, 4341–4349.

Holmes, N. (2006). CD45: All is not yet crystal clear. *Immunology* **117**, 145–155.

Horgan, K. J., Tanaka, Y., Luce, G. E., van Seventer, G. A., Nutman, T. B., and Shaw, S. (1994). CD45RB expression defines two interconvertible subsets of human CD4+ T cells with memory function. *Eur. J. Immunol.* **24**, 1240–1243.

Hsi, E. D., Kopecky, K. J., Appelbaum, F. R., Boldt, D., Frey, T., Loftus, M., and Hussein, M. A. (2003). Prognostic significance of CD38 and CD20 expression as assessed by quantitative flow cytometry in chronic lymphocytic leukaemia. *Br. J. Haematol.* **120**, 1017–1025.

Huang, W., Sinha, J., Newman, J., Reddy, B., Budhai, L., Furie, R., Vaishnaw, A., and Davidson, A. (2002). The effect of anti-CD40 ligand antibody on B cells in human systemic lupus erythematosus. *Arthritis Rheum.* **46**, 1554–1562.

Huck, S., Jamin, C., Youinou, P., and Zouali, M. (1998). High-density expression of CD95 on B cells and underrepresentation of the B-1 cell subset in human lupus. *J. Autoimmun.* **11**, 449–455.

Huntington, N. D., and Tarlinton, D. M. (2004). CD45: direct and indirect government of immune regulation. *Immunol. Lett.* **94**, 167–174.

Huntington, N. D., Xu, Y., Puthalakath, H., Light, A., Willis, S. N., Strasser, A., and Tarlinton, D. M. (2006). CD45 links the B cell receptor with cell survival and is required for the persistence of germinal centers. *Nat. Immunol.* **7,** 190–198.

Hus, I., Podhorecka, M., Bojarska-Junak, A., Rolinski, J., Schmitt, M., Sieklucka, M., Wasik-Szczepanek, E., and Dmoszynska, A. (2006). The clinical significance of ZAP-70 and CD38 expression in B-cell chronic lymphocytic leukaemia. *Ann. Oncol.* **17,** 683–690.

Hystad, M. E., Myklebust, J. H., Bo, T. H., Sivertsen, E. A., Rian, E., Forfang, L., Munthe, E., Rosenwald, A., Chiorazzi, M., Jonassen, I., Staudt, L. M., and Smeland, E. B. (2007). Characterization of early stages of human B cell development by gene expression profiling. *J. Immunol.* **179,** 3662–3671.

Ibrahim, S., Keating, M., Do, K. A., O'Brien, S., Huh, Y. O., Jilani, I., Lerner, S., Kantarjian, H. M., and Albitar, M. (2001). CD38 expression as an important prognostic factor in B-cell chronic lymphocytic leukemia. *Blood* **98,** 181–186.

Jackson, S. M., and Capra, J. D. (2005). IgH V-region sequence does not predict the survival fate of human germinal center B cells. *J. Immunol.* **174,** 2805–2813.

Jackson, S. M., Harp, N., Patel, D., Henderson, M., Roy, N. M., Courtney, M. A., Johnson, A., and Capra, J. D. (2007a). CD45RO: A marker for BCR-mediated selection. *Scand. J. Immunol.* **66,** 249–260.

Jackson, S. M., Harp, N., Patel, D., Wulf, J., Spaeth, E. D., Dike, U. K., James, J. A., and Capra, J. D. (2008). CD45RB inversely correlates with somatic mutation and enriches for terminally differentiating B cells in humans. *Submitted for Publication.*

Jackson, S. M., Harp, N., Patel, D., Zhang, J., Willson, S., Kim, Y. J., Clanton, C., and Capra, J. D. (2007b). CD45RO enriches for activated, highly mutated human germinal center B cells. *Blood* **110,** 3917–3925.

Jacobi, A. M., and Diamond, B. (2005). Balancing diversity and tolerance: lessons from patients with systemic lupus erythematosus. *J. Exp. Med.* **202,** 341–344.

Jacobi, A. M., Odendahl, M., Reiter, K., Bruns, A., Burmester, G. R., Radbruch, A., Valet, G., Lipsky, P. E., and Dorner, T. (2003). Correlation between circulating CD27high plasma cells and disease activity in patients with systemic lupus erythematosus. *Arthritis Rheum.* **48,** 1332–1342.

Jego, G., Robillard, N., Puthier, D., Amiot, M., Accard, F., Pineau, D., Harousseau, J. L., Bataille, R., and Pellat-Deceunynck, C. (1999). Reactive plasmacytoses are expansions of plasmablasts retaining the capacity to differentiate into plasma cells. *Blood* **94,** 701–712.

Kelsoe, G. (1996). Life and death in germinal centers (redux). *Immunity.* **4,** 107–111.

Kelsoe, G. (1995). *In situ* studies of the germinal center reaction. *Adv. Immunol.* **60,** 267–288.

Kikuchi, Y., Koarada, S., Tada, Y., Ushiyama, O., Morito, F., Suzuki, N., Ohta, A., Miyake, K., Kimoto, M., Horiuchi, T., and Nagasawa, K. (2002). RP105-lacking B cells from lupus patients are responsible for the production of immunoglobulins and autoantibodies. *Arthritis Rheum.* **46,** 3259–3265.

Kimoto, H., Nagaoka, H., Adachi, Y., Mizuochi, T., Azuma, T., Yagi, T., Sata, T., Yonehara, S., Tsunetsugu-Yokota, Y., Taniguchi, M., and Takemori, T. (1997). Accumulation of somatic hypermutation and antigen-driven selection in rapidly cycling surface Ig+ germinal center (GC) B cells which occupy GC at a high frequency during the primary anti-hapten response in mice. *Eur. J. Immunol.* **27,** 268–279.

Kitanaka, A., Ito, C., Coustan-Smith, E., and Campana, D. (1997). CD38 ligation in human B cell progenitors triggers tyrosine phosphorylation of CD19 and association of CD19 with lyn and phosphatidylinositol 3-kinase. *J. Immunol.* **159,** 184–192.

Kitanaka, A., Ito, C., Nishigaki, H., and Campana, D. (1996). CD38-mediated growth suppression of B-cell progenitors requires activation of phosphatidylinositol 3-kinase and involves its association with the protein product of the c-cbl proto-oncogene. *Blood* **88,** 590–598.

Klapper, W., Moosig, F., Sotnikova, A., Qian, W., Schroder, J. O., and Parwaresch, R. (2004). Telomerase activity in B and T lymphocytes of patients with systemic lupus erythematosus. *Ann. Rheum. Dis.* **63**, 1681–1683.

Klein, U., Kuppers, R., and Rajewsky, K. (1993). Human IgM+IgD+ B cells, the major B cell subset in the peripheral blood, express V kappa genes with no or little somatic mutation throughout life. *Eur. J. Immunol.* **23**, 3272–3277.

Klein, U., Rajewsky, K., and Kuppers, R. (1998). Human immunoglobulin (Ig)M+IgD+ peripheral blood B cells expressing the CD27 cell surface antigen carry somatically mutated variable region genes: CD27 as a general marker for somatically mutated (memory) B cells. *J. Exp. Med.* **188**, 1679–1689.

Klonowski, K. D., Primiano, L. L., and Monestier, M. (1999). Atypical VH-D-JH rearrangements in newborn autoimmune MRL mice. *J. Immunol.* **162**, 1566–1572.

Koelsch, K., Zheng, N. Y., Zhang, Q., Duty, A., Helms, C., Mathias, M. D., Jared, M., Smith, K., Capra, J. D., and Wilson, P. C. (2007). Mature B cells class switched to IgD are autoreactive in healthy individuals. *J. Clin. Invest* **117**, 1558–1565.

Koffler, D., Carr, R. I., Agnello, V., Fiezi, T., and Kunkel, H. G. (1969). Antibodies to polynucleotides: Distribution in human serums. *Science* **166**, 1648–1649.

Kohler, G., and Milstein, C. (1975). Continuous cultures of fused cells secreting antibody of predefined specificity. *Nature* **256**, 495–497.

Kolar, G. R., and Capra, J. D. (2004a). Ig V region restrictions in human chronic lymphocytic leukemia suggest some cases have a common origin. *J. Clin. Invest* **113**, 952–954.

Kolar, G. R., and Capra, J. D. (2004b). Immunoglobulin heavy-chain receptor editing is observed in the NOD/SCID model of human B-cell development. *Scand. J. Immunol.* **60**, 108–111.

Kolar, G. R., Mehta, D., Pelayo, R., and Capra, J. D. (2007). A novel human B cell subpopulation representing the initial germinal center population to express AID. *Blood* **109**, 2545–2552.

Kolar, G. R., Mehta, D., Wilson, P. C., and Capra, J. D. (2006). Diversity of the Ig repertoire is maintained with age in spite of reduced germinal centre cells in human tonsil lymphoid tissue. *Scand. J. Immunol.* **64**, 314–324.

Kolar, G. R., Yokota, T., Rossi, M. I., Nath, S. K., and Capra, J. D. (2004). Human fetal, cord blood, and adult lymphocyte progenitors have similar potential for generating B cells with a diverse immunoglobulin repertoire. *Blood* **104**, 2981–2987.

Kosco-Vilbois, M. H., Zentgraf, H., Gerdes, J., and Bonnefoy, J. Y. (1997). To 'B' or not to 'B' a germinal center? *Immunol. Today* **18**, 225–230.

Kristensson, K., Borrebaeck, C. A., and Carlsson, R. (1992). Human CD4+ T cells expressing CD45RA acquire the lymphokine gene expression of CD45RO+ T-helper cells after activation *in vitro*. *Immunology* **76**, 103–109.

Krober, A., Bloehdorn, J., Hafner, S., Buhler, A., Seiler, T., Kienle, D., Winkler, D., Bangerter, M., Schlenk, R. F., Benner, A., Lichter, P., Dohner, H., *et al.* (2006). Additional genetic high-risk features such as 11q deletion, 17p deletion, and V3-21 usage characterize discordance of ZAP-70 and VH mutation status in chronic lymphocytic leukemia. *J. Clin. Oncol.* **24**, 969–975.

Krober, A., Seiler, T., Benner, A., Bullinger, L., Bruckle, E., Lichter, P., Dohner, H., and Stilgenbauer, S. (2002). V(H) mutation status, CD38 expression level, genomic aberrations, and survival in chronic lymphocytic leukemia. *Blood* **100**, 1410–1416.

Kruetzmann, S., Rosado, M. M., Weber, H., Germing, U., Tournilhac, O., Peter, H. H., Berner, R., Peters, A., Boehm, T., Plebani, A., Quinti, I., and Carsetti, R. (2003). Human immunoglobulin M memory B cells controlling Streptococcus pneumoniae infections are generated in the spleen. *J. Exp. Med.* **197**, 939–945.

Kuliszkiewicz-Janus, M., Zimny, A., Sokolska, V., Sasiadek, M., and Kuliczkowski, K. (2005). Immunoglobulin D myeloma—problems with diagnosing and staging (own experience and literature review). *Leuk. Lymphoma* **46**, 1029–1037.

Kumagai, M., Coustan-Smith, E., Murray, D. J., Silvennoinen, O., Murti, K. G., Evans, W. E., Malavasi, F., and Campana, D. (1995). Ligation of CD38 suppresses human B lymphopoiesis. *J. Exp. Med.* **181,** 1101–1110.

Kwan, D. K., and Norman, A. (1974). Letter: Identification of two populations of human lymphocytes. *Acta Cytol.* **18,** 189–191.

Kyle, R. A., and Rajkumar, S. V. (2004). Multiple myeloma. *N. Engl. J. Med.* **351,** 1860–1873.

Lemaire, R., Winne, A., Sarkissian, M., and Lafyatis, R. (1999). SF2 and SRp55 regulation of CD45 exon 4 skipping during T cell activation. *Eur. J. Immunol.* **29,** 823–837.

Li, Y., Spellerberg, M. B., Stevenson, F. K., Capra, J. D., and Potter, K. N. (1996). The I binding specificity of human VH 4-34 (VH 4-21) encoded antibodies is determined by both VH framework region 1 and complementarity determining region 3. *J. Mol. Biol.* **256,** 577–589.

Link, J. M., and Schroeder, H. W., Jr. (2002). Clues to the etiology of autoimmune diseases through analysis of immunoglobulin genes. *Arthritis Res.* **4,** 80–83.

Lister, J., Miklos, J. A., Swerdlow, S. H., and Bahler, D. W. (1996). A clonally distinct recurrence of Burkitt's lymphoma at 15 years. *Blood* **88,** 1407–1410.

Liu, Y. J., and Arpin, C. (1997). Germinal center development. *Immunol. Rev.* **156,** 111–126.

Liu, Y. J., Arpin, C., de, B. O., Guret, C., Banchereau, J., Martinez-Valdez, H., and Lebecque, S. (1996a). Sequential triggering of apoptosis, somatic mutation and isotype switch during germinal center development. *Semin. Immunol.* **8,** 169–177.

Liu, Y. J., and Banchereau, J. (1996). The paths and molecular controls of peripheral B-cell development. *Immunologist* **4,** 55–66.

Liu, Y. J., de, B. O., Arpin, C., Briere, F., Galibert, L., Ho, S., Martinez-Valdez, H., Banchereau, J., and Lebecque, S. (1996b). Normal human IgD+IgM- germinal center B cells can express up to 80 mutations in the variable region of their IgD transcripts. *Immunity.* **4,** 603–613.

Liu, Y. J., de, B. O., and Fugier-Vivier, I. (1997). Mechanisms of selection and differentiation in germinal centers. *Curr. Opin. Immunol.* **9,** 256–262.

Liu, Y. J., Malisan, F., de, B. O., Guret, C., Lebecque, S., Banchereau, J., Mills, F. C., Max, E. E., and Martinez-Valdez, H. (1996c). Within germinal centers, isotype switching of immunoglobulin genes occurs after the onset of somatic mutation. *Immunity.* **4,** 241–250.

Llewelyn, M. B., Hawkins, R. E., and Russell, S. J. (1992). Discovery of antibodies. *BMJ* **305,** 1269–1272.

Looney, R. J. (2005). B cells as a therapeutic target in autoimmune diseases other than rheumatoid arthritis. *Rheumatology. (Oxford)* **44**(Suppl 2), ii13–ii17.

Looney, R. J., Anolik, J., and Sanz, I. (2004). B lymphocytes in systemic lupus erythematosus: lessons from therapy targeting B cells. *Lupus* **13,** 381–390.

Looney, R. J., Anolik, J. H., Campbell, D., Felgar, R. E., Young, F., Arend, L. J., Sloand, J. A., Rosenblatt, J., and Sanz, I. (2004). B cell depletion as a novel treatment for systemic lupus erythematosus: A phase I/II dose-escalation trial of rituximab. *Arthritis Rheum.* **50,** 2580–2589.

Lucas, B., Vasseur, F., and Penit, C. (1994). Production, selection, and maturation of thymocytes with high surface density of TCR. *J. Immunol.* **153,** 53–62.

Lund, F. E., Yu, N., Kim, K. M., Reth, M., and Howard, M. C. (1996). Signaling through CD38 augments B cell antigen receptor (BCR) responses and is dependent on BCR expression. *J. Immunol.* **157,** 1455–1467.

MacLennan, I. C. (1994a). Germinal centers. *Annu. Rev. Immunol.* **12,** 117–139.

MacLennan, I. C. (1994b). Somatic mutation. From the dark zone to the light. *Curr. Biol.* **4,** 70–72.

Malavasi, F., Deaglio, S., Ferrero, E., Funaro, A., Sancho, J., Ausiello, C. M., Ortolan, E., Vaisitti, T., Zubiaur, M., Fedele, G., Aydin, S., Tibaldi, E. V., *et al.* (2006). CD38 and CD157 as receptors of the immune system: A bridge between innate and adaptive immunity. *Mol. Med.* **12,** 334–341.

Manz, R. A., Thiel, A., and Radbruch, A. (1997). Lifetime of plasma cells in the bone marrow. *Nature* **388**, 133–134.

Martinez-Valdez, H., Guret, C., de, B. O., Fugier, I., Banchereau, J., and Liu, Y. J. (1996). Human germinal center B cells express the apoptosis-inducing genes Fas, c-myc, P53, and Bax but not the survival gene bcl-2. *J. Exp. Med.* **183**, 971–977.

Matolcsy, A., Casali, P., Warnke, R. A., and Knowles, D. M. (1996). Morphologic transformation of follicular lymphoma is associated with somatic mutation of the translocated Bcl-2 gene. *Blood* **88**, 3937–3944.

Matrai, Z., Lin, K., Dennis, M., Sherrington, P., Zuzel, M., Pettitt, A. R., and Cawley, J. C. (2001). CD38 expression and Ig VH gene mutation in B-cell chronic lymphocytic leukemia. *Blood* **97**, 1902–1903.

Matsuda, F., Ishii, K., Bourvagnet, P., Kuma, K., Hayashida, H., Miyata, T., and Honjo, T. (1998). The complete nucleotide sequence of the human immunoglobulin heavy chain variable region locus. *J. Exp. Med.* **188**, 2151–2162.

Matto, M., Nuutinen, U. M., Ropponen, A., Myllykangas, K., and Pelkonen, J. (2005). CD45RA and RO isoforms have distinct effects on cytokine- and B-cell-receptor-mediated signalling in human B cells. *Scand. J. Immunol.* **61**, 520–528.

Maurer, D., Fischer, G. F., Fae, I., Majdic, O., Stuhlmeier, K., Von, J. N., Holter, W., and Knapp, W. (1992). IgM and IgG but not cytokine secretion is restricted to the CD27+ B lymphocyte subset. *J. Immunol.* **148**, 3700–3705.

Maurer, D., Holter, W., Majdic, O., Fischer, G. F., and Knapp, W. (1990). CD27 expression by a distinct subpopulation of human B lymphocytes. *Eur. J. Immunol.* **20**, 2679–2684.

McNeill, L., Cassady, R. L., Sarkardei, S., Cooper, J. C., Morgan, G., and Alexander, D. R. (2004). CD45 isoforms in T cell signalling and development. *Immunol. Lett.* **92**, 125–134.

Meffre, E., Davis, E., Schiff, C., Cunningham-Rundles, C., Ivashkiv, L. B., Staudt, L. M., Young, J. W., and Nussenzweig, M. C. (2000). Circulating human B cells that express surrogate light chains and edited receptors. *Nat. Immunol.* **1**, 207–213.

Meffre, E., Schaefer, A., Wardemann, H., Wilson, P., Davis, E., and Nussenzweig, M. C. (2004). Surrogate light chain expressing human peripheral B cells produce self-reactive antibodies. *J. Exp. Med.* **199**, 145–150.

Mehta, K., Shahid, U., and Malavasi, F. (1996). Human CD38, a cell-surface protein with multiple functions. *FASEB J.* **10**, 1408–1417.

Mhes, L., Simon, A., Rejto, L., Kiss, A., Remenyi, G., Batar, P., Telek, B., and Udvardy, M. (2003). [Prognositc value of CD38 cell surface marker in chronic lymphocytic leukemia]. *Orv. Hetil.* **144**, 1531–1535.

Miller, M. J., Wei, S. H., Parker, I., and Cahalan, M. D. (2002). Two-photon imaging of lymphocyte motility and antigen response in intact lymph node. *Science* **296**, 1869–1873.

Mockridge, C. I., Potter, K. N., Wheatley, I., Neville, L. A., Packham, G., and Stevenson, F. K. (2007). Reversible anergy of sIgM-mediated signaling in the two subsets of CLL defined by VH-gene mutational status. *Blood* **109**, 4424–4431.

Morabito, F., Damle, R. N., Deaglio, S., Keating, M., Ferrarini, M., and Chiorazzi, N. (2006). The CD38 ectoenzyme family: Advances in basic science and clinical practice. *Mol. Med.* **12**, 342–344.

Moskowitz, c. (2006). Is it time to stop treating subset of DLBCL with R-CHOP. *Blood* **107**, 4197–4198.

Muller-Hermelink, H. K., and Rudiger, T. (2006). Oncogenic pathways in distinct DLBCL subsgroups. *Blood* **107**, 3818.

Munakata, Y., Saito, S., Hoshino, A., Muryoi, T., Hirabayashi, Y., Shibata, S., Miura, T., Ishii, T., Funato, T., and Sasaki, T. (1998). Somatic mutation in autoantibody-associated VH genes of circulating IgM+IgD+ B cells. *Eur. J. Immunol.* **28**, 1435–1444.

Muramatsu, M. (2006). [AID controls immunoglobulin class switch recombination]. *Seikagaku* **78**, 719–724.

Muramatsu, M., Kinoshita, K., Fagarasan, S., Yamada, S., Shinkai, Y., and Honjo, T. (2000). Class switch recombination and hypermutation require activation-induced cytidine deaminase (AID), a potential RNA editing enzyme. *Cell* **102,** 553–563.

Muramatsu, M., Sankaranand, V. S., Anant, S., Sugai, M., Kinoshita, K., Davidson, N. O., and Honjo, T. (1999). Specific expression of activation-induced cytidine deaminase (AID), a novel member of the RNA-editing deaminase family in germinal center B cells. *J. Biol. Chem.* **274,** 18470–18476.

Natkunam, Y., Hsi, E. D., Aoun, P., Zhao, S., Elson, P., Pohlman, B., Naushad, H., Bast, M., Levy, R., and Lossos, I. S. (2007). Expression of the human germinal center-associated lymphoma (HGAL) protein identifies a subset of classic Hodgkin lymphoma of germinal center derivation and improved survival. *Blood* **109,** 298–305.

Naylor, M., and Capra, J. D. (1999). Mutational status of Ig V(H) genes provides clinically valuable information in B-cell chronic lymphocytic leukemia. *Blood* **94,** 1837–1839.

Nedellec, S., Renaudineau, Y., Bordron, A., Berthou, C., Porakishvili, N., Lydyard, P. M., Pers, J. O., and Youinou, P. (2005). B cell response to surface IgM cross-linking identifies different prognostic groups of B-chronic lymphocytic leukemia patients. *J. Immunol.* **174,** 3749–3756.

Ng, Y. S., Wardemann, H., Chelnis, J., Cunningham-Rundles, C., and Meffre, E. (2004). Bruton's tyrosine kinase is essential for human B cell tolerance. *J. Exp. Med.* **200,** 927–934.

Nieminen, T., Kayhty, H., Virolainen, A., and Eskola, J. (1998). Circulating antibody secreting cell response to parenteral pneumococcal vaccines as an indicator of a salivary IgA antibody response. *Vaccine* **16,** 313–319.

Nossal, G. J. (1994a). Differentiation of the secondary B-lymphocyte repertoire: The germinal center reaction. *Immunol. Rev.* **137,** 173–183.

Nossal, G. J. (1994b). Twenty-five years of germinal centre physiology: Implications for tolerance in the secondary B cell repertoire. *Scand. J. Immunol.* **40,** 575–578.

Novak, T. J., Farber, D., Leitenberg, D., Hong, S. C., Johnson, P., and Bottomly, K. (1994). Isoforms of the transmembrane tyrosine phosphatase CD45 differentially affect T cell recognition. *Immunity.* **1,** 109–119.

Nowak, J. (1983). [Determination of B and T lymphocytes and their subpopulations with the use of monoclonal antibodies]. *Pol. Arch. Med. Wewn.* **69,** 393–399.

Ocana, E., gado-Perez, L., Campos-Caro, A., Munoz, J., Paz, A., Franco, R., and Brieva, J. A. (2007). The prognostic role of CXCR3 expression by chronic lymphocytic leukemia B cells. *Haematologica* **92,** 349–356.

Odendahl, M., Jacobi, A., Hansen, A., Feist, E., Hiepe, F., Burmester, G. R., Lipsky, P. E., Radbruch, A., and Dorner, T. (2000). Disturbed peripheral B lymphocyte homeostasis in systemic lupus erythematosus. *J. Immunol.* **165,** 5970–5979.

Odendahl, M., Keitzer, R., Wahn, U., Hiepe, F., Radbruch, A., Dorner, T., and Bunikowski, R. (2003). Perturbations of peripheral B lymphocyte homoeostasis in children with systemic lupus erythematosus. *Ann. Rheum. Dis.* **62,** 851–858.

Odendahl, M., Mei, H., Hoyer, B. F., Jacobi, A. M., Hansen, A., Muehlinghaus, G., Berek, C., Hiepe, F., Manz, R., Radbruch, A., and Dorner, T. (2005). Generation of migratory antigen-specific plasma blasts and mobilization of resident plasma cells in a secondary immune response. *Blood* **105,** 1614–1621.

Ody, C., Jungblut-Ruault, S., Cossali, D., Barnet, M., urrand-Lions, M., Imhof, B. A., and Matthes, T. (2007). Junctional adhesion molecule C (JAM-C) distinguishes CD27[+] germinal center B lymphocytes from non-germinal center cells and constitutes a new diagnostic tool for B-cell malignancies. *Leukemia* **21,** 1285–1293.

Ogilvy, S., Louis-Dit-Sully, C., Cooper, J., Cassady, R. L., Alexander, D. R., and Holmes, N. (2003). Either of the CD45RB and CD45RO isoforms are effective in restoring T cell, but not B cell, development and function in CD45-null mice. *J. Immunol.* **171,** 1792–1800.

Okazaki, I., Yoshikawa, K., Kinoshita, K., Muramatsu, M., Nagaoka, H., and Honjo, T. (2003). Activation-induced cytidine deaminase links class switch recombination and somatic hypermutation. *Ann. N. Y. Acad. Sci.* **987,** 1–8.

Paramithiotis, E., and Cooper, M. D. (1997). Memory B lymphocytes migrate to bone marrow in humans. *Proc. Natl. Acad. Sci. USA* **94,** 208–212.

Pascual, V., and Capra, J. D. (1992). VH4-21, a human VH gene segment overrepresented in the autoimmune repertoire. *Arthritis Rheum.* **35,** 11–18.

Pascual, V., Liu, Y. J., Magalski, A., de, B. O., Banchereau, J., and Capra, J. D. (1994). Analysis of somatic mutation in five B cell subsets of human tonsil. *J. Exp. Med.* **180,** 329–339.

Pascual, V., Victor, K., Lelsz, D., Spellerberg, M. B., Hamblin, T. J., Thompson, K. M., Randen, I., Natvig, J., Capra, J. D., and Stevenson, F. K. (1991). Nucleotide sequence analysis of the V regions of two IgM cold agglutinins. Evidence that the VH4-21 gene segment is responsible for the major cross-reactive idiotype. *J. Immunol.* **146,** 4385–4391.

Pascual, V., Victor, K., Randen, I., Thompson, K., Steinitz, M., Forre, O., Fu, S. M., Natvig, J. B., and Capra, J. D. (1992a). Nucleotide sequence analysis of rheumatoid factors and polyreactive antibodies derived from patients with rheumatoid arthritis reveals diverse use of VH and VL gene segments and extensive variability in CDR-3. *Scand. J. Immunol.* **36,** 349–362.

Pascual, V., Victor, K., Spellerberg, M., Hamblin, T. J., Stevenson, F. K., and Capra, J. D. (1992b). VH restriction among human cold agglutinins. The VH4-21 gene segment is required to encode anti-I and anti-i specificities. *J. Immunol.* **149,** 2337–2344.

Pascual, V., Wilson, P., Liu, Y. J., Banchereau, J., and Capra, J. D. (1997). Biased VH4 gene segment repertoire in the human tonsil. *Chem. Immunol.* **67,** 45–57.

Pers, J. O., Devauchelle, V., Daridon, C., Bendaoud, B., Le, B. R., Bordron, A., Hutin, P., Renaudineau, Y., Dueymes, M., Loisel, S., Berthou, C., Saraux, A., et al. (2007). BAFF-modulated repopulation of B lymphocytes in the blood and salivary glands of rituximab-treated patients with Sjogren's syndrome. *Arthritis Rheum.* **56,** 1464–1477.

Phan, T. G., Paus, D., Chan, T. D., Turner, M. L., Nutt, S. L., Basten, A., and Brink, R. (2006). High affinity germinal center B cells are actively selected into the plasma cell compartment. *J. Exp. Med.* **203,** 2419–2424.

Pileri, S. A., Leoncini, L., and Falini, B. (1995). Revised European-American Lymphoma Classification. *Curr. Opin. Oncol.* **7,** 401–407.

Portanova, J. P., Creadon, G., Zhang, X., Smith, D. S., Kotzin, B. L., and Wysocki, L. J. (1995). An early post-mutational selection event directs expansion of autoreactive B cells in murine lupus. *Mol. Immunol.* **32,** 117–135.

Potter, K. N., Li, Y., Mageed, R. A., Jefferis, R., and Capra, J. D. (1999). Molecular characterization of the VH1-specific variable region determinants recognized by anti-idiotypic monoclonal antibodies G6 and G8. *Scand. J. Immunol.* **50,** 14–20.

Potter, K. N., Li, Y., Pascual, V., Williams, R. C., Jr., Byres, L. C., Spellerberg, M., Stevenson, F. K., and Capra, J. D. (1993). Molecular characterization of a cross-reactive idiotope on human immunoglobulins utilizing the VH4-21 gene segment. *J. Exp. Med.* **178,** 1419–1428.

Prak, E. L., and Weigert, M. (1995). Light chain replacement: A new model for antibody gene rearrangement. *J. Exp. Med.* **182,** 541–548.

Preud'homme, J. L., Petit, I., Barra, A., Morel, F., Lecron, J. C., and Lelievre, E. (2000). Structural and functional properties of membrane and secreted IgD. *Mol. Immunol.* **37,** 871–887.

Pritsch, O., Maloum, K., Magnac, C., Davi, F., Binet, J. L., Merle-Beral, H., and Dighiero, G. (1997). What do chronic B cell malignancies teach us about B cell subsets? *Chem. Immunol.* **67,** 85–101.

Pruzanski, W., and Keystone, E. C. (1977). Biologic role of lymphocytes. *Can. Med. Assoc. J.* **117,** 114–116.

Pugh-Bernard, A. E., Silverman, G. J., Cappione, A. J., Villano, M. E., Ryan, D. H., Insel, R. A., and Sanz, I. (2001). Regulation of inherently autoreactive VH4-34 B cells in the maintenance of human B cell tolerance. *J. Clin. Invest* **108,** 1061–1070.

Putterman, C., Deocharan, B., and Diamond, B. (2000). Molecular analysis of the autoantibody response in peptide-induced autoimmunity. *J. Immunol.* **164,** 2542–2549.

Radic, M. Z., and Weigert, M. (2004). Intricacies of anti-DNA autoantibodies. *J. Immunol.* **172,** 3367–3368.

Radic, M. Z., and Weigert, M. (1995). Origins of anti-DNA antibodies and their implications for B-cell tolerance. *Ann. N. Y. Acad. Sci.* **764,** 384–396.

Rasmussen, T., Lodahl, M., Hancke, S., and Johnsen, H. E. (2004). In multiple myeloma clonotypic. *Leuk. Lymphoma* **45,** 1413–1417.

Razanajaona, D., Denepoux, S., Blanchard, D., de, B. O., Liu, Y. J., Banchereau, J., and Lebecque, S. (1997). *In vitro* triggering of somatic mutation in human naive B cells. *J. Immunol.* **159,** 3347–3353.

Rettig, M. B., Vescio, R. A., Cao, J., Wu, C. H., Lee, J. C., Han, E., DerDanielian, M., Newman, R., Hong, C., Lichtenstein, A. K., and Berenson, J. R. (1996). VH gene usage is multiple myeloma: Complete absence of the VH4.21 (VH4-34) gene. *Blood* **87,** 2846–2852.

Riboldi, P., Gaidano, G., Schettino, E. W., Steger, T. G., Knowles, D. M., la-Favera, R., and Casali, P. (1994). Two acquired immunodeficiency syndrome-associated Burkitt's lymphomas produce specific anti-i IgM cold agglutinins using somatically mutated VH4-21 segments. *Blood* **83,** 2952–2961.

Richter, W., Jury, K. M., Loeffler, D., Manfras, B. J., Eiermann, T. H., and Boehm, B. O. (1995). Immunoglobulin variable gene analysis of human autoantibodies reveals antigen-driven immune response to glutamate decarboxylase in type 1 diabetes mellitus. *Eur. J. Immunol.* **25,** 1703–1712.

Ridderstad, A., and Tarlinton, D. M. (1998). Kinetics of establishing the memory B cell population as revealed by CD38 expression. *J. Immunol.* **160,** 4688–4695.

Robillard, N., vet-Loiseau, H., Garand, R., Moreau, P., Pineau, D., Rapp, M. J., Harousseau, J. L., and Bataille, R. (2003). CD20 is associated with a small mature plasma cell morphology and t(11;14) in multiple myeloma. *Blood* **102,** 1070–1071.

Rodig, S. J., Shahsafaei, A., Li, B., and Dorfman, D. M. (2005). The CD45 isoform B220 identifies select subsets of human B cells and B-cell lymphoproliferative disorders. *Hum. Pathol.* **36,** 51–57.

Roes, J., and Rajewsky, K. (1993). Immunoglobulin D (IgD)-deficient mice reveal an auxiliary receptor function for IgD in antigen-mediated recruitment of B cells. *J. Exp. Med.* **177,** 45–55.

Sagaert, X., Sprangers, B., and De Wolf-Peeters, C. (2007). The dynamics of the B follicle: Understanding the normal counterpart of B-cell-derived malignancies. *Leukemia* **21,** 1378–1386.

Samuels, J., Ng, Y. S., Coupillaud, C., Paget, D., and Meffre, E. (2005). Impaired early B cell tolerance in patients with rheumatoid arthritis. *J. Exp. Med.* **201,** 1659–1667.

Sarkissian, M., Winne, A., and Lafyatis, R. (1996). The mammalian homolog of suppressor-of-white-apricot regulates alternative mRNA splicing of CD45 exon 4 and fibronectin IIICS. *J. Biol. Chem.* **271,** 31106–31114.

Sasaki, S., Jaimes, M. C., Holmes, T. H., Dekker, C. L., Mahmood, K., Kemble, G. W., Arvin, A. M., and Greenberg, H. B. (2007). Comparison of the influenza virus-specific effector and memory B-cell responses to immunization of children and adults with live attenuated or inactivated influenza virus vaccines. *J. Virol.* **81,** 215–228.

Schroers, R., Griesinger, F., Trumper, L., Haase, D., Kulle, B., Klein-Hitpass, L., Sellmann, L., Duhrsen, U., and Durig, J. (2005). Combined analysis of ZAP-70 and CD38 expression as a predictor of disease progression in B-cell chronic lymphocytic leukemia. *Leukemia* **19,** 750–758.

Seyama, K., Nonoyama, S., Gangsaas, I., Hollenbaugh, D., Pabst, H. F., Aruffo, A., and Ochs, H. D. (1998). Mutations of the CD40 ligand gene and its effect on CD40 ligand expression in patients with X-linked hyper IgM syndrome. *Blood* **92,** 2421–2434.

Shen, Y., Iqbal, J., Xiao, L., Lynch, R. C., Rosenwald, A., Staudt, L. M., Sherman, S., Dybkaer, K., Zhou, G., Eudy, J. D., Delabie, J., McKeithan, T. W., *et al.* (2004). Distinct gene expression profiles in different B-cell compartments in human peripheral lymphoid organs. *BMC. Immunol.* **5,** 20.

Shi, Y., Agematsu, K., Ochs, H. D., and Sugane, K. (2003). Functional analysis of human memory B-cell subpopulations: IgD+CD27+ B cells are crucial in secondary immune response by producing high affinity IgM. *Clin. Immunol.* **108,** 128–137.

Shlomchik, M., Mascelli, M., Shan, H., Radic, M. Z., Pisetsky, D., Marshak-Rothstein, A., and Weigert, M. (1990). Anti-DNA antibodies from autoimmune mice arise by clonal expansion and somatic mutation. *J. Exp. Med.* **171,** 265–292.

Shlomchik, M. J., Cooke, A., and Weigert, M. (2004). Autoimmunity: The genes and phenotypes of autoimmunity. *Curr. Opin. Immunol.* **16,** 738–740.

Shlomchik, M. J., Marshak-Rothstein, A., Wolfowicz, C. B., Rothstein, T. L., and Weigert, M. G. (1987). The role of clonal selection and somatic mutation in autoimmunity. *Nature* **328,** 805–811.

Silvennoinen, O., Nishigaki, H., Kitanaka, A., Kumagai, M., Ito, C., Malavasi, F., Lin, Q., Conley, M. E., and Campana, D. (1996). CD38 signal transduction in human B cell precursors. Rapid induction of tyrosine phosphorylation, activation of syk tyrosine kinase, and phosphorylation of phospholipase C-gamma and phosphatidylinositol 3-kinase. *J. Immunol.* **156,** 100–107.

Silverman, G. J. (2006). Targeting of B cells in SLE: Rationale and therapeutic opportunities. *Bull. NYU. Hosp. Jt. Dis.* **64,** 51–56.

Sims, G. P., Ettinger, R., Shirota, Y., Yarboro, C. H., Illei, G. G., and Lipsky, P. E. (2005). Identification and characterization of circulating human transitional B cells. *Blood* **105,** 4390–4398.

Slifka, M. K., Antia, R., Whitmire, J. K., and Ahmed, R. (1998). Humoral immunity due to long-lived plasma cells. *Immunity.* **8,** 363–372.

Snider, D. P., Uppenkamp, I. K., Titus, J. A., and Segal, D. M. (1991). Processing fate of protein antigen attached to IgD or MHC molecules on normal B lymphocytes using heterocrosslinked bispecific antibodies. *Mol. Immunol.* **28,** 779–788.

Stein, K., Hummel, M., Korbjuhn, P., Foss, H. D., Anagnostopoulos, I., Marafioti, T., and Stein, H. (1999). Monocytoid B cells are distinct from splenic marginal zone cells and commonly derive from unmutated naive B cells and less frequently from postgerminal center B cells by polyclonal transformation. *Blood* **94,** 2800–2808.

Steiniger, B., Timphus, E. M., Jacob, R., and Barth, P. J. (2005). CD27+ B cells in human lymphatic organs: re-evaluating the splenic marginal zone. *Immunology* **116,** 429–442.

Steinitz, M., Klein, G., Koskimies, S., and Makel, O. (1977). EB virus-induced B lymphocyte cell lines producing specific antibody. *Nature* **269,** 420–422.

Stollar, B. D., Zon, G., and Pastor, R. W. (1986). A recognition site on synthetic helical oligonucleotides for monoclonal anti-native DNA autoantibody. *Proc. Natl. Acad. Sci. USA* **83,** 4469–4473.

Strelkauskas, A. J., Schauf, V., Wilson, B. S., Chess, L., and Schlossman, S. F. (1978). Isolation and characterization of naturally occurring subclasses of human peripheral blood T cells with regulatory functions. *J. Immunol.* **120,** 1278–1282.

Szczepek, A. J., Seeberger, K., Wizniak, J., Mant, M. J., Belch, A. R., and Pilarski, L. M. (1998). A high frequency of circulating B cells share clonotypic Ig heavy-chain VDJ rearrangements with autologous bone marrow plasma cells in multiple myeloma, as measured by single-cell and in situ reverse transcriptase-polymerase chain reaction. *Blood* **92,** 2844–2855.

Tan, E. M., Cohen, A. S., Fries, J. F., Masi, A. T., McShane, D. J., Rothfield, N. F., Schaller, J. G., Talal, N., and Winchester, R. J. (1982). The 1982 revised criteria for the classification of systemic lupus erythematosus. *Arthritis Rheum.* **25,** 1271–1277.

Tangye, S. G., and Good, K. L. (2007). Human IgM⁺CD27⁺ B cells: memory B cells or "memory" B cells? *J. Immunol.* **179,** 13–19.

Tarlinton, D. M., and Hodgkin, P. D. (2004). Targeting plasma cells in autoimmune diseases. *J. Exp. Med.* **199,** 1451–1454.

Tchilian, E. Z., and Beverley, P. C. (2006). Altered CD45 expression and disease. *Trends Immunol.* **27,** 146–153.

Tchilian, E. Z., Dawes, R., Hyland, L., Montoya, M., Le, B. A., Borrow, P., Hou, S., Tough, D., and Beverley, P. C. (2004). Altered CD45 isoform expression affects lymphocyte function in CD45 Tg mice. *Int. Immunol.* **16,** 1323–1332.

Ten, B. E., Siegert, C. E., Vrielink, G. J., Van, D., Ceelen, A., and De, K. W. (2007). Analyses of CD27⁺⁺ plasma cells in peripheral blood from patients with bacterial infections and patients with serum antinuclear antibodies. *J. Clin. Immunol.* **27,** p. 467–476.

Thomas, M. D., Srivastava, B., and Allman, D. (2006). Regulation of peripheral B cell maturation. *Cell Immunol.* **239,** 92–102.

Thompson, K. M., Sutherland, J., Barden, G., Melamed, M. D., Randen, I., Natvig, J. B., Pascual, V., Capra, J. D., and Stevenson, F. K. (1991). Human monoclonal antibodies against blood group antigens preferentially express a VH4-21 variable region gene-associated epitope. *Scand. J. Immunol.* **34,** 509–518.

Tiller, T., Tsuiji, M., Yurasov, S., Velinzon, K., Nussenzweig, M. C., and Wardemann, H. (2007). Autoreactivity in human IgG⁺ memory B cells. *Immunity.* **26,** 205–213.

Truffinet, V., Pinaud, E., Cogne, N., Petit, B., Guglielmi, L., Cogne, M., and Denizot, Y. (2007). The 3′ IgH locus control region is sufficient to deregulate a c-myc transgene and promote mature B cell malignancies with a predominant Burkitt-like phenotype. *J. Immunol.* **179,** 6033–6042.

Tsokos, G. C., and Balow, J. E. (1981). Spontaneous and pokeweed mitogen-induced plaque-forming cells in systemic lupus erythematosus. *Clin. Immunol. Immunopathol.* **21,** 172–183.

Tsuiji, M., Yurasov, S., Velinzon, K., Thomas, S., Nussenzweig, M. C., and Wardemann, H. (2006). A checkpoint for autoreactivity in human IgM⁺ memory B cell development. *J. Exp. Med.* **203,** 393–400.

van Zelm, M. C., Szczepanski, T., van der, B. M., and van Dongen, J. J. (2007). Replication history of B lymphocytes reveals homeostatic proliferation and extensive antigen-induced B cell expansion. *J. Exp. Med.* **204,** 645–655.

Vorob'ev, I. A., Gorgidze, L. A., Gretsov, E. M., Korneva, E. P., Kharazishvili, D. V., Khudoleeva, O. A., and Churakova, Z. (2006). [Flow fluorimetry in differential diagnosis of diffuse large B-cell lymphoma]. *Ter. Arkh.* **78,** 46–51.

Wallace, V. A., Fung-Leung, W. P., Timms, E., Gray, D., Kishihara, K., Loh, D. Y., Penninger, J., and Mak, T. W. (1992). CD45RA and CD45RB high expression induced by thymic selection events. *J. Exp. Med.* **176,** 1657–1663.

Wang, J., and Boxer, L. M. (2005). Regulatory elements in the immunoglobulin heavy chain gene 3′-enhancers induce c-myc deregulation and lymphomagenesis in murine B cells. *J. Biol. Chem.* **280,** 12766–12773.

Wardemann, H., Hammersen, J., and Nussenzweig, M. C. (2004). Human autoantibody silencing by immunoglobulin light chains. *J. Exp. Med.* **200,** 191–199.

Wardemann, H., Yurasov, S., Schaefer, A., Young, J. W., Meffre, E., and Nussenzweig, M. C. (2003). Predominant autoantibody production by early human B cell precursors. *Science* **301,** 1374–1377.

Warner, N. L. (1976). Differentiation and Ontogeny of Lymphoid Cells. *Adv. Exp. Med. Biol.* **66,** 3–11.

Wehr, C., Eibel, H., Masilamani, M., Illges, H., Schlesier, M., Peter, H. H., and Warnatz, K. (2004). A new CD21low B cell population in the peripheral blood of patients with SLE. *Clin. Immunol.* **113,** 161–171.

Wei, C., Anolik, J., Cappione, A., Zheng, B., Pugh-Bernard, A., Brooks, J., Lee, E. H., Milner, E. C., and Sanz, I. (2007). A new population of cells lacking expression of CD27 represents a notable component of the B cell memory compartment in systemic lupus erythematosus. *J. Immunol.* **178,** 6624–6633.

Weill, J. C., Weller, S., and Reynaud, C. A. (2004). A bird's eye view on human B cells. *Semin. Immunol.* **16,** 277–281.

Weller, S., Braun, M. C., Tan, B. K., Rosenwald, A., Cordier, C., Conley, M. E., Plebani, A., Kumararatne, D. S., Bonnet, D., Tournilhac, O., Tchernia, G., Steiniger, B., *et al.* (2004). Human blood IgM "memory" B cells are circulating splenic marginal zone B cells harboring a prediversified immunoglobulin repertoire. *Blood* **104,** 3647–3654.

Weller, S., Faili, A., Garcia, C., Braun, M. C., Le Deist, F. F., de Saint Basile, G. G., Hermine, O., Fischer, A., Reynaud, C. A., and Weill, J. C. (2001). CD40-CD40L independent Ig gene hypermutation suggests a second B cell diversification pathway in humans. *Proc. Natl. Acad. Sci. USA* **98,** 1166–1170.

Werner-Favre, C., Bovia, F., Schneider, P., Holler, N., Barnet, M., Kindler, V., Tschopp, J., and Zubler, R. H. (2001). IgG subclass switch capacity is low in switched and in IgM-only, but high in IgD$^+$IgM$^+$, post-germinal center (CD27$^+$) human B cells. *Eur. J. Immunol.* **31,** 243–249.

White, M. B., Word, C. J., Humphries, C. G., Blattner, F. R., and Tucker, P. W. (1990). Immunoglobulin D switching can occur through homologous recombination in human B cells. *Mol. Cell Biol.* **10,** 3690–3699.

Wilson, P. C., de, B. O., Liu, Y. J., Potter, K., Banchereau, J., Capra, J. D., and Pascual, V. (1998). Somatic hypermutation introduces insertions and deletions into immunoglobulin V genes. *J. Exp. Med.* **187,** 59–70.

Wilson, P. C., and Donald, C. J. (1998). The super-information age of immunoglobulin genetics. *J. Exp. Med.* **188,** 1973–1975.

Wilson, P. C., Wilson, K., Liu, Y. J., Banchereau, J., Pascual, V., and Capra, J. D. (2000). Receptor revision of immunoglobulin heavy chain variable region genes in normal human B lymphocytes. *J. Exp. Med.* **191,** 1881–1894.

Wouters, C. H., Diegenant, C., Ceuppens, J. L., Degreef, H., and Stevens, E. A. (2004). The circulating lymphocyte profiles in patients with discoid lupus erythematosus and systemic lupus erythematosus suggest a pathogenetic relationship. *Br. J. Dermatol.* **150,** 693–700.

Wrammert, J., Smith, K., Miller, J., Langely, T., Kokko, K., Larsen, C., Zheng, N. Y., Mays, I., Garman, L., Helms, C., James, J., and Air, G. M. (2008). Rapid cloning of high affinity human monoclonal antibodies against influenza virus. *Nature* .

Wu, H. (1995). Somatic Hypermutations Generated in the Immunoglobulin Genes of Human Follicular Lymphoma Cells *in vivo* and in vitro Department of Bacteriology and Immunology, Haartman Institute, University of Helsinki, Finland, Helsinki, Finland.

Xiao, Y., Hendriks, J., Langerak, P., Jacobs, H., and Borst, J. (2004). CD27 is acquired by primed B cells at the centroblast stage and promotes germinal center formation. *J. Immunol.* **172,** 7432–7441.

Yasue, T., Nishizumi, H., Aizawa, S., Yamamoto, T., Miyake, K., Mizoguchi, C., Uehara, S., Kikuchi, Y., and Takatsu, K. (1997). A critical role of Lyn and Fyn for B cell responses to CD38 ligation and interleukin 5. *Proc. Natl. Acad. Sci. USA* **94,** 10307–10312.

Yasui, H., Akahori, Y., Hirano, M., Yamada, K., and Kurosawa, Y. (1989). Class switch from mu to delta is mediated by homologous recombination between sigma mu and sigma mu sequences in human immunoglobulin gene loci. *Eur. J. Immunol.* **19,** 1399–1403.

Yokoi, H., Seki, M., Okazoe, S., Okumura, K., and Yoshikawa, H. (2003). The role and expression of CD27 and CD70 lymphocytes in the human tonsil. *Lymphology* **36**, 74–83.

Yokoyama, M. (1999). [Regulatory role of CD38 in humoral immune responses]. *Seikagaku* **71**, 438–442.

Youinou, P. (2007). B cell conducts the lymphocyte orchestra. *J. Autoimmun.* **28**, 143–151.

Youinou, P., Jamin, C., and Lydyard, P. M. (1999). CD5 expression in human B-cell populations. *Immunol. Today* **20**, 312–316.

Young, J. L., Ramage, J. M., Gaston, J. S., and Beverley, P. C. (1997). *In vitro* responses of human CD45R0brightRA- and CD45R0-RAbright T cell subsets and their relationship to memory and naive T cells. *Eur. J. Immunol.* **27**, 2383–2390.

Yu, C. C., Mamchak, A. A., and DeFranco, A. L. (2003). Signaling mutations and autoimmunity. *Curr. Dir. Autoimmun.* **6**, 61–88.

Yu, Y., Rabinowitz, R., Polliack, A., Ben-Bassat, H., and Schlesinger, M. (2002). Hyposialated 185 kDa CD45RA$^+$ molecules attain a high concentration in B lymphoma cells and in activated human B cells. *Eur. J. Haematol.* **68**, 22–30.

Yurasov, S., and Nussenzweig, M. C. (2007). Regulation of autoreactive antibodies. *Curr. Opin. Rheumatol.* **19**, 421–426.

Yurasov, S., Tiller, T., Tsuiji, M., Velinzon, K., Pascual, V., Wardemann, H., and Nussenzweig, M. C. (2006). Persistent expression of autoantibodies in SLE patients in remission. *J. Exp. Med.* **203**, 2255–2261.

Yurasov, S., Wardemann, H., Hammersen, J., Tsuiji, M., Meffre, E., Pascual, V., and Nussenzweig, M. C. (2005). Defective B cell tolerance checkpoints in systemic lupus erythematosus. *J. Exp. Med.* **201**, 703–711.

Zanetti, M. (2007). Gating on germinal center B cells. *Blood* **110**, 3816–3817.

Zheng, N. Y., Wilson, K., Jared, M., and Wilson, P. C. (2005). Intricate targeting of immunoglobulin somatic hypermutation maximizes the efficiency of affinity maturation. *J. Exp. Med.* **201**, 1467–1478.

Zheng, N. Y., Wilson, K., Wang, X., Boston, A., Kolar, G., Jackson, S. M., Liu, Y. J., Pascual, V., Capra, J. D., and Wilson, P. C. (2004). Human immunoglobulin selection associated with class switch and possible tolerogenic origins for C delta class-switched B cells. *J. Clin. Invest* **113**, 1188–1201.

Zhou, C., Saxon, A., and Zhang, K. (2003). Human activation-induced cytidine deaminase is induced by IL-4 and negatively regulated by CD45: implication of CD45 as a Janus kinase phosphatase in antibody diversification. *J. Immunol.* **170**, 1887–1893.

Zouali, M. (1992). Development of human antibody variable genes in systemic autoimmunity. *Immunol. Rev.* **128**, 73–99.

Zucchetto, A., Bomben, R., Dal, B. M., Bulian, P., Benedetti, D., Nanni, P., Del, P. G., Degan, M., and Gattei, V. (2006a). CD49d in B-cell chronic lymphocytic leukemia: correlated expression with CD38 and prognostic relevance. *Leukemia* **20**, 523–525.

Zucchetto, A., Bomben, R., Dal, B. M., Sonego, P., Nanni, P., Rupolo, M., Bulian, P., Dal, M. L., Del, P. G., Del Principe, M. I., Degan, M., and Gattei, V. (2006b). A scoring system based on the expression of six surface molecules allows the identification of three prognostic risk groups in B-cell chronic lymphocytic leukemia. *J. Cell Physiol.* **207**, 354–363.

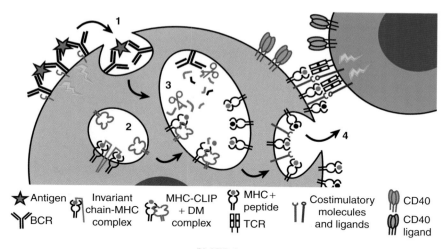

Antigen **Invariant chain-MHC complex** **MHC-CLIP + DM complex** **MHC + peptide** **Costimulatory molecules and ligands** **CD40**
BCR **TCR** **CD40 ligand**

PLATE 4

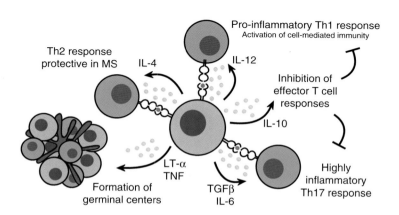

Th2 response protective in MS

IL-4

IL-12

Pro-inflammatory Th1 response
Activation of cell-mediated immunity

Inhibition of effector T cell responses

IL-10

LT-α
TNF

TGFβ
IL-6

Highly inflammatory Th17 response

Formation of germinal centers

PLATE 5

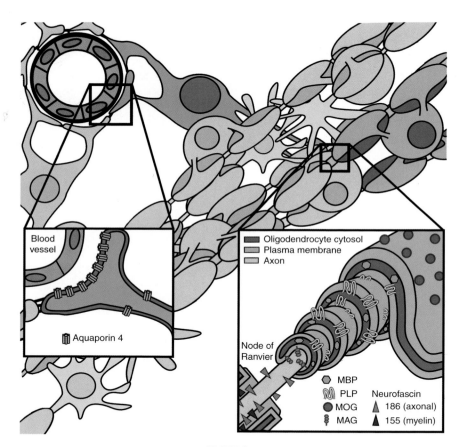

PLATE 6

A

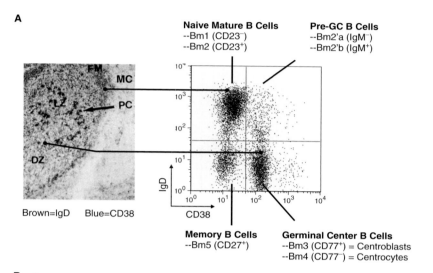

Naive Mature B Cells
--Bm1 (CD23⁻)
--Bm2 (CD23⁺)

Pre-GC B Cells
--Bm2'a (IgM⁻)
--Bm2'b (IgM⁺)

Brown=IgD   Blue=CD38

Memory B Cells
--Bm5 (CD27⁺)

Germinal Center B Cells
--Bm3 (CD77⁺) = Centroblasts
--Bm4 (CD77⁻) = Centrocytes

B

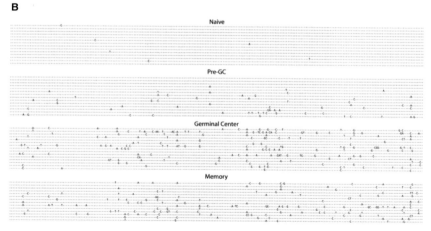

Naive

Pre-GC

Germinal Center

Memory

PLATE 7

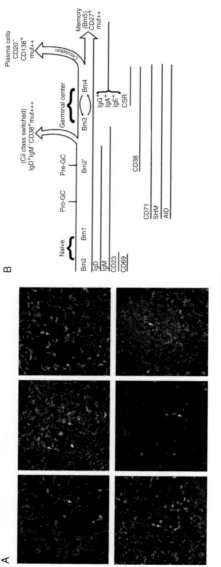

B

Plasma cells
CD20⁻
CD138⁺
mut++

Memory
(Bm5)
CD27⁺
mut++

Germinal center

(Cδ class switched)
IgD⁺IgM⁻CD38⁺mut+++

Bm2 Bm1 Pro-GC Pre-GC Bm2' Bm3 Bm4

Naive

IgG⁺ʱ
IgA⁺
IgE⁺
CSR

IgD
IgM
CD23
CD69

CD38

CD71
SHM
AID

PLATE 8

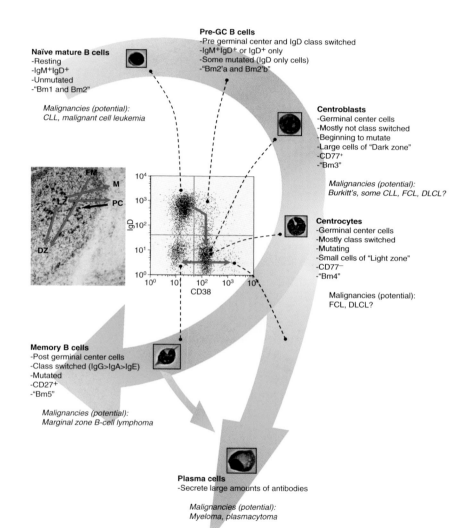

Naïve mature B cells
-Resting
-IgM+IgD+
-Unmutated
-"Bm1 and Bm2"

Malignancies (potential):
CLL, malignant cell leukemia

Pre-GC B cells
-Pre germinal center and IgD class switched
-IgM+IgD+ or IgD+ only
-Some mutated (IgD only cells)
-"Bm2'a and Bm2'b"

Centroblasts
-Germinal center cells
-Mostly not class switched
-Beginning to mutate
-Large cells of "Dark zone"
-CD77+
-"Bm3"

Malignancies (potential):
Burkitt's, some CLL, FCL, DLCL?

Centrocytes
-Germinal center cells
-Mostly class switched
-Mutating
-Small cells of "Light zone"
-CD77−
-"Bm4"

Malignancies (potential):
FCL, DLCL?

Memory B cells
-Post germinal center cells
-Class switched (IgG>IgA>IgE)
-Mutated
-CD27+
-"Bm5"

Malignancies (potential):
Marginal zone B-cell lymphoma

Plasma cells
-Secrete large amounts of antibodies

Malignancies (potential):
Myeloma, plasmacytoma

PLATE 9

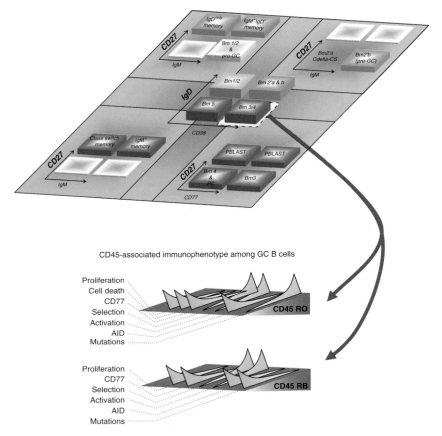

CD45-associated immunophenotype among GC B cells

PLATE 10

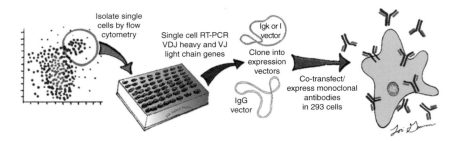

PLATE 11

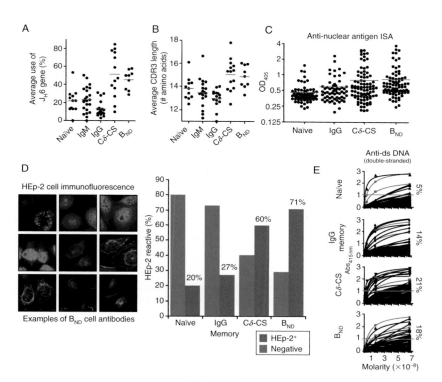

PLATE 12

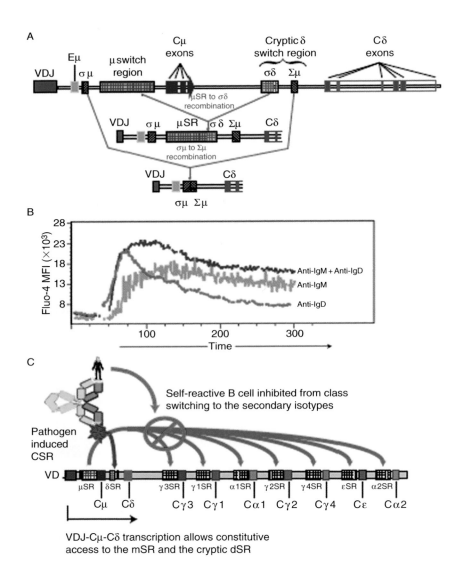

A

PLATE 13